Studies in Logic

Studies in Mathematical Logic and Foundations

Volume 65

Elementary Logic with Applications

A Procedural Perspective for Computer Scientists

Volume 54
Proof Theory of N4-related Paraconsistent Logics
Norihiro Kamide and Heinrich Wansing

Volume 55
All about Proofs, Proofs for All
Bruno Woltzenlogel Paleo and David Delahaye, eds

Volume 56
Dualities for Structures of Applied Logics
Ewa Orłowska, Anna Maria Radzikowska and Ingrid Rewitzky

Volume 57
Proof-theoretic Semantics
Nissim Francez

Volume 58
Handbook of Mathematical Fuzzy Logic, Volume 3
Petr Cintula, Petr Hajek and Carles Noguera, eds.

Volume 59
The Psychology of Argument. Cognitive Approaches to Argumentation and Persuasion
Fabio Paglieri, Laura Bonelli and Silvia Felletti, eds

Volume 60
Absract Algebraic Logic. An Introductory Textbook
Josep Maria Font

Volume 61
Philosophical Applications of Modal Logic
Lloyd Humberstone

Volume 62
Argumentation and Reasoned Action. Proceedings of the 1st European Conference on Argumentation, Lisbon 2015. Volume I
Dima Mohammed and Marcin Lewiński, eds

Volume 63
Argumentation and Reasoned Action. Proceedings of the 1st European Conference on Argumentation, Lisbon 2015. Volume II
Dima Mohammed and Marcin Lewiński, eds

Volume 64
Logic of Questions in the Wild. Inferential Erotetic Logic in Information Seeking Dialogue Modelling
Paweł Łupkowski

Volume 65
Elementary Logic with Applications. A Procedural Perspective for Computer Scientists
D. M. Gabbay and O. T. Rodrigues

Studies in Logic Series Editor
Dov Gabbay   dov.gabbay@kcl.ac.uk

Elementary Logic with Applications
A Procedural Perspective for Computer Scientists

D. M. Gabbay

O. T. Rodrigues

© Individual author and College Publications 2016
All rights reserved.

ISBN 978-1-84890-225-1

College Publications
Scientific Director: Dov Gabbay
Managing Director: Jane Spurr

http://www.collegepublications.co.uk

Printed by Lightning Source, Milton Keynes, UK

All rights reserved. No part of this publication may be reproduced, stored in a retrieval system or transmitted in any form, or by any means, electronic, mechanical, photocopying, recording or otherwise without prior permission, in writing, from the publisher.

CONTENTS

Preface ix

Introduction xiii

1 **Propositional logic** 1
   1.1 Introducing classical propositional logic 3
       1.1.1 Formal language 4
       1.1.2 Computing truth tables 8
   1.2 Notions of truth and validity 12
       1.2.1 Equivalences 13
       1.2.2 Arguments and consequence relations 16
   1.3 Worked examples 23

2 **Some non-classical logics** 39
   2.1 Many-valued logics 40
   2.2 Other logics 43
       2.2.1 Temporal logic 43
       2.2.2 Intuitionistic logic 46
       2.2.3 Resource logics 48
   2.3 Another look at intuitionistic logic 50
   2.4 Worked examples 54

3 Introducing forward rules 69
3.1 Natural deduction rules 72
3.1.1 Rules for conjunction 72
3.1.2 Rules for disjunction 73
3.1.3 Rules for implication 74
3.1.4 Rules for negation 75
3.2 Using subcomputations 77
3.3 Example proofs 81
3.3.1 Proofs using $\land$, $\lor$ and $\to$ rules 82
3.3.2 Proofs involving negation rules 84
3.4 Formal descriptions of proofs 89

4 From forward to backward rules 95
4.1 Introductory heuristics 96
4.1.1 Adding positive and negative assumptions 97
4.1.2 Restart rule 103
4.1.3 A heuristic for negation 105
4.1.4 Rewriting assumptions 106
4.1.5 Summary of computation rules 109
4.2 Worked examples 110
4.3 A deterministic goal-directed algorithm 122

5 Methodology and metatheorems 129
5.1 Consequence relations and Hilbert systems 130
5.2 Hilbert formulation of a logic 132
5.3 Metatheorems about classical logic consequence 136
5.4 Propositional soundness and completeness 138
5.5 Proof of soundness 142
5.6 Proof of completeness 143
5.7 Fragment of $\{\land, \to, \bot\}$ 155
5.8 Summary of systems and completeness theorems 158
5.9 Worked examples 158

6 Introducing predicate logic 167
6.1 Simple sentences 168
6.1.1 Building simple sentences 169
6.1.2 Truth of simple sentences 171
6.2 Variables and quantifiers 173
6.2.1 Scope of quantifiers 176
6.2.2 Quantifier rules and equivalences 178

		6.2.3 Prenex normal form	182
		6.2.4 Equality	184
	6.3	Formal definitions	184
	6.4	Worked examples	189

7 Forward and backward predicate rules — 205
    7.1 Reasoning with variables 206
        7.1.1 Introduction to unification 210
        7.1.2 A unification algorithm 212
        7.1.3 The meaning of variables in clauses 215
        7.1.4 Skolemization 217
    7.2 Reasoning forwards 221
    7.3 Reasoning backward 227
        7.3.1 Connection with resolution 235
    7.4 Decidability 237
    7.5 Worked examples 238

8 Introduction to PROLOG — 243
    8.1 Getting started with PROLOG's syntax 243
    8.2 Trying it all out 248
    8.3 PROLOG's strategy for proving goals 253
    8.4 Working with Lists 256
        8.4.1 Terminology and Notation 257
        8.4.2 Unifying lists 258
        8.4.3 Useful predicates about lists 259
    8.5 Operators in PROLOG 261
        8.5.1 Operator Precedence and Associativity 262
        8.5.2 Some built-in PROLOG operators 263
        8.5.3 Declaring your own operators 267
    8.6 Backtracking 268
    8.7 Cut 270
    8.8 PROLOG and classical logic 272
    8.9 Clauses as first-order logic formulae 273
    8.10 Theorem proving in PROLOG 276
    8.11 From PROLOG to classical logic 278

Answers to the exercises — 281

Bibliography — 335

Index 342

PREFACE

Logic is now widely recognized to be one of the foundation disciplines for information technology. It has been said that logic may come to play a role in information technology similar to that played by calculus in physics. Like calculus, logic has found applications in virtually all aspects of the subject, from software engineering and hardware design to programming and artificial intelligence. Moreover, it has served to stimulate the search for clear conceptual foundations, a quest which has recently acquired a certain urgency. While progress in the past has been enhanced by theoretical development, it has to a significant degree proceeded independently of theory. In the future this is unlikely to be the case. Many of the current aspirations are seen to require major theoretical advances, and many lie in areas where logic is one of the basic disciplines. If these aspirations are to be fulfilled, logic and information technology must enter a more intimate and enduring relationship. Logic must be adapted and changed to suit its new applications.

The paradigm of logic as an executable language is currently based on the Horn clause fragment of classical logic, and yet there is no existing formulation of classical logic which leads naturally to and supports this paradigm, from the point of view of either teaching or research. Furthermore, existing theorem provers are based on various resolution methods which reformulate and rewrite well-formed formulae in normal forms which are quite alien to the way we perceive the meaning of the classical connectives.

In recent years many extensions of classical logic such as temporal, modal, relevance and fuzzy logics have been used extensively in computer science. Resolution becomes even less intuitive when extended, where possible, to handle these logics. A new formulation of classical logic is needed which can naturally be modified to yield the effect of non-classical logics.

In this book we present classical and non-classical logic and develop a theorem prover for classical logic, in a way which tries to maintain a procedural point of view and has the following main characteristics:

1. The intuitive meaning of the connectives will be retained and no unintuitive translation of well-formed formulae will be used unless it has a natural and intuitive meaning beyond mere logical equivalence.

2. The reasoning rules and the theorem prover will follow as closely as possible the intuitive reasoning procedures employed by humans.

3. The principles involved in the reasoning of the logic and the theorem prover are layered and structured. We show a range of stronger and stronger fragments from intuitionistic logic to full classical logic.

4. Variations on the principles involved in the proof procedures and the theorem prover allow for variations of classical logic in a natural way. The methods can be extended to handle modal and temporal logics.

5. The above set-up and presentation of classical logic has theoretical value for pure logic itself. The trade-off between logical principles and restrictions on the proof procedures and theorem prover yields (owing to points 3 and 4) a better understanding of classical logic and its variations.

This will show that the role of logic in computing is not only that of an essential tool but that of a true partnership.

This book is taken from several logic courses first taught by Dov Gabbay in the Department of Computing at Imperial College, London, since autumn 1984 and then by Odinaldo Rodrigues in the Department of Informatics at King's College London, since 1998.

The courses were given to a mixture of undergraduate and postgraduate students, in classes varying in size from 20 to 350. The format is usually that of 22 hours of lectures over 11 weeks, plus weekly large group tutorials involving the whole class and small group tutorials with roughly 30 students each. The material is covered according to the level of the degree programme in which they are taught. Typical undegraduate coverage includes Chapters 1, 3, 6, elements of Chapter 7 and Chapter 8.

Chapter 5 is not usually taught, even to MSc students. Computer science programmes tend to allow little time for more theoretical topics such as completeness theorems and others of a more mathematical nature. However, students seem to feel more secure to have full mathematical backup to the material they are taught.

We would like to dedicate this book to all of the logic teachers, researchers and scholars, who advance the understanding of logic.

Acknowledgements

We are grateful to many colleagues who participated in one way or another in the realisation of this book. In particular, Jane Spurr helped in all production aspects including the production of the camera-ready version of the book. Valentin Shehtman supplied many of the worked examples. Other examples were extracted from [Lavrov and Maksimova, 1995, Mendelson, 1964, Bizam and Herczeg, 1978, Spivak, 1995]. We would also like to thank the many students along the years who helped shape and refine the way this material has been taught.

Dov Gabbay and Odinaldo Rodrigues
London
Summer 2016

INTRODUCTION

There are many logics in existence, and more seem to be devised each year. They range from the rules of argumentation experimented with by Ancient Greek philosophers to arcane and highly abstract formalisms used by computer scientists to give meaning to the more complex operations that computers carry out. At first, logics from these extremes seem to have little in common, yet we claim that they are all logics. What common properties do they have that allow such a claim?

Essentially, logics are systems for reasoning that some statement is true, given that some statements are true. (We put aside the question of what it means for a statement to be true, for now.) Logics differ in the kinds of statements which can be described, and the truth of statements relative to other statements. In other words, in one logic we might be able to show that John is sometimes at home, given that John is always at home. In a second logic, this may not be the case, and in a third logic we may not be able even to state that John is sometimes at home, let alone decide whether it is true or not! As for most tools, there are various logics, some of which are useful for some problems, and not for others. Therein lies the reason for this book's existence.

This book is devised to introduce classical logic in such a way that we can easily deviate into discussing non-classical logics as well. Rather than be bemused by the variety of logics displayed, we would have you concentrate on their similarities. Seemingly large differences between two logics often arise from a small difference in their definition. Understanding these small differences and their magnified effects is necessary for you to comprehend the concepts common to all logics.

Overview of the book

Chapter 1 introduces the most common logic, *classical propositional logic*, and with it the basic notions which we shall transfer to each of the logics we examine. We immediately start to interfere with the definition of classical propositional logic to reach a number of *non-classical* logics in Chapter 2, before returning to the classical world in Chapter 3 to present a natural deduction method for determining the truth of statements from other statements. While this method is historically important, and useful in certain circumstances, it is difficult to automate. Therefore, in Chapter 4 we tinker with the method to produce a goal-directed automated version. In Chapter 5 we take a step back from reasoning within logic, to consider methodological properties of the logic itself, and of the reasoning methods we use. We show that our reasoning method is both sound and complete. We then move on to predicate logics, which are introduced in Chapter 6, with the concepts being presented informally before a mathematical definition is given. Chapter 7 continues with classical predicate logic, showing how reasoning can be carried out with the additional features that predicate logic offers. Chapter 8 contains a basic introduction to PROLOG. The book contains many exercises, the answers to which are to be found at its back. We urge the reader to attempt all the exercises, as they have been included to support the main body of the text by illustrating and sometimes expanding upon points raised in the preceding material. There are also worked examples which should be attempted without looking at the answers.

The following diagram gives the schematic dependence of the chapters:

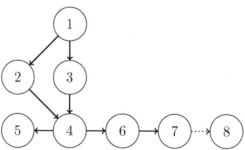

PROPOSITIONAL LOGIC

This book is about logical systems. Our view of such systems is that they consist of three major components:

1. a formal language,

2. an intended semantic interpretation,

3. a family of reasoning rules.

The formal language is used to represent knowledge, i.e. collections of assumptions. The intended interpretation, which could be either purely mathematical or just intuitive, represents our perception of the connection between the formal language and the knowledge. The reasoning rules allow us to draw logical conclusions from the representation of the knowledge. The most common and most widely used logical system is known as *classical first-order predicate calculus*. The power of the predicate calculus means that it (or its extensions) can be used for most application areas.

The general field of logic has many applications. It started in *philosophy* for the analysis of human arguments and of various concepts we use, such as causality, necessity, time, concepts in the philosophy of religion, ethics, etc. Logic is used to analyze the foundations of *mathematics*. Some parts of logic such as set theory, model theory, recursion theory, constructive mathematics and topoi are now considered as mathematics subjects in their own right. In *linguistics* suitable logics and logical grammars are constructed to analyze our use of language; for example, how we use pronouns, quantifiers, tenses and so on. The use of logic in linguistics is important for natural language processing in computer science. The logical analysis of the language must, in many cases, be done first, in order to enable us to apply computer science to language. In addition, logic is used to study cognition in *psychology*, and special quantum logics have been constructed for the study of the foundations of quantum mechanics in *physics*. However, our formulation of logic in this book is motivated by its use in computing. Logic is used directly in computer science, especially in the areas of:

- *specification and verification of programs*
  Logic sentences are used to describe the desired behaviour of a program, and the program itself; manipulating the sentences can show whether the program implements the specification.

- *distributed computing, concurrency and process control*
  Certain forms of logic are devised specifically to reason about physical phenomena, such as time and space, and these are used to reason about interacting processes.

- *database management*
  Querying databases is essentially a matter of testing to see if one sentence is a logical consequence of some other sentences; similarly, updating and consistency maintenance can also be viewed as logical operations.

- *circuit design and VLSI*
  Logic has been applied to the problem of describing and verifying the behaviour of electronic circuits.

- *expert systems, planning and artificial intelligence*
  These areas are substantially based on the application of rules to data, which is the *raison d'être* for logic.

- *natural language processing*
  The semantics of sentences such as 'Alun ate the cake that Bethan had bought that morning' can be represented by logic formulae. These formulae can then be used to reason about the consequences of the sentences.

- *logic programming*
  This is the most obvious connection between computing and logic,

where a special formulation of logic (called Horn clause logic) is used as a basis for the programming language Prolog.

1.1 Introducing classical propositional logic

We begin with the simplest of logics: the *classical propositional logic*. The rest of this chapter will describe the language of classical propositional logic, which essentially can contain all the characteristic elements of many major logical systems. Many different logics, formulations and styles can all be recognized through their propositional parts.[1]

The first assumption of classical logic is that we are dealing with atomic propositions (statements), which can receive a truth value of true ($\top$) or false ($\bot$), in any given situation. By situation we mean a local snapshot, a description of what is happening at a certain place at a certain time from a fixed and agreed point of observation. The classical logic is two valued, i.e. propositions can only have a truth value of true or false. Here we are already making a very serious assumption with far-reaching logical consequences. Statements like

    John loves Mary.

may not in reality have a clear-cut truth value. In situations where it may not be possible for propositions to have a precise truth value, a *non-classical* logic may be required. We present a logic for such situations in Chapter 2.

The second assumption of classical logic is that we are dealing with a single situation, or snapshot. A notable different assumption could be that we are dealing with a sequence of snapshots, a changing situation over time. This gives rise to the so-called *temporal logics*.

The third assumption for classical logic, as for most other logics, is that the truth value of a complex structured sentence depends only on the truth value of its parts. Thus the value of the complex sentence

    John and June got married and John and June had a child.

---

[1]This footnote can be appreciated after reading Chapters 7 and 8: different propositional logics can be presented through their own characterisitic propositional axioms and rules. They can be turned into predicate logics by enriching the language and adding more axioms for the quantifiers. Up to very recently the quantificational component of a logic was more or less a standard addition (with some variation) to its propositional part. (See Remark 8.3.8 below.) It is only recently that general theories of quantifiers are being developed.

depends on whether the two sentences

> John and June got married.
> John and June had a child.

are true or false.

There are many intensional sentences whose truth value does not depend on the truth values of the components. For example,

> Tom believes that John and June had a child.

may be true or false independently of whether John and June did indeed have a child or not.

1.1.1 Formal language

The formal language of propositional logic consists of a set of symbols (typically $\{p, q, r, \ldots\}$) which are the atomic propositions, and the symbols $\wedge$ for conjunction (pronounced 'and'), $\vee$ for disjunction ('or'), $\rightarrow$ for implication ('implies') and $\neg$ for negation ('not'), to build up more complex sentences.

If A and B are the components, we can build up the formulae $(A)\wedge(B)$, $(A)\vee(B)$, $(A)\rightarrow(B)$ and $\neg(A)$ from them. The truth values of the more complex sentences are determined by the truth values of the atomic propositions and the logical symbols in the complex sentences. Take the sentence $\neg(p)$, and suppose that the proposition p is given the value $\bot$ (false). We find the value of $\neg(p)$ by looking up the value given by $\neg$ when applied to $\bot$, in the table

A	$\neg A$
$\top$	$\bot$
$\bot$	$\top$

For an argument which is $\bot$, $\neg$ produces $\top$ (true), so $\neg(p)$ is true when p is false. Now consider the sentence $(p)\wedge(q)$. The table for $\wedge$ is

A	B	$A \wedge B$
$\top$	$\top$	$\top$
$\bot$	$\top$	$\bot$
$\top$	$\bot$	$\bot$
$\bot$	$\bot$	$\bot$

1.1. INTRODUCING CLASSICAL PROPOSITIONAL LOGIC

Suppose that p is given the value $\top$ and q is given the value $\bot$; then we can see that $(p)\wedge(q)$ has the value $\bot$. Similarly if both p and q are given the value $\top$, then $(p)\wedge(q)$ has the value $\top$ as well, which is why $\wedge$ is pronounced 'and'.

The $\vee$ operator has the table

A	B	A∨B
$\top$	$\top$	$\top$
$\bot$	$\top$	$\top$
$\top$	$\bot$	$\top$
$\bot$	$\bot$	$\bot$

and thus $(p)\vee(q)$ has the value $\top$ whenever either p or q is given the value $\top$; hence $\vee$ is called 'or'. Finally, the table for $\to$ is

A	B	A→B
$\top$	$\top$	$\top$
$\top$	$\bot$	$\bot$
$\bot$	$\top$	$\top$
$\bot$	$\bot$	$\top$

so that when p is $\top$ and q is $\bot$, $(q)\to(p)$ is $\top$.

The table for $\to$ is less straightforward than the previous ones, so let us check the rationale behind it. Suppose we see an advertisement for a shop which claims: 'For payment in cash we give a 25% reduction on the price of all television sets.' We decide to check the truth of this advertisement and go into the shop to buy a TV set. There are four possible outcomes:

1. We pay cash and get a 25% reduction.

2. We pay cash, but do not get a 25% reduction. (The shop clerk tells us 'this is our last set, take it or leave it!')

3. We do not pay cash (i.e. we get credit), but nevertheless we get the 25% reduction. (The shop clerk is very nice and says 'Well, if you can't pay now, pay when your next salary cheque comes in'.)

4. We do not pay cash and we do not get the 25% reduction.

We summarize these outcomes in the following table:

	Pay Cash	Get 25%	Advert
1	True	True	
2	True	False	
3	False	True	
4	False	False	

What do we put in the right-hand column of this table? When is the advertisement misleading? We cannot complain about outcomes 1 and 3 since in both cases the 25% reduction was given. Outcome 2 would, if true, clearly make the advertisement false, so the only case which may leave us in doubt about what to put in the table is outcome 4. We cannot say with conviction that the advertisement is misleading, because we did not pay cash. On the other hand, neither can we confirm that the advertisement is true, since we did not test the condition of paying by cash. We need to place some value there, if we are to continue to use truth tables, so we must decide on some convention. Classical logic is content with assuming truth for outcome 4.

	Pay Cash	Get 25%	Advert
1	True	True	True
2	True	False	False
3	False	True	True
4	False	False	???

Later in the book we will introduce other views of $\rightarrow$.

If we have nested complex sentences, such as $\neg((p)\vee(q))$, we work out the value of the innermost sentences first, working outwards, so that we evaluate $(p)\vee(q)$ and use the result of that to evaluate the $\neg$. If p and q are both $\bot$, then $(p)\vee(q)$ is also $\bot$, and thus $\neg((p)\vee(q))$ is $\top$.

Exercise 1.1.1 *Work out the truth values of the following formulae when p has the value $\top$, and q has the value $\bot$:*

1. $\neg((p)\wedge(q))$
2. $(p)\rightarrow(p)$
3. $(p)\rightarrow(\neg(p))$
4. $((q)\vee(\neg(q)))\wedge(p)$

We now formally define the notion of a well-formed formula (wff) of classical propositional logic and the truth values given to the various operators.

1.1. INTRODUCING CLASSICAL PROPOSITIONAL LOGIC

Definition 1.1.2 (Classical propositional logic)

1. *The language of classical propositional logic contains a set $\mathcal{L}_p$ of atomic propositions with typical members $\{p, q, r, s, q_1, q_2, q_3, \ldots\}$ and a set of connectives $\{\wedge, \vee, \neg, \rightarrow\}$.*

2. *The notion of a well-formed formula (wff) is defined inductively by the following conditions:*

   (a) *Any atom $p \in \mathcal{L}_p$ is a wff. We say it is built up from $\{p\}$.*

   (b) *If A and B are wffs then so are: $\neg(A)$, $(A)\wedge(B)$, $(A)\vee(B)$, $(A)\rightarrow(B)$. If A is built up from the atoms $\{p_1, \ldots, p_n\}$ and B is built up from the atoms $\{q_1, \ldots, q_k\}$, then $\neg(A)$ is built up from the same atoms as A and $(A) \wedge (B)$, $(A) \vee (B)$ and $(A) \rightarrow (B)$ are built up from $\{p_1, \ldots, p_n, q_1, \ldots, q_k\}$.*

   (c) *Let A be a formula built up from $q_1, \ldots, q_k$. We indicate this fact by writing $A(q_1, \ldots, q_k)$. Let $B_1, \ldots, B_k$ be wffs. We define by structural induction the result of substituting in A the formulae B_i for the atom q_i, for $i = 1, \ldots, k$. We denote this substitution by $A(q_1/B_1, \ldots, q_k/B_k)$ and refer to it as a substitution instance of A.*

   - *For atomic $A(q) = q$, we let $A(q/B) = B$.*
   - *$(\neg(A))(q_1/B_1, \ldots, q_k/B_k) = \neg(A)(q_1/B_1, \ldots, q_k/B_k)$*
   - *$((A) \rightarrow (B))(q_1/B_1, \ldots, q_k/B_k) = A(q_1/B_1, \ldots, q_k/B_k) \rightarrow B(q_1/B_1, \ldots, q_k/B_k)$ and similarly for $(A) \wedge (B)$ and $(A) \vee (B)$.*

3. *An interpretation (or assignment) is any function h assigning truth values to the atomic propositions. h is a function from $\mathcal{L}_p$ into $\{\top, \bot\}$, i.e. $h : \mathcal{L}_p \mapsto \{\top, \bot\}$.*

4. *We can extend the definition of h from atomic propositions to any wff A by induction, as follows:*

   (a) *$h(\neg(A)) = \top$ if $h(A) = \bot$, otherwise the value is $\bot$.*

   (b) *$h((A)\wedge(B)) = \top$ if $h(A) = \top$ and $h(B) = \top$, otherwise the value is $\bot$.*

   (c) *$h((A)\vee(B)) = \top$ if $h(A) = \top$ or $h(B) = \top$ or both, otherwise the value is $\bot$.*

(d) $h((A){\to}(B)) = \top$ if $h(A) = \bot$ or $h(B) = \top$ or both, otherwise the value is $\bot$.

The above definition of $h(A)$ agrees with our understanding of the meaning of the connectives as presented in the truth tables.

5. We find it convenient to assume that our language contains as atomic propositions the constant atoms $\top$ and $\bot$. $\top$ is always true and $\bot$ is always false.

By defining a precedence hierarchy for each of our connectives, we can omit many of the brackets which clutter our formulae. Let us agree that the precedence of the connectives is as follows: $\neg$ is stronger than $\wedge$ which is stronger than $\vee$ which is stronger than $\to$. This is similar to the situation we have in algebra, where $-$ (unary minus) is stronger than $\times$, which is stronger than $+$. Thus $-2 + 5 \times 3$ is equivalent to $(-2) + (5 \times 3) = 13$. Similarly $\neg q \vee r \wedge s \to q$ is $((\neg q) \vee (r \wedge s)) \to q$. To clarify the structure of written formulae further, we will sometimes use [and] instead of (and), e.g. as in $[(\neg q) \vee (r \wedge s)] \to q$.

In general $A \wedge B \wedge C$ means $(((A) \wedge (B)) \wedge (C))$ and similarly for $A \vee B \vee C$.

1.1.2 Computing truth tables

To calculate the truth values of complex wffs for arbitrary truth values of propositions, we build a truth table for the sentence. This is how to compute a truth table for the wff $A = (\neg p \to q) \vee r$.

Step 1 Count the number of atoms in A. In this case it is 3 (p, q and r).

Step 2 Form a table with $2^3 = 8$ rows and with as many columns as there are subformulae (i.e. nested sentences) of A. In our case the table is shown below, where r is the first atom, q is the second atom and p is the third atom (from the right).

1.1. INTRODUCING CLASSICAL PROPOSITIONAL LOGIC   9

	p	q	r	$\neg p$	$(\neg p \to q)$	$(\neg p \to q) \vee r$
1						
2						
3						
4						
5						
6						
7						
8						

Step 3 Fill in the columns of the atoms with $\top$ and $\bot$ by alternating downwards $\top\bot\top\bot\ldots$ for the first atoms, $\top\top\bot\bot\ldots$ for the second atoms and so on in powers of 2:

	p	q	r	$\neg p$	$(\neg p \to q)$	$(\neg p \to q) \vee r$
1	$\top$	$\top$	$\top$			
2	$\top$	$\top$	$\bot$			
3	$\top$	$\bot$	$\top$			
4	$\top$	$\bot$	$\bot$			
5	$\bot$	$\top$	$\top$			
6	$\bot$	$\top$	$\bot$			
7	$\bot$	$\bot$	$\top$			
8	$\bot$	$\bot$	$\bot$			

Step 4 Compute the values in the other columns row by row using the basic truth tables for the connectives:

	p	q	r	$\neg p$	$(\neg p \to q)$	$(\neg p \to q) \vee r$
1	$\top$	$\top$	$\top$	$\bot$	$\top$	$\top$
2	$\top$	$\top$	$\bot$	$\bot$	$\top$	$\top$
3	$\top$	$\bot$	$\top$	$\bot$	$\top$	$\top$
4	$\top$	$\bot$	$\bot$	$\bot$	$\top$	$\top$
5	$\bot$	$\top$	$\top$	$\top$	$\top$	$\top$
6	$\bot$	$\top$	$\bot$	$\top$	$\top$	$\top$
7	$\bot$	$\bot$	$\top$	$\top$	$\bot$	$\top$
8	$\bot$	$\bot$	$\bot$	$\top$	$\bot$	$\bot$

Exercise 1.1.3 *Compute the truth tables of the following sentences:*

1. $\neg A \vee B$

2. $\neg(\neg A \wedge \neg B)$

3. $A \vee \neg A$

4. $((A \wedge B) \to C) \to (A \to (B \to C))$

5. $(A \to B) \wedge (B \to A)$; this formula is abbreviated $(A \leftrightarrow B)$

6. $(A \to (B \wedge C)) \to ((A \to B) \wedge (A \to C))$

Most wffs have contingent truth tables, meaning that they become true or false depending on the truth values of their atoms. For example, the following can be made either true or false depending on the values of p and q:

$$[(p \to q) \wedge q] \to p$$

If p is intended to mean 'you are a heavy smoker' and q 'you cough', then the above might be read (more or less) as

> [if you are a heavy smoker then you cough, and you cough, then you are a heavy smoker].

Now if you cough, and yet are not a heavy smoker, then the above formula is false. On the other hand, if you are a heavy smoker, the formula is true regardless of whether you cough or not.

We saw that every wff has a truth table, but is the converse true? For a given truth table for, say, p, q and r, is there a corresponding formula $A(p,q,r)$ with the given truth table? The answer is yes. We demonstrate this by means of an example. Can we find a formula $A(p,q,r)$ with the following truth table?

	p	q	r	A
1	T	T	T	T
2	⊥	T	T	T
3	T	⊥	T	T
4	⊥	⊥	T	⊥
5	T	T	⊥	⊥
6	⊥	T	⊥	⊥
7	T	⊥	⊥	⊥
8	⊥	⊥	⊥	⊥

The wff with the above table is

$$A = [p \wedge q \wedge r] \vee [\neg p \wedge q \wedge r] \vee [p \wedge \neg q \wedge r]$$

1.1. INTRODUCING CLASSICAL PROPOSITIONAL LOGIC

How do we find A? We go over the table and take out the rows for which the table for A gives a value of $\top$ to A. In this example there are three such rows:

	p	q	r	A
1	$\top$	$\top$	$\top$	$\top$
2	$\bot$	$\top$	$\top$	$\top$
3	$\top$	$\bot$	$\top$	$\top$

For each such row we describe what the values of p, q and r are as indicated:

- For $\top\ \top\ \top$ write $[p \wedge q \wedge r]$
- For $\bot\ \top\ \top$ write $[\neg p \wedge q \wedge r]$
- For $\top\ \bot\ \top$ write $[p \wedge \neg q \wedge r]$.

A is the disjunction (a combination using $\vee$) of all the wffs for the rows, i.e.
$$A = [p \wedge q \wedge r] \vee [\neg p \wedge q \wedge r] \vee [p \wedge \neg q \wedge r]$$

What we said in symbols means the following in English: A is $\top$ exactly when p, q and r are all true, or when p is false and q and r are true, or when p is true, q is false and r is true. We should show that A has exactly the required table. In other words, that it is impossible for our formula A to be true under an assignment of truth values to p, q and r for which the truth table records a value of $\bot$ for A. No such situation can arise, since for the wff A to be true, one of its disjuncts must be true. Each of its disjuncts uniquely describes one row in the truth table, in particular a row which gives a value of $\top$ to A.

Exercise 1.1.4

1. Find a wff A for each of the following tables:

p	q	$A=?$
$\top$	$\top$	$\top$
$\top$	$\bot$	$\bot$
$\bot$	$\top$	$\bot$
$\bot$	$\bot$	$\top$

p	q	r	$A=?$
$\top$	$\top$	$\top$	$\top$
$\top$	$\top$	$\bot$	$\bot$
$\top$	$\bot$	$\top$	$\bot$
$\top$	$\bot$	$\bot$	$\top$
$\bot$	$\top$	$\top$	$\bot$
$\bot$	$\top$	$\bot$	$\top$
$\bot$	$\bot$	$\top$	$\top$
$\bot$	$\bot$	$\bot$	$\bot$

2. Can the following be all true?

> The television is switched on. If the television is switched on and has power then there is a picture. There is no picture.

(Hint: translate into symbols and check the truth table for a row in which the conjunction of the sentences is true.)

1.2 Notions of truth and validity

We have seen in the exercises that some wffs have truth tables with the value $\top$ throughout in the right-hand column. Such wffs are called *tautologies*. They are always true no matter what truth values their atoms have. We can also define a tautology in terms of the truth assignment h.

Definition 1.2.1 *A wff A is a* tautology *iff for every assignment h, $h(A) = \top$. The symmetric concept is that A is a contradiction iff for every h, $h(A) = \bot$. If A is a tautology we denote it by $\models A$.*[2]

Some commonly encountered tautologies include

$p \vee \neg p$ $p \to p$
$[p \wedge (p \to q)] \to q$ $p \wedge q \to p$
$p \wedge q \to q$ $\neg\neg p \to p$
$p \to \neg\neg p$ $p \to p \vee q$
$q \to p \vee q$ $[(p \to q) \wedge \neg q] \to \neg p$

Tautologies are logical truths. Contradictions, on the other hand, are logical falsities, and can never be true. Here are some example contradictions:

$p \wedge \neg p$
$\neg(p \to p)$
$p \wedge (p \to q) \wedge \neg q$

Notice that if A is a tautology, $\neg A$ is a contradiction. If a formula A is not a contradiction, then there is at least one assignment h to its atoms

---
[2] Tautologies are also called *theorems* of logic, especially in the context of some other ways of generating them. See next chapter.

1.2. NOTIONS OF TRUTH AND VALIDITY

which gives $h(A) = \top$. When this is the case, the formula A is said to be *consistent* or *satisfiable*. Some examples of consistent (but not tautological) sentences are

$$p$$
$$\neg q \vee (q \rightarrow p)$$
$$q \wedge (p \rightarrow p)$$

1.2.1 Equivalences

Two formulae A, B which have the same truth table (i.e. $h(A) = h(B)$ for all h) are said to be *equivalent*. Logically they 'say' the same thing. We indicate equivalence by writing $\equiv$ (pronounced 'if and only if', or 'iff'). Thus $A \equiv B$ means that A and B have the same truth table. For example,

$$\neg\neg p \equiv p$$
$$\neg(p \vee q) \equiv \neg p \wedge \neg q$$
$$\neg(p \wedge q) \equiv \neg p \vee \neg q$$

These first three are called the De Morgan laws. Others include

$$\neg(p \rightarrow q) \equiv p \wedge \neg q$$
$$p \wedge (q \vee r) \equiv (p \wedge q) \vee (p \wedge r)$$
$$p \rightarrow q \equiv \neg q \rightarrow \neg p$$
$$p \vee (q \wedge r) \equiv (p \vee q) \wedge (p \vee r)$$
$$p \wedge p \equiv p$$
$$p \wedge q \equiv q \wedge p$$
$$(p \wedge q) \wedge r \equiv p \wedge (q \wedge r)$$
$$p \vee p \equiv p$$
$$p \vee q \equiv q \vee p$$
$$(p \vee q) \vee r \equiv p \vee (q \vee r)$$
$$p \wedge \neg p \equiv \bot$$
$$p \vee \neg p \equiv \top$$

Using equivalences, we can define some of the truth connectives in terms of each other:

$$p \rightarrow q \equiv \neg p \vee q$$

$$p \wedge q \equiv \neg(\neg p \vee \neg q)$$
$$p \rightarrow q \equiv \neg(p \wedge \neg q)$$
$$p \vee q \equiv \neg(\neg p \wedge \neg q)$$
$$p \wedge q \equiv \neg(p \rightarrow \neg q)$$
$$p \vee q \equiv \neg p \rightarrow q$$

We can also use the equivalences to transform formulae to other equivalent formulae like we do in school algebra. See Exercise 1.2.5 and its solution.

Exercise 1.2.2

1. Let $p|q$ be defined as the wff with p, q having the table below:

p	q	$p\|q$
T	T	⊥
⊥	T	T
T	⊥	T
⊥	⊥	T

Define $\wedge, \vee, \neg$ and $\rightarrow$ using $|$. (Hint: let $\neg p$ be $p|p$.) The '$|$' operator is known as the Sheffer stroke.

2. Which of the following are tautologies?

   (a) $(p \vee q) \wedge \neg p \rightarrow q$
   (b) $(\neg q \rightarrow \neg p) \rightarrow (p \rightarrow q)$
   (c) $(p \wedge q \vee r) \rightarrow (p \wedge q \wedge r)$
   (d) $((p \rightarrow q) \rightarrow p) \rightarrow p$

Definition 1.2.3 (Normal forms) *There are two basic normal forms in which sentences can be written, based on disjunctions (combinations using $\vee$) and conjunctions (combinations using $\wedge$). Such normal forms are sometimes used for automatic methods of checking whether a sentence is a tautology.*

1.2. NOTIONS OF TRUTH AND VALIDITY

1. Disjunctive normal form—*a formula of the form*

$$A_1 \vee A_2 \vee \cdots \vee A_k$$

   where each A_i is a conjunction of either atoms or their negations, is called a wff in a disjunctive normal form. These are formulae we get out of tables in the manner we described in the previous section.

   The formula $\bot$ is considered to be in a disjunctive normal form in which the number of disjuncts is $k = 0$.

2. Conjunctive normal form—*a formula of the form*

$$A_1 \wedge A_2 \wedge \cdots \wedge A_k$$

   where each A_i is a disjunction of atoms or their negations, is called a wff in a conjunctive normal form. This is the dual to disjunctive normal form where $\wedge$ and $\vee$ are interchanged.

   The formula $\top$ is considered to be in a conjunctive normal form in which the number of conjuncts is 0.

3. One can give a more strict definition of normal forms. First we do not allow repetition; thus $p \wedge p$ is not allowed. Second we want each atom p appearing in the formula to appear either as p or $\neg p$ in each basic component of the normal form. Thus for example $p \vee q$ is not acceptable. It should be $(p \wedge q) \vee (p \wedge \neg q) \vee (\neg p \wedge q)$ which is equivalent. Such a normal form we shall call a strict normal form.

We can prove that all wff can be rewritten into either of the normal forms by using the equivalences we have presented above. In fact every formula can be put into an equivalent strict normal form if we first build the full truth table for it and then construct the corresponding formula for this truth table as done in Section 1.1.2.

Theorem 1.2.4 *For any wff A, there are equivalent formulae A^c and A^d where A^c is in conjunctive normal form and A^d is in disjunctive normal form.*

Proof. We prove the theorem by induction over the structure of the formula A.

1. The base case is when A is an atomic proposition, so that $A = A^c = A^d$.

2. When A is $\neg B$, by induction we can rewrite B into the normal form formulae B^c and B^d. By applying the De Morgan laws, negating B^c gives a formula in disjunctive normal form, and negating B^d gives a formula in conjunctive normal form. Hence if $B^c = B_1^c \wedge \cdots \wedge B_n^c$ where the B_i^c are conjunctions of propositions or their negations, then each B_i^c can be negated to become a disjunction $(\neg B_i^c)^d$ so that $A^d = (\neg B_1^c)^d \vee \cdots \vee (\neg B_n^c)^d$. Similarly $A^c = (\neg B_1^d)^c \vee \cdots \vee (\neg B_n^d)^c$.

3. When A is $B \vee C$, by induction we can rewrite both B and C into the normal form formulae B^c, B^d, C^c and C^d. Clearly $B^d \vee C^d$ is in disjunctive normal form. Now $B^c \vee C^c$ can be transformed into conjunctive normal form by repeated use of the equivalence

$$p \vee (q \wedge r) \equiv (p \vee q) \wedge (p \vee r)$$

Assume that $B^c = B_1^c \wedge \cdots \wedge B_m^c$ and $C^c = C_1^c \wedge \cdots \wedge C_n^c$. Now

$$\begin{aligned}&(B_1^c \wedge \cdots \wedge B_m^c) \vee (C_1^c \wedge \cdots \wedge C_n^c) \\ \equiv\ &((B_1^c \wedge \cdots \wedge B_m^c) \vee C_1^c) \wedge \cdots \wedge ((B_1^c \wedge \cdots \wedge B_m^c) \vee C_n^c) \\ \equiv\ &((B_1^c \vee C_1^c) \wedge \cdots \wedge (B_m^c \vee C_1^c)) \wedge \cdots \wedge ((B_1^c \vee C_n^c) \wedge \cdots \wedge (B_m^c \vee C_n^c))\end{aligned}$$

The B_i^c and C_j^c are all disjunctions of propositions or their negations; hence the entire formula is in conjunctive normal form.

4. A similar argument holds for when A is $B \wedge C$.

5. When A is $B \to C$, we can rewrite A to be $\neg B \vee C$, and then the proof continues as for disjunction above.

∎

Exercise 1.2.5 *Transform the following to both disjunctive normal form and conjunctive normal form:*

$$[(p \to q) \vee r] \wedge \neg q \to p$$
$$(p \vee q) \wedge r$$
$$p \wedge q \wedge \neg r$$

1.2.2 Arguments and consequence relations

An *argument* is a claim that one formula B (called the *goal* or the *conclusion*) follows logically from a set of formulae $A_1, \ldots, A_n$ (called

1.2. NOTIONS OF TRUTH AND VALIDITY

the *data* or the *assumptions*). The set of assumptions may be empty (i.e. no assumption). In such a case $A_1, \ldots, A_n$ do not appear. We can formally use the trick of saying this is the case of $n = 0$. Representing the assumptions by $A_1, \ldots, A_n$ and the goal by B, we represent the argument by

$$\frac{A_1, \ldots, A_n}{B}$$

or alternatively $A_1, \ldots, A_n \vdash B$, which has the virtue of being a more compact notation. Here is an example of an argument:

> Either John is at home or John is in his office.
> John is not in the office.
> ―――――――――――――
> Therefore John is at home.

Writing this formally with propositional symbols h for John being at home, and and o for John being at the office, we have

$$\frac{h \vee o}{\neg o}$$
$$\frac{}{h}$$

or $h \vee o, \neg o \vdash h$. Perhaps the most well-known argument is

$$\frac{A, A \to B}{B}$$

or $A, A \to B \vdash B$. This is known as *modus ponens* and will be used extensively in future chapters.

Definition 1.2.6 (Validity of arguments) *An argument with assumptions $A_1, \ldots, A_n$ and conclusion B is logically valid when the formula $[(A_1) \wedge \ldots \wedge (A_n)] \to B$ is a tautology, otherwise the argument is invalid. Hence an argument is valid if, whenever the assumptions of the argument are true, the conclusion is true.*

Given the following information,

Assumptions	Conclusion
$(p \to q)$	$\neg q$
$\neg p$	

does the conclusion follow? To answer, we check if $[(p{\to}q){\wedge}\neg p]{\to}\neg q$ is a tautology. We can write $A_1,\ldots,A_n \vdash ? B$ when we are not sure whether the argument is valid. Otherwise the question mark is omitted. We leave the reader to check whether the argument is valid or not.

A valid argument from a set of assumptions to a conclusion can be considered to form a relationship between the assumptions and the conclusion, namely that the conclusion is true as a consequence of the assumptions being true. Thus the $\vdash$ symbol is often referred to as the *consequence relation*. Consequence relations satisfy the following properties:

- *Reflexivity*
  If the conclusion is also one of the assumptions then the argument is valid, i.e. $A_1,\ldots,A_n, B \vdash B$.

- *Monotonicity*
  If a conclusion from a set of assumptions is valid, then the same conclusion is valid from the set of assumptions with some additional formulae added, i.e. if $A_1,\ldots,A_n \vdash B$ then $A_1,\ldots,A_n, C \vdash B$ for any C.

- *Cut*[3]
  If a conclusion from a set of assumptions is added to that set of assumptions, and this larger set is used to make a second conclusion, then the second conclusion can be made from the original assumptions, i.e. if $A_1,\ldots,A_n \vdash B$ and $A_1,\ldots,A_n, B \vdash C$ then $A_1,\ldots,A_n \vdash C$.

In fact one can define the notion of a logical system on the set of wffs to be any relation satisfying the above three properties. Different logical systems satisfy different *additional properties*. For example, the above $\vdash$ for classical logic (defined using truth tables) also satisfies

$$A_1,\ldots,A_n, A \vdash B \vee C \quad \text{iff} \quad A_1,\ldots,A_n \vdash (A{\to}B) \vee C$$

Note that the presence of C is important. In some other logics (such as intuitionistic logic, which we will discuss in Chapter 2), C is not allowed. Further discussion is postponed to Chapter 5.

---

[3] The notion of cut is central in logic. A full discussion is given in Chapter 3. Monotonicity or non-monotonicity is also very central. This will be discussed in Chapter 10.

1.2. NOTIONS OF TRUTH AND VALIDITY

Here is a list of valid arguments. They are so basic that they are called *rules*.

List of valid rules

1. $\dfrac{A \wedge B}{A}$ and $\dfrac{A \wedge B}{B}$ ($\wedge$ *elimination*)

2. $\dfrac{A, B}{A \wedge B}$ ($\wedge$ *introduction*)

3. $\dfrac{A}{A \vee B}$ and $\dfrac{B}{A \vee B}$ ($\vee$ *introduction*)

4. $\dfrac{A \to C,\ B \to C,\ A \vee B}{C}$ ($\vee$ *elimination*)

5. $\dfrac{A,\ A \to B}{B}$ This *modus ponens* rule has a related form:

$\dfrac{\neg A,\ A \vee B}{B}$ ($\to$ *elimination*)

6. $\dfrac{B,\ \neg B}{C}$ ($\neg$ *elimination*)

7. $\dfrac{B \to C,\ B \to \neg C}{\neg B}$ ($\neg$ *introduction*)

8. If we want to show that $\dfrac{B_1, \ldots, B_k}{A \to C}$ is a valid argument we can show instead that $\dfrac{B_1, \ldots, B_k,\ A}{C}$ is valid.

Exercise 1.2.7

1. Show that A logically implies B, i.e. $A \vdash B$, is a valid argument if and only if $A \wedge \neg B$ is not consistent.

2. Is $A \to B, B \to C \vdash A \to C$ valid?

Definition 1.2.8 (Consistency of a set of formulae) *A set of wffs*

$$\{A_1, \ldots, A_n, C\}$$

is consistent if there is a row in the truth table of their conjunction which makes the conjunction true. In other words, if values can be given to all the atomic sentences in $\{A_1, \ldots, A_n, C\}$ which make all of $A_1, \ldots, A_n, C$ true.

For example, consider the set $\{p \to q, \neg p, \neg q\}$, which is consistent if we can find values for p and q which will make all members of the set true. Assigning falsity to both p and q achieves this. Now consider the set $\{r \to s, r, \neg s\}$. In this case we cannot find any such values for r and s (try drawing a truth table for the formula $r \to s \wedge r \wedge \neg s$).

Exercise 1.2.9

1. There are n people standing in a queue. The first person says

    'The last person in the line is lying.'

    The other people in the line say

    'The person in front of me is lying.'

    We can write one sentence of the form 'person 1 is telling the truth iff person n is not telling the truth' for the first person in the queue, and $n-1$ sentences of the form 'person i is telling the truth iff person $i - 1$ is not telling the truth', for the other people in the queue. We use p_i to represent the sentence 'person i is telling the truth'

$$\begin{array}{rcl} A_1 & is & p_1 \leftrightarrow \neg p_n \\ A_2 & is & p_2 \leftrightarrow \neg p_1 \\ & \vdots & \\ A_n & is & p_n \leftrightarrow \neg p_{n-1} \end{array}$$

    Is the set $\{A_1, \ldots, A_n\}$ consistent, i.e. can you give values to $p_1, \ldots, p_n$ so that all of $A_1, \ldots, A_n$ are satisfied? Does it matter whether n is odd or even?

We have learnt the notion of contradiction. A is a contradiction if $\neg A$ is a tautology, i.e. if the truth table for A is all the $\bot$ in the right-hand column. A is not consistent if and only if A is a contradiction. This gives us the connection of consistency with arguments. If $\{A_1, \ldots, A_n, C\}$

1.2. NOTIONS OF TRUTH AND VALIDITY

is not consistent, and if $\{A_1, \ldots, A_n\}$ is true (in some interpretation) then $\neg C$ is true (in that interpretation). Thus $A_1 \wedge \cdots \wedge A_n \to \neg C$ is a tautology. The other direction is also true. If $A_1 \wedge \cdots \wedge A_n \to \neg C$ is a tautology, then any row in the truth table which makes $A_1, \ldots, A_n$ true must also make $\neg C$ true, i.e. make C false. Thus we cannot make all of $\{A_1, \ldots, A_n, C\}$ all true.

The practical conclusion from the above is that checking whether $A_1, \ldots, A_n \vdash B$ is a logically valid argument is the same as checking whether $\{A_1, \ldots, A_n, \neg B\}$ is inconsistent. Another view of the argument

$$\frac{A_1, \ldots, A_n}{B}$$

is to consider $A_1, \ldots, A_n$ as *data* and to consider B as a *query* to be asked of the data. B follows logically from $A_1, \ldots, A_n$ if the answer to the query B should be yes.

Thus we have the following three ways to look at the formula (i.e. 1, 2 and 3 below are equivalent).

$$A_1 \wedge \cdots \wedge A_n \to B$$

1. as a tautology,

2. as a valid logical argument

$$\frac{A_1, \ldots, A_n}{B}$$

   with assumptions $A_1, \ldots, A_n$ and conclusion B,

3. as a statement that if we query a database with goal B and the database contains $A_1, \ldots, A_n$ among its data, then the answer is yes.

$$\frac{\text{Data}}{A_1, \ldots, A_n} \ ? \ \frac{\text{Query}}{B}$$

In Chapters 3 and 4 we shall look at some rule-based methods for answering the question 'does the query logically follow from the data?' In the next chapter, we broaden our horizons by beginning to look at so-called *non-classical* logics where components of the classical logical system have been changed, offering in some cases more expressive logics, and in others more restrictive logics.

Exercise 1.2.10

1. *Let p stand for 'John loves Mary', q stand for 'Mary loves John', r stand for 'John is tall', and s stand for 'Mary is tall'. Translate the following into symbols (you may need to introduce other atoms):*

   (a) *John loves Mary only if she loves him and she is not tall.*

   (b) *If John is tall and Mary loves him then Mary is not tall.*

   (c) *If John and Mary love each other then Mary loves herself.*

2. *Use the equivalence rules (see Section 1.2.1) to push all occurrences of the negation symbol '¬' next to the atoms in the following expressions:*

$$\neg((a\rightarrow b)\vee(a\rightarrow c)\wedge\neg a)$$
$$\neg(a\wedge\neg b)\rightarrow a$$

3. *Show the equivalences below:*

$$(A\rightarrow\top) \equiv \top \qquad (A\rightarrow\bot) \equiv \neg A$$
$$(\bot\rightarrow A) \equiv \top \qquad (\top\rightarrow A) \equiv A$$
$$(A\vee\top) \equiv \top \qquad (A\vee\bot) \equiv A$$
$$(A\wedge\top) \equiv A \qquad (A\wedge\bot) \equiv \bot$$

4. *Below are several arguments in English. For each argument:*

   - *define a dictionary using the atoms p, q, r, s,..., i.e. assign one of these symbols to stand for each relevant proposition in the argument,*
   - *translate the argument into logic using the dictionary,*
   - *check whether the argument is logically valid.*

   (a) *Either it is warm or it is raining. Unless it is not warm, we go outside. Therefore we go outside.*

   (b) *If the king is in the room, then the courtiers laugh only if he laughs. The courtiers always laugh when the jester is in the room. The king never laughs when the jester is in the room. Therefore either the king or the jester is not in the room.*

   (c) *If we are not hungry, and the food is very hot, we eat slowly. If we are not hungry, either we eat slowly or the food is very hot. The food is not very hot. Therefore we are hungry.*

(d) If Jones did not meet Smith last night, then either Smith was a murderer, or Jones is telling a lie. If Smith was not a murderer, then Jones did not meet Smith last night, and the murder happened after midnight. If the murder happened after midnight, then either Smith was a murderer, or Jones is telling a lie, but not both. Therefore, Smith was a murderer.

1.3 Worked examples

Example 1.3.1 (Worked examples for Section 1.1.1)
Exercises

1. Following the definition, show that in every formula the numbers of left and right brackets are equal.

2. Explain why the following expression is not a formula:
$$\neg((p) \wedge (q)) \vee (p)$$

3. According to the definitions, show that if A is a formula, then $(A)(q/B)$ is a formula.

4. Let $l(A)$ be the length of a formula A (i.e. the total number of characters in the expression (A)). Prove that $l((A)(q/B)) \geq l(A)$.

5. Show that for any formula A, 3 divides $(l(A) - 1)$.

6. Show that if A, B are formulae and q is an atom occurring n times in A, then $l((A)(q/B)) = l(A) + n \cdot (l(B) - 1)$.

7. Find an example showing that formulae $(A)(p, q/B, C)$ and $((A)(p/B))(q/C)$ may be distinct.

Solutions

1. For atomic propositions there is nothing to prove. If the claim is true for A, then it is true for $\neg(A)$, where the numbers of left and right brackets increase by one. If it is true for A and B, then it is true for $(A) \wedge (B), (A) \vee (B), (A) \rightarrow (B)$, because the total numbers of left and right brackets increase by two.

2. Suppose it is. This expression does not begin with '(' and is not an atomic proposition. So it is $\neg(B)$ for some formula B. But the expression $B = (p) \wedge (q)) \vee (p$ is not a formula, because every non-atomic formula ends with ')'.

3. By induction. If A is atomic then $(A)(q/B)$ is either A or B. If $A = \neg(C)$ then $(A)(q/B) = \neg((C)(q/B))$. $(C)(q/B)$ is a formula by the induction hypothesis. Hence $(A)(q/B)$ is a formula by 1.1.2(b). The cases when A is an implication, disjunction or conjunction are similar.

4. By induction. If A is atomic then $l(A) = 1$, and the claim is trivial.
If $A = (C) \vee (D)$ then $l(A) = l(C) + l(D) + 5$.
$(A)(q/B) = ((C)(q/B)) \vee ((D)(q/B))$, and so $l((A)(q/B)) = l((C)(q/B)) + l((D)(q/B)) + 5$. By the induction hypothesis, $l((C)(q/B)) \geq l(C)$, and $l((D)(q/B)) \geq l(D)$. Therefore $l((A)(q/B)) \geq l(A)$.
Other cases are similar.

5. By induction. The case when A is atomic is clear.
If $A = \neg(B)$, and $l(B) = 3n+1$, then $l(A) = l(B)+3 = 3(n+1)+1$. If $A = (B) \triangledown (C)$, $\triangledown$ being $\rightarrow, \vee$ or $\wedge$, and $l(B) = 3m+1, l(C) = 3n+1$ then $l(A) = l(B) + l(C) + 5 = 3m + 3n + 7 = 3(m+n+2) + 1$.

6. Again by induction. We use the notation $A' = (A)(q/B)$.
If A is atomic, $A \neq q$, then $A' = A$, $n = 0$, and thus $l(A') = l(A) + n(l(B) - 1)$.
If $A = q$, then $A' = B, n = 1$, and so $l(A') = l(B) = 1 + (l(B) - 1) = l(A) + n(l(B) - 1)$.
If $A = \neg(C)$ then $A' = \neg(C')$ by 1.1.2(c), and $l(A) = l(C)+3, l(A') = l(C') + 3$. If q occurs n times in A, then it occurs n times in C, and so by the induction hypothesis, $l(C') = l(C) + n(l(B) - 1)$. Hence $l(A') = l(C') + 3 = l(C) + n(l(B) - 1) + 3 = l(A) + n(l(B) - 1)$.
If $A = (C) \triangledown (D)$ then $A' = (C') \triangledown (D')$, and $l(A) = l(C) + l(D) + 5, l(A') = l(C') + l(D') + 5$. If q occurs n times in A, x times in C, y times in D, then $n = x + y$. By the induction hypothesis, $l(C') = l(C) + x(l(B) - 1), l(D') = l(D) + y(l(B) - 1)$. Thus
$$\begin{aligned} l(A') &= l(C') + l(D') + 5 \\ &= l(C) + x(l(B) - 1) + l(D) + y(l(B) - 1) + 5 \\ &= l(A) + n(l(B) - 1) \end{aligned}$$

7. This may happen when q occurs in B. For instance, take $A = (p) \lor (q)$, $B = q, C = p$. Then $(A)(p, q/B, C) = (q) \lor (p)$, but $((A)(p/B))(q/C) = ((q) \lor (q))(q/C) = (p) \lor (p)$.

Example 1.3.2 (Worked examples for Section 1.1.2)
Exercises

1. Find all values of the atoms p, q, r, s, t, u for which the following formula is false:

   (a) $((p \to q \land r) \to (\neg q \to \neg p)) \to \neg p$
   (b) $(p \land q) \lor (p \land r) \lor (q \land r) \lor (s \land t) \lor (s \land u) \lor (t \land u) \lor (\neg p \land \neg s)$
   (c) $p \lor q \lor r \to (p \lor q) \land (p \lor r)$
   (d) $(p \lor q) \land (q \lor r) \land (r \lor p) \to p \land q \land r$
   (e) $p \lor q \to (\neg p \land q) \lor (p \land \neg q)$

2. Consider the following propositions about a natural number n:

$$n < 100$$
$$n > 35$$
$$n > 9$$
$$n > 10$$
$$n > 5$$

   Find n, for which two of these propositions are false, and three others are true. Is such n unique?

3. m, n are natural numbers. Of the following four propositions

   (a) n divides $(m + 1)$
   (b) $m = 2n + 5$
   (c) 3 divides $(m + n)$
   (d) $(m + 7n)$ is prime

   three are true, and one is false. Find all possible pairs a, b.

4. Three runners A, B, C had a race and finished almost at the same time. Three sports commentators gave the following immediate reports:

   (a) A has won.

(b) *B has not won.*

(c) *C was not the last.*

It turned out that exactly one of the commentators was right. What was the result of the race?

5. *After racing, four jockeys (A, B, C, D) made the following statements about their results:*

   (a) *'I was neither the first, nor the last.'*

   (b) *'I was not the last.'*

   (c) *'I was the first.'*

   (d) *'I was the last.'*

   Three of these statements are true, and one is false. Find, who was telling a lie and who came first.

6. *A professor of logic meets 10 of his former students, Albert, Alice, Bob, Bertha, Clifford, Connie, David, Dora, Edgar and Edith, who have become five married couples. When asked about their husbands, the ladies gave the following answers:*
   Alice: My husband is Clifford, and Bob has married Dora.
   Bertha: My husband is Albert, and Bob has married Connie.
   Connie: Clifford is my husband, Bertha's husband is Edgar.
   Dora: My husband is Bob, and David has married Edith.
   Edith: Yes, David is my husband. And Albert's wife is Alice.
   Additional true information coming from the men was that every lady gave one correct and one wrong answer. This was sufficient to find out the truth. Reproduce the professor's argument.

7. *Each of four dwarfs, Ben, Ken, Len and Vin, is either always telling the truth or always lying. Here is their chat:*
   Ben (to Ken). You are a liar.
   Len (to Ben). You yourself are a liar!
   Vin (to Len). They are both liars. And you too.
   Who is telling the truth?

Solutions

1. (a) The implication $((p \to q \land r) \to (\neg q \to p)) \to \neg p$ is false iff $(p \to q \land r) \to (\neg q \to \neg p)$ is $\top$, and $\neg p$ is $\bot$. Then p is $\top$ and it remains to consider only q, r:

1.3. WORKED EXAMPLES

q	r	$q \wedge r$	$p \to q \wedge r$	$\neg q \to \neg p$	$(p \to q \wedge r) \to (\neg q \to \neg p)$
T	T	T	T	T	T
T	⊥	⊥	⊥	T	T
⊥	T	⊥	⊥	⊥	T
⊥	⊥	⊥	⊥	⊥	T

So the answer is: p is T; q, r are arbitrary.

(b) *All possible combinations of truth values are listed in the table:*

p	q	r	s	t	u
T	⊥	⊥	T	⊥	⊥
T	⊥	⊥	⊥	T	⊥
T	⊥	⊥	⊥	⊥	T
T	⊥	⊥	⊥	⊥	⊥
⊥	T	⊥	T	⊥	⊥
⊥	⊥	T	T	⊥	⊥
⊥	⊥	⊥	T	⊥	⊥

Comment. The formulae $p \wedge q, p \wedge r, q \wedge r, s \wedge t, s \wedge u, t \wedge u, \neg p \wedge \neg s$ must be false, and thus $(p \vee s)$ must be true. So we notice that at least one of the following combinations appears:

- *p is T, and q, r are ⊥*
- *s is T, and t, u are ⊥.*

Having three values fixed, we compute the truth table for the other three.

(c)

p	q	r
⊥	⊥	T
⊥	T	⊥

Comment. $p \vee q \vee r$ is T and ($p \vee q$ is ⊥ or $p \vee r$ is ⊥).

(d)

p	q	r
⊥	T	T
T	⊥	T
T	T	⊥

Comment. $p \vee q, q \vee r, p \vee r$ must be true, and $p \wedge q \wedge r$ false (i.e. at least one of the atoms is $\bot$). If p is $\bot$ then q, r are $\top$. Two other cases are analogous.

(e) p, q must both be true.

2. Denote these propositions by $p_1, \ldots, p_5$. Then the following formulae are true:
$$\neg p_1 \to p_2, p_2 \to p_4, p_4 \to p_3, p_3 \to p_5$$
We notice that if p_2 is $\top$ then the three other propositions (p_3, p_4, p_5) are $\top$. Hence p_2 is $\bot$.

Since $\neg p_1 \to p_2$ is $\top$ we get that p_1 is $\top$. But then p_4 is false (otherwise, four propositions are true).

Therefore p_1, p_3, p_5 are true; p_2, p_4 are false.

That is, we have: $n < 100, n \leq 35, n > 9, n \leq 10, n > 5$. Thus $n = 10$ is the unique possibility.

3. The argument is similar to the previous exercise. Let $p_1, \ldots, p_4$ be the propositions in question. Then one can check that the following formulae are true:
$$p_3 \to \neg p_2, p_3 \to \neg p_4$$
Since exactly one of the atoms is false, it follows easily that p_1, p_2, p_4 are $\top$ and p_3 is $\bot$. Then the following is true:

(a) n divides $(m+1)$.

(b) $m = 2n + 5$.

(c) $(m + 7n)$ is prime.

By (a), (b), n divides $2n + 6$, and thus n divides 6, i.e. $n = 1, 2, 3$ or 6. Taking (c) into account, we get the two possibilities:

(i) $n = 2, m = 9$.

(ii) $n = 6, m = 17$.

4. Consider the statements a, b, c as atoms. Clearly, $a \to b$ is true. So there are the following two possibilities:

(i) a, b are $\top$, c is $\bot$.

Then the positions of A, B, C are *1,2,3*.

1.3. WORKED EXAMPLES

(ii) c, b are $\top$, a is $\bot$.

Then A has not won, B has not won either, and C was the last. This is impossible. Therefore only the first possibility remains.

5. Consider the statements of A, B, C, D respectively as atoms a, b, c, d. The following two propositions are obviously true:

$$d \to b, c \land d \to a$$

Since exactly one of a, b, c must be false, there are two possibilities for their truth values:

(i) a, b, d are $\top$, c is $\bot$.

(ii) a, b, c are $\top$, d is $\bot$.

In the case (i) only B might be the first. (Note that D was the last, and the positions of A, C cannot be found exactly.) In the case (ii) we obtain a contradiction, because nobody can be the last. Therefore, B has won, and C was lying.

6. There are 25 atomic propositions of the form 'a man whose name begins with X and a woman whose name begins with Y are married', which we denote by $p_{AA}, p_{AB}, \ldots, p_{EE}$. The information at the professor's disposal is expressed by a rather long conjunction including clauses $(p_{XY} \to \neg p_{XZ})$ and $(p_{YX} \to \neg p_{ZX})$ for any triple X, Y, Z where $Y \neq Z$, and also the following ones:

(a) $p_{CA} \leftrightarrow \neg p_{BD}$

(b) $p_{AB} \leftrightarrow \neg p_{BC}$

(c) $p_{CC} \leftrightarrow \neg p_{EB}$

(d) $p_{BD} \leftrightarrow \neg p_{DE}$

(e) $p_{DE} \leftrightarrow \neg p_{AA}$

It will be convenient to show truth values of the atoms in a 5×5 table. The first part of our conjunction means that $\top$ occurs exactly once at each row and each column of the table.

Now assume that p_{CA} is $\top$ and start filling in the table:

	A	B	C	D	E
A	⊥				
B	⊥				
C	⊤	⊥	⊥	⊥	⊥
D	⊥				
E	⊥				

Since (a) and (c) are true, we have that p_{BD} is $\bot$, p_{DE} is $\top$, and

	A	B	C	D	E
A	⊥				⊥
B	⊥			⊥	⊥
C	⊤	⊥	⊥	⊥	⊥
D	⊥	⊥	⊥	⊥	⊤
E	⊥				⊥

Using (c) we get

	A	B	C	D	E
A	⊥	⊥			⊥
B	⊥	⊥		⊥	⊥
C	⊤	⊥	⊥	⊥	⊥
D	⊥	⊥	⊥	⊥	⊤
E	⊥	⊤	⊥	⊥	⊥

and eventually

	A	B	C	D	E
A	⊥	⊥	⊥	⊤	⊥
B	⊥	⊥	⊤	⊥	⊥
C	⊤	⊥	⊥	⊥	⊥
D	⊥	⊥	⊥	⊥	⊤
E	⊥	⊤	⊥	⊥	⊥

We have to check also another assumption: p_{CA} is $\bot$. Then if we fill

1.3. WORKED EXAMPLES 31

in the table according to (a), (d), (e), we have

	A	B	C	D	E
A	$\top$	$\bot$	$\bot$	$\bot$	$\bot$
B	$\bot$	$\bot$	$\bot$	$\top$	$\bot$
C	$\bot$			$\bot$	
D	$\bot$			$\bot$	$\bot$
E	$\bot$			$\bot$	

and we observe that (b) becomes false.

Therefore the solution is unique.

7. Let B, K, L, V be the atoms stating that corresponding dwarfs are truthful. The following formulae are known to be true:

   (a) $B \leftrightarrow \neg K$

   (b) $L \leftrightarrow \neg B$

   (c) $V \rightarrow \neg B \wedge \neg K \wedge \neg L$

   (d) $\neg V \rightarrow \neg(\neg B \wedge \neg K) \wedge L$

Now assume that V is true. Then B, K must be $\bot$ by (c), and so (a) becomes false.

Hence V is $\bot$. By (d) L is $\top$, and by (b), (a) we get that B is $\bot$, K is $\top$. (a)-(d) are true for these values of atoms. Therefore only Len is telling the truth.

Example 1.3.3 (Worked examples for Section 1.2.1)
Exercises

1. Show that $A \equiv B$ iff $(A \leftrightarrow B)$ is a tautology.

2. Show that if $A_1 \equiv B_1$ and $A_2 \equiv B_2$ then $(A_1 \rightarrow A_2) \equiv (B_1 \rightarrow B_2)$.

3. Prove that if $A \equiv B$ then $(C)(p/A) \equiv (C)(p/B)$.

4. Show that every substitution instance of a tautology is also a tautology.

5. Prove that if $A \equiv B$ then $(A)(p/C) \equiv (B)(p/C)$.

6. Check the following equivalences:

(a) $A \vee (A \wedge B) \equiv A$

(b) $A \wedge (A \to B) \equiv A \wedge B$

(c) $A \vee (B \wedge \neg A) \equiv A \vee B$

(d) $A \to (B \to C) \equiv A \wedge B \to C$

(e) $A \to (B \to C) \equiv B \to (A \to C)$

(f) $(A \to B) \to A \equiv A$

(g) $(A \to B) \to B \equiv A \vee B$

(h) $(A \leftrightarrow B) \leftrightarrow C \equiv A \leftrightarrow (B \leftrightarrow C)$.

7. Find disjunctive normal forms for the formulae

   (a) $p \leftrightarrow q$

   (b) $\neg(p \leftrightarrow q)$

   (c) $(p \leftrightarrow q) \leftrightarrow r$

8. Explain how to find $(\neg A)^d$ provided A^d is given as a strict normal form.

9. Find a formula A built of three atomic propositions, such that the strict normal forms A^d and $(\neg A)^d$ have an equal number of disjuncts.

10. Let X be the set of all formulae built of three atomic propositions. Find a consistent formula $A \in X$ with the following property:

    for any $B \in X$, if $B \to A$ is a tautology, then either $A \equiv B$ or B is a contradiction.

11. Find a formula A, such that the formula

    $$(A \wedge q \to \neg p) \to ((p \to \neg q) \to A)$$

    is a tautology.

Solutions

1. *Exercise 1.1.3.5 shows that for any assignment h, $h(A \leftrightarrow B) = \top$ iff $h(A) = h(B)$. Hence the claim follows.*

2. *Immediately by definitions: given that $h(A_1) = h(B_1)$ and $h(A_2) = h(B_2)$ we obtain that $h(A_1 \to A_2) = h(B_1 \to B_2)$.*

1.3. WORKED EXAMPLES

3. By induction. For example, if $C = D \wedge E$ then

$$(C)(p/A) = (D)(p/A) \wedge (E)(p/A),$$
$$(C)(p/B) = (D)(p/B) \wedge (E)(p/B).$$

If the claim holds for D, E we obtain that $(C)(p/A) \equiv (C)(p/B)$ as in the previous exercise.

4. Let A be a tautology. To show that $A' = (A)(q_1/B_1, \ldots, q_k/B_k)$ is a tautology take any assignment h and prove that $h(A') = \top$. Let h_1 be an assignment such that $h_1(q_1) = h(B_1), \ldots, h_1(q_k) = h(B_k), h_1(r) = h(r)$ for any other atom r. It suffices to prove that $h(C') = h_1(C)$ for any formula C (because then we have: $h(A') = h_1(A) = \top$). This is easily done by induction on C. For example, if $C = D \wedge E$ and $h(D') = h_1(D), h(E') = h_1(E)$ then $C' = D' \wedge E'$, and so $h(C') = \top$ iff $h(D') = h(E') = \top$ iff $h_1(D) = h_1(E) = \top$ iff $h_1(C) = \top$.

5. If $A \equiv B$ then $(A \leftrightarrow B)$ is a tautology (Exercise 1 above). Then $(A \leftrightarrow B)(p/C)$ is also a tautology (Exercise 4 above). But this is the same as $(A)(p/C) \leftrightarrow (B)(p/C)$. Now apply Exercise 1 above again.

6. Sometimes it is easier to compute truth tables, and sometimes straightforward arguments work better.

    (a) For any assignment h, $h(A \vee (A \wedge B)) = \top$
    iff $h(A) = \top$ or $h(A \wedge B) = \top$
    iff $h(A) = \top$ or $h(A) = h(B) = \top$
    iff $h(A) = \top$. Hence $h(A \vee (A \wedge B)) = h(A)$.

    Another method: compute the truth table of $A \vee (A \wedge B)$:

A	B	$A \wedge B$	$A \vee (A \wedge B)$
$\top$	$\top$	$\top$	$\top$
$\top$	$\bot$	$\bot$	$\top$
$\bot$	$\top$	$\bot$	$\bot$
$\bot$	$\bot$	$\bot$	$\bot$

    (b) $h(A \wedge (A \to B)) = \top$ iff $h(A) = \top$ and $h(A \to B) = \top$
    iff $h(A) = \top$ and $(h(A) = \bot$ or $h(B) = \top)$
    iff $h(A) = h(B) = \top$
    iff $h(A \wedge B) = \top$.

(c) $h(A \vee (B \wedge \neg A)) = \top$ iff $h(A) = \top$ or $h(B \wedge \neg A) = \top$
iff $h(A) = \top$ or $h(B) = h(\neg A) = \top$
iff $h(A) = \top$ or $(h(B) = \top$ and $h(A) = \bot)$
iff $h(A) = \top$ or $h(B) = \top$.

(d) The easiest way is to use equivalences mentioned in the text:

$$\begin{aligned} A \to (B \to C) &\equiv \neg A \vee (B \to C) \\ &\equiv \neg A \vee (\neg B \vee C) \\ &\equiv (\neg A \vee \neg B) \vee C \\ &\equiv \neg(A \wedge B) \vee C \\ &\equiv A \wedge B \to C \end{aligned}$$

(e) Use the previous equivalence:

$$A \to (B \to C) \equiv A \wedge B \to C$$
$$B \to (A \to C) \equiv B \wedge A \to C$$

(f) Compute the truth table:

A	B	$A \to B$	$(A \to B) \to A$
$\top$	$\top$	$\top$	$\top$
$\top$	$\bot$	$\bot$	$\top$
$\bot$	$\top$	$\top$	$\bot$
$\bot$	$\bot$	$\top$	$\bot$

(g) Compute the truth table:

A	B	$A \to B$	$(A \to B) \to B$	$A \vee B$
$\top$	$\top$	$\top$	$\top$	$\top$
$\top$	$\bot$	$\bot$	$\top$	$\top$
$\bot$	$\top$	$\top$	$\top$	$\top$
$\bot$	$\bot$	$\top$	$\bot$	$\bot$

(h) One way is to compute truth tables. Another way is to observe that $(p \leftrightarrow q)$ has the same table as addition in the two-element Abelian group in which $\top$ is the null element. Group addition is always associative.

7. (a) $(p \wedge q) \vee (\neg p \wedge \neg q)$
   (b) $(p \wedge \neg q) \vee (\neg p \wedge q)$

1.3. WORKED EXAMPLES   35

  (c) $(p \wedge q \wedge r) \vee (\neg p \wedge \neg q \wedge r) \vee (p \wedge \neg q \wedge \neg r) \vee (\neg p \wedge q \wedge \neg r)$

8. $(\neg A)^d$ consists exactly of those disjuncts built of atoms and their negations which are missing in A^d. This follows immediately if we remember correspondence between truth tables and normal forms.

9. Use the previous exercise. A^d should have four disjuncts. For example,

$$(p \wedge q \wedge r) \vee (p \wedge \neg q \wedge r) \vee (p \wedge \neg q \wedge \neg r) \vee (\neg p \wedge q \wedge r)$$

10. Take an A such that in its truth table only one line gives the value $\top$. If $B \to A$ is a tautology, B can be true only at the same line.
    So, for example, we can take $A = p \wedge q \wedge r$ or $A = \neg p \wedge q \wedge \neg r$, etc.

11. It is sufficient to make $(p \to \neg q) \to A$ a tautology. And this happens if $A = (p \to \neg q)$.

Example 1.3.4 (Worked examples for Section 1.2.2)
Exercises

1. (a) Find formulae A, B, C such that $\{A, B\}, \{A, C\}, \{B, C\}$ are consistent, while $\{A, B, C\}$ is not.
   (b) For any n find an inconsistent set of n formulae, of which every $(n-1)$ formulae are consistent.

2. Find four pairwise inconsistent non-contradictory formulae.

3. What is the maximal number of pairwise inconsistent non-contradictory formulae with two atomic propositions p, q?

4. Check validity of the rules 1–8 of page 19.

5. Given that $A_1, \ldots, A_n, B \vdash C$ and $A_1, \ldots, A_n, C \vdash B$ show that $A_1, \ldots, A_n \vdash B \leftrightarrow C$.

Solutions

1. (a) For example, if $A = p, B = q, C = \neg(p \wedge q)$.
   (b) For example, $A_1 = p_1, \ldots, A_{n-1} = p_{n-1}, A_n = \neg(p_1 \wedge \cdots \wedge p_{n-1})$. The set $\{A_1, \ldots, A_{i-1}, A_{i+1}, \ldots, A_n\}$ is consistent because all these formulae are true when every atom but p_i is $\top$.

2. $p \wedge q, p \wedge \neg q, \neg p \wedge q, \neg p \wedge \neg q$.

3. Four, because for every formula there is at least one line in the truth table when it is true, and no line fits for two distinct formulae.

4. Rules 1, 2, 3 are straightforward.

   Rule 4. We have to show that $(A \to C) \wedge (B \to C) \wedge (A \vee B) \to C$ is a tautology. Instead of drawing the truth table we can do it as follows.

   Assume that $A \to C, B \to C, A \vee B$ are true. Then A is true or B is true (or both). If A is $\top$ and $A \to C$ is $\top$ then C is $\top$. If B is $\top$ and $B \to C$ is $\top$ then C is $\top$. Therefore C is true under our assumption.

   Rules 5, 6 are checked easily.

   Rule 7. Assume that $B \to C, B \to \neg C$ are both true. If B were true then $C, \neg C$ would be both true. Thus B is false, i.e. $\neg B$ is true.

   Rule 8. Suppose $\dfrac{B_1, \ldots, B_k, A}{C}$ is valid. Then $\dfrac{B_1, \ldots, B_k}{A \to C}$ is valid, because if $B_1, \ldots, B_k$ are true then the truth of A implies the truth of C.

5. By rule 8 we obtain $A_1, \ldots, A_n \vdash B \to C$ and $A_1, \ldots, A_n \vdash C \to B$. Hence $A_1, \ldots, A_n \vdash B \leftrightarrow C$, because if $A_1, \ldots, A_n$ are true then both $B \to C, C \to B$ are true.

Summary

In this chapter we learnt the following main ideas:

- A logical system consists of a formal language, a semantic interpretation and a set of reasoning rules.

- The language of classical propositional logic uses letters such as p, q, r, etc. to represent atomic sentences, and connectives $\wedge$, $\vee$, $\neg$ and $\to$ to combine sentences into more complex sentences.

- Truth tables can be used to interpret the sentences of classical propositional logic.

- Sentences which are always true are called tautologies; those which are always false are contradictions. If a sentence is not a contradiction then it is said to be consistent.

1.3. WORKED EXAMPLES

- Sentences can be put in a convenient normal form.

- A valid argument is a relationship (known as the consequence) between a set of assumptions and a conclusion, and holds provided the conclusion is true whenever the assumptions are all true.

- The consequence relation for any logical system possesses three properties: reflexivity, monotonicity and transitivity. Differing additional properties distinguish differing logical systems.

- If a set of assumptions and a negated formula are not consistent, then the formula is a conclusion in a valid argument from the assumptions.

2
SOME NON-CLASSICAL LOGICS

Classical logic is a powerful language for certain forms of reasoning, and much can be done with it. However, it does not capture all that the natural reasoning methods of humans do. For example, we might not know for certain that an atomic sentence was true, but that there was a 75% chance that it was—this cannot be captured by classical propositional logic. As a further illustration, let the atoms p and q represent two different coins which can be exchanged for an item r. This could be easily stated by $p \land q \to r$. However, to state that the amount represented by p and q is used up in the exchange cannot be done by classical reasoning. When we use an implication $A \to B$, we assume that no change is undergone by A when it is used to conclude B. This does not carry over into the chemical example, nor for a number of problems in computer science. (Note that attempts to express the change to p and q are doomed to failure: $p \land q \to r \land \neg p \land \neg q$ reduces to $\neg(p \land q)$.)

Many non-classical logics have been devised to handle these and other problems which exceed the capabilities of classical logic. In this chapter, we present a selection of variations of classical propositional logic to illustrate the possibilities. There are alternatives to classical logic which add more connectives to the language, and other alternatives which vary the meaning of the existing connectives. Amongst the most important of the latter are *many-valued* logics, *intuitionistic* logic, *modal* and *temporal* logics and *resource* (*relevance* and *linear*) logics.

2.1 Many-valued logics

The first assumption that we made in Chapter 1 was that atomic sentences were either true or false, and that no other values were permitted. In the first alternative to classical logic that we shall describe, Łukasiewicz many-valued logics, we withdraw that assumption and instead we assume that we have a range of possible values and any atom gets one of these values. Depending on the number of values in the range, different many-valued logics can be defined.

The language of propositional many-valued logic is the same as the language of the classical propositional calculus; we have a set $Q = \{q_1, q_2, \ldots\}$ of atomic sentences and a set of connectives $\{\wedge, \vee, \neg, \rightarrow\}$. Let L_n be the set $\{0, 1/n, 2/n, \ldots, (n-1)/n, 1\}$. L_n is the set of truth values: 0 represents false (0 degree of truth), 1 represents true and the other values represent degrees of truth between the two extremes. For $n = 1$ we get $\{0, 1\}$, which is classical logic. For $n = 2$, we get $\{0, 1/2, 1\}$, a three-valued logic. The truth table for the connectives for the case $n = 2$ is defined in Table 2.1. The table also contains the general equation for computing the truth values for any value of n.

Table 2.1: The truth table for $n = 2$.

A	B	$A \wedge B$	$A \vee B$	$A \rightarrow B$	$\neg A$
x	y	$\min(x, y)$	$\max(x, y)$	$\min(1, 1 - x + y)$	$1 - x$
1	1	1	1	1	0
1	1/2	1/2	1	1/2	0
1	0	0	1	0	0
1/2	1	1/2	1	1	1/2
1/2	1/2	1/2	1/2	1	1/2
1/2	0	0	1/2	1/2	1/2
0	1	0	1	1	1
0	1/2	0	1/2	1	1
0	0	0	0	1	1

The rationale behind the truth table is as follows. When we claim $A \rightarrow B$, then our claim is true if B is 'at least as true as' A. This corresponds to $\bot \rightarrow \bot$ is $\top$, $\bot \rightarrow \top$ is $\top$ and $\top \rightarrow \top$ is $\top$ in the classical two-valued truth tables. Our attitude towards $A \rightarrow B$ where A is 'more true' than B might be that it is false, but this judgement is too harsh. A

2.1. MANY-VALUED LOGICS

may be just slightly more true than B. If x is the truth value of A, and y is the truth value of B, then $x - y$ is how far $A \to B$ is from the truth. We can thus give $A \to B$ the value $1 - (x - y) = 1 - x + y$. Combining both cases for $x \leq y$ and $x < y$ gives the formula $\min(1, 1 - x + y)$. The functions of x and y for the other connectives, $A \land B$ has truth value $\min(x, y)$, $A \lor B$ has $\max(x, y)$, and $\neg A$ has $(1-x)$, reflect the traditional way of understanding these connectives.

The value 1 is considered to be *truth*, and so any formula A is an L_n-tautology if it always gets the value 1 in the L_n truth table. For example, $A \to A$ is an L_2-tautology, while $\neg A \lor A$ is not. When the truth value of A, i.e. x is $1/2$, $\neg A \lor A$ is $\max(1 - 1/2, 1/2) = 1/2$ and hence $\neg A \lor A$ does not always get the value 1. (Contrast this with classical logic, where $A \to A$ is equivalent to $\neg A \lor A$.) As for classical logic, A is an L_n-contradiction if it always gets the value 0 in the L_n truth table. The negation of an L_n-tautology is an L_n-contradiction, and vice versa.

Łukasiewicz also introduced the infinite-valued logic L_∞, where the truth values are all rational numbers x such that $0 \leq x \leq 1$. The same formulae given in Table 2.1 for calculating the meaning of the connectives apply. The following formulae are all L_∞-tautologies.[1]

1. $A \to A$

2. $A \to (B \to A)$

3. $(A \to B) \to ((B \to C) \to (A \to C))$

4. $(A \to (B \to C)) \to (B \to (A \to C))$

5. $((A \to B) \to B) \to ((B \to A) \to A)$

6. $((A \to B) \to (B \to A)) \to (B \to A)$

Exercise 2.1.1

1. For the three-valued logic, can one find wffs $d_1(x), d_2(x), d_3(x)$ for atomic x, such that d_1 always gets the value 1, d_2 always gets the value $1/2$, and d_3 always gets the value 0?

2. Compute the truth table in the three-valued logic $\{0, 1/2, 1\}$ of the formulae:

---

[1]The same L_∞-tautologies are obtained also when we let the truth values be all real numbers $0 \leq x \leq 1$. This is so because of the continuity of the truth table functions.

(a) $\neg A \vee B$

(b) $\neg A \wedge \neg B$

(c) $A \vee \neg A$

(d) $((A \wedge B) \to C) \to (A \to (B \to C))$

(e) $(A \to B) \wedge (B \to A)$

(f) $(A \to (B \wedge C)) \to ((A \to B) \wedge (A \to C))$

3. Let $A^0 \to B$ be B and $A^{n+1} \to B$ be $A \to (A^n \to B)$.

   (a) Show that (for any assignment) the truth value of $(\neg A)^k \to A$ is either 0 or 1 in L_n, provided $k \geq n-1$.

   (b) Find a formula which always gets the value 1 in the truth table of $\{0, 1/n, 2/n, \ldots, 1\}$ but does not always get the value 1 in the table for $\{0, 1/(n+1), 2/(n+1), \ldots, 1\}$.

4. Show that $((A^n \to B) \to A) \to A$ is a tautology of L_n but not of L_{n+1}.

5. Find a formula A_k of L_n with a single atom x such that

$$A_k(x) = 1 \text{ iff } x = 1/k, \text{ for } k \geq 2$$

6. Let us denote by L_n also the set of tautologies (theorems) of the logic L_n.

   (a) Show that $L_n \subseteq L_m$ if m is a divisor of n.[2]

   (b) Show that for wffs without negation, $L_n \subseteq L_m$ iff $m \leq n$.

Definition 2.1.2

1. We can define a consequence relation for the logic L_n and L_∞ as follows:

   $A_1, \ldots, A_k \vdash_{L_n} B$ iff for any assignment h

$$h(A_1) + \cdots + h(A_k) - h(B) \leq k - 1$$

   In particular, we have $\vdash_{L_n} B$ iff for all h, $h(B) = 1$.

---

[2] Actually it is also true that if m is not a divisor of n then $L_n \not\subseteq L_m$. The proof is not as easy.

2. $\vdash_{L_n}$ is the consequence relation which satisfies the equivalence

$$A_1, \ldots, A_k, B \vdash_{L_n} C \text{ iff } A_1, \ldots, A_k \vdash_{L_n} B \to C$$

3. There are other consequence relations definable using the many-valued truth table. For example, we can let $A_1, \ldots, A_n \vdash_n B$ iff for all h

$$\min_i \{h(A_i)\} \leq h(B)$$

The above equivalence is also known as the deduction theorem.

Exercise 2.1.3 Verify the deduction theorem mentioned in item 2 of the previous definition.

2.2 Other logics

Classical logic and the many-valued logics are static systems, which rely on fixed sets of truth values which can be assigned to the atomic sentences, and the truth values of complex sentences are computed inductively according to a prescribed truth table. Thus if A is a formula built up from the propositions $q_1, q_2, \ldots, q_n$, we can evaluate the truth value of A provided we know the truth values of $q_1, q_2, \ldots, q_n$. The function giving these values is the assignment h, which for example is given by

$$\begin{aligned} h(p \to q) &= \min(1, 1 - h(p) + h(q)) \\ h(\neg p) &= 1 - h(p) \end{aligned}$$

for the classical and many-valued logics. To introduce and explain other logics, such as intuitionistic and linear logics, we need to consider the above slightly differently. We make h context dependent on indices t from structured set T. We get different logics for different structures.

2.2.1 Temporal logic

We assume we have a flow of time of the form $(T, \leq)$, where T is the set of moments of time and $\leq$ is a reflexive, transitive and antisymmetric earlier–later relation on T, i.e. $\leq$ satisfies the following:

- $t \leq t$

- $t \leq s$ and $s \leq r$ imply $t \leq r$

- $t \leq s$ and $s \leq t$ imply $t = s$.

Note that we do not require that $(T, \leq)$ be linear. Thus we are allowing for a branching future in our time flow.

Let us regard the assignment h of truth values to the propositions as being a description of the world or situation with respect to a particular time t. Thus h_t might say that the value of q_1 at time t is $\top$ and the value of q_2 is $\bot$ and so forth. According to the truth tables, we can compute a truth value for the formula A at time t, written $h_t(A)$. A wff is considered to be a tautology if, no matter what h_t is, the value of A, namely $h_t(A)$, is true.

The difference between classical logic and other many-valued logics is in how they compute the value of $h_t(A)$ at time t, given the values for the propositions, and in what values one can give to the propositions. Suppose that at time t we want to say that, always in the future, A implies B. We cannot write '$A \rightarrow B$ holds at t' or compute $h_t(A \rightarrow B)$ to get the required meaning because $\rightarrow$ behaves according to the agreed truth table—we check at time t the value of A and the value of B and return the value true provided A is not true at the same time as B is false. We need a special connective, say $\Box A$, which will say that A holds at all moments from t onwards. Thus we have $h_t(\Box A) = \top$ if for all $s \geq t$ we have $h_s(A) = \top$.

This connective still satisfies the principle that the value of $\Box A$ at time t depends on the values of A (its subformula) at all other times.

Let us now consider the combination $A \twoheadrightarrow B$ which we define to be $\Box(A \rightarrow B)$.

This is the connective we want. $h_t(A \twoheadrightarrow B)$ has to go to any future time $s \geq t$ and check the values of A and B at s and find that $A \rightarrow B$ holds at s. Therefore

$$h_t(A \twoheadrightarrow B) = \top \quad \text{iff for all } s \geq t, h_s(A) = \top \text{ implies } h_s(B) = \top$$

Let us agree from now on that $\twoheadrightarrow$ is understood to be interpreted as above. What kind of logic do we get? We present the definition of the meaning of the logic below.

Definition 2.2.1 (Temporal many-valued logic) *Let $(T, \leq)$ be a flow of time and let L be a set of truth values (e.g. L can be L_n or L_∞). An interpretation (or a model) for the language with $\{\neg, \wedge, \vee, \rightarrow, \Box\}$ is any set of functions $\{H_t \mid t \in T\}$ assigning truth values to the atomic propositions for each moment of time t. We can extend the definition of H_t*

2.2. OTHER LOGICS

from atomic propositions to any complex formulae by induction, as follows:

$H_t(\neg A) = 1 - H_t(A)$

$H_t(A \wedge B) = \min(H_t(A), H_t(B))$

$H_t(A \vee B) = \max(H_t(A), H_t(B))$

$H_t(A \to B) = \min(1, 1 - H_t(A) + H_t(B))$

$H_t(\Box A) = \min_{s \geq t} H_s(A)$

A model is presented as $(T, \leq, H)$, where H is the family of functions H_t. We say that A is a temporal L-tautology, and write $\models_L A$, iff $H_t(A) = 1$ for all possible H_t over all possible models of time, where L is the set of truth values..

We can define a temporal many-valued consequence relation by letting

$$A_1, \ldots, A_k \models_L B$$

iff for all $(T, \leq, H)$ and all $t \in T$, we have

$$H_t(A_1) + \cdots + H_t(A_k) - H_t(B) \leq k - 1$$

We assume the model of time and H_t are such that different values are given to propositions at different times, and we understand $A \twoheadrightarrow B$ to mean that at all future times A implies B. Each $A \twoheadrightarrow B$ is evaluated as for the many-valued logics at each $s \geq t$, and takes the lowest value that any of these times assigns it.

The new temporal tautologies are those formulae A for which $H_t(A) = 1$ for all times t. For example, $p \twoheadrightarrow p$ is both a classical (when $\twoheadrightarrow$ is replaced by $\to$) and a temporal tautology, whereas $p \vee (p \twoheadrightarrow q)$ is not a temporal tautology although it is a classical tautology. A counterexample is to have p false at time t, but to have p true and q false at some time $s > t$.

To say that A will always be true, one can write $(A \twoheadrightarrow A) \twoheadrightarrow A$, which is what we denoted $\Box A$. Furthermore, to say that A will be true at some (unknown) future moment is the same as saying that its negation will not always be true, i.e. $\neg \Box \neg A$, which is denoted $\Diamond A$. The $\Box$ and $\Diamond$ are known as *modalities*.

2.2.2 Intuitionistic logic

So far the future allows for truth values to change in any way we want. Thus if $t < s$, we may have $H_t(q) = 1$ and $H_s(q) = 0$. Propositions can become false as time passes. In the case of many-valued logics, atoms can fluctuate in value over time, rising and falling through a series of values. A restriction on this leads naturally to intuitionistic logic. Suppose we take the two-valued $L_1 = \{0, 1\}$ and insist that truth values or propositions can only go up (or remain the same) as time goes on. In other words, propositions cannot become 'falser' over time. Formally we might write

$$t \leq s \text{ implies } H_t(q) \leq H_s(q) \tag{2.1}$$

What impact does this restriction have? At first sight, it does not seem to be a practical restriction because we know that in reality, truth values can change either way. Is there a plausible interpretation for this restriction? Consider the following situation. We have an event at a fixed time t_0, giving values to all wffs; for example, suppose that a murder has been committed. An investigating team is set up which gathers information as time goes on. To assert A at time t means in this case that at time t the team discovered that A happened at time t_0, i.e. at the time of the murder. Thus information accumulates over time. Each formula A has three possible values:

- We record A as true if A is definitely confirmed (by some physical evidence, say).

- We record $\neg A$ as true if $\neg A$ is definitely confirmed.

- Otherwise, we record A as false.

Thus $H_t(A) = 1$ means that at time t we have definite evidence that A was true at t_0. $H_t(\neg A) = 1$ means that at time t we have definite evidence that $\neg A$ was true at t_0, and $H_t(A) = 0$ means that at time t we are unable to confirm or deny the truth of A. Because of the 'increasing truth' condition, we know that once we have ascertained that either A or $\neg A$ was true at t_0, that information will remain.

Now if $H_t(A \twoheadrightarrow B) = 1$, we know that we have firm evidence to establish that if A was true at t_0, then B was also true at t_0. This is evidence of a *link* between evidence of the truth of A which can be used as evidence of the truth of B. For example, if we can prove that the

2.2. OTHER LOGICS

person with a particular fingerprint was the murderer, and we can prove that Alun possesses the particular fingerprint, then we can prove that Alun was the murderer.

The above interpretation yields the intuitionistic meaning of $\rightarrow$ and $\neg$. We keep the meaning of $\wedge$ and $\vee$ as before. Thus $\vDash_{L_1} A$ gives the set of all intuitionistic tautologies. If we let $L = L_n$ or $L = L_\infty$, we get a general notion of intuitionistic many-valued logic. The following is a definition of an intuitionistic many-valued logic.

Definition 2.2.2 (Intuitionistic many-valued logic) *We define the language as in Definition 2.2.1 with the following changes:*

1. $t \leq s$ implies $H_t(q) \leq H_s(q)$ for all propositions q

2. $H_t(\neg A) = \min_{s \geq t}(1, 1 - H_s(A))$.

Note that if we use the constant $\bot$ so that $H_t(\bot) = 0$ for all values of t, then $H_t(\neg A) = H_t(A \rightarrow \bot)$ in the intuitionistic many-valued interpretation.

Definition 2.2.3 (Kripke models for intuitionistic logic) *We give separately the version of the previous definition for the case of two truth values $\{0, 1\}$, and the language $\{\wedge, \vee, \rightarrow, \neg, \bot\}$. A Kripke structure, or a Kripke model, has the form $(T, \leq, h)$ where $(T, \leq)$ is a partially ordered set and h is an assignment giving for each $t \in T$ and atom q a value $h(t, q) \in \{0, 1\}$. We require persistence for all q, namely*

- $t \leq s$ and $h(t, q) = 1$ imply $h(s, q) = 1$.

We extend h to all wffs as follows:

- $h(t, A \wedge B) = 1$ iff $h(t, A) = 1$ and $h(t, B) = 1$.

- $h(t, A \vee B) = 1$ iff $h(t, A) = 1$ or $h(t, B) = 1$.

- $h(t, A \rightarrow B) = 1$ iff for all s such that $t \leq s$ and $h(s, A) = 1$ we have $h(s, B) = 1$.

- $h(t, \bot) = 0$.

- $h(t, \neg A) = 1$ iff for all s such that $t \leq s$ we have $h(s, A) = 0$.

- We define a semantic consequence relation $\vDash_I$ by $A_1, \ldots, A_n \vDash_I B$ iff for all models $(T, \leq, h)$ and all $t \in T$ we have: if $h(t, A_i) = 1$, for $i = 1, \ldots, n$, then $h(t, B) = 1$.

- B is an intuitionistic tautology iff $\varnothing \vDash_I B$.

Exercise 2.2.4 *Show that in the previous definition the following persistence property holds:*

- $t \leq s$ *and* $h(t, A) = 1$ *imply* $h(s, A) = 1$.

2.2.3 Resource logics

We are now in a position to discuss the family of logics which are grouped under the heading of *resource* logics. The intuitionistic interpretation was not really temporal—it had to do with the accumulation of evidence, with $t \leq s$ meaning s is a time at which more evidence has been obtained than at time t. The temporal component was used as a means of ordering the increasing evidence. However, it is possible to deal directly with the additional information and do without the concept of time.

Let α_t be the information available at time t, and α_s be the information available at time s. Now $\alpha_s = \alpha_t \otimes \beta$ where β is the additional information which combined (or added) to α_t yields α_s. We can thus replace the flow of time by a flow of information. Let I be a set of pieces of information, and let $\otimes$ be a 'combining' operator for putting together members of I to obtain further members of I. $H_t(q)$ gives the truth value of q relative to the piece of information t. We can define $t \leq s$ iff for some x, $t \otimes x = s$. We also need $\varnothing$ to represent no information, so that $\varnothing \otimes t = t \otimes \varnothing = t$ for all t. We now have a new way of understanding $\twoheadrightarrow$:

$H_t(A \twoheadrightarrow B) = 1$ iff for all s such that $H_s(A) = 1$ we get $H_{t \otimes s}(B) = 1$

This definition states that $A \twoheadrightarrow B$ is true in a situation provided that whenever a situation in which A is true is added, a situation in which B is true is reached. The definition corresponds to natural reasoning.

There are several options for understanding $\neg A$ in this context, the simplest being

$$H_t(\neg A) = 1 - H_t(A)$$

which corresponds to earlier interpretations of $\neg$. We could also use

$$H_t(\neg A) = H_t(A \twoheadrightarrow \bot)$$

as above.

2.2. OTHER LOGICS

Validity can be defined by $\models A$ iff $H_\varnothing(A) = 1$ for all information sets and definitions of the $\otimes$ operator. These different definitions of $\otimes$ yield different logics. There is a basic requirement of $\otimes$ that it be associative, i.e. $(x \otimes y) \otimes z = x \otimes (y \otimes z)$. We can require that $\otimes$ be commutative ($t \otimes s = s \otimes t$) meaning that the order in which information is combined does not matter. A further requirement might be that information cannot be reinforced by repetition, so that $t \otimes t = t$.

We summarize the discussion in the following definition.

Definition 2.2.5 (Many-valued resource logic) *A many-valued substructural logic has the same connectives as before, namely $\{\wedge, \vee, \twoheadrightarrow, \neg, \bot\}$, but is interpreted with respect to $\langle I, \otimes, \varnothing \rangle$, where I is a set with an associative binary operator $\otimes$ and identity $\varnothing$ such that $\varnothing \otimes t = t$ for all t. H is an assignment associating with each $t \in I$ and proposition q, a truth value $H_t(q)$. H can be inductively extended to all wffs by*

$$H_t(A \wedge B) = \min(H_t(A), H_t(B))$$

$$H_t(A \vee B) = \max(H_t(A), H_t(B))$$

$$H_t(A \twoheadrightarrow B) = \min_s \min(1, 1 - H_s(A) + H_{t \otimes s}(B))$$

$$H_t(\neg A) = H_t(A \twoheadrightarrow \bot)$$

where $\bot$ is a constant which may be interpreted by $H_t(\bot) = 0$ for all t (although logics exist which do not require this).
$\models A$ *(A is valid) iff for all $\langle I, \otimes, \varnothing \rangle$ and H we have $H_\varnothing(A) = 1$.*

Exercise 2.2.6 *Check the validity of the following formulae when (a) $\otimes$ is associative and commutative, and when (b) $\otimes$ is associative and $t \otimes t = t$, and when in both cases $\varnothing \otimes t = t$.*

1. $A \twoheadrightarrow A$
2. $A \twoheadrightarrow (A \twoheadrightarrow A)$
3. $(A \twoheadrightarrow (B \twoheadrightarrow C)) \twoheadrightarrow (B \twoheadrightarrow (A \twoheadrightarrow C))$
4. $(A \twoheadrightarrow (A \twoheadrightarrow B)) \twoheadrightarrow (A \twoheadrightarrow B)$
5. $A \twoheadrightarrow ((A \twoheadrightarrow A) \twoheadrightarrow A)$

Example 2.2.7 (Linear logic and relevance logic) *These are very well known resource logics and this example will give a flavour of their nature. Imagine several assumptions, say*

1. A

2. $A \to (A \to B)$

3. C

We can use modus ponens and derive B by using A twice, i.e.

$$\frac{A, A \to (A \to B)}{A \to B}$$

and again

$$\frac{A, A \to B}{B}$$

So, in deriving B, we used assumption (1) twice, assumption (2) once and assumption (3) not at all.

In general, let $\alpha : D$ be a pair with α a list of exactly how many times each assumption was used in deriving D. We can maintain our lists by letting

$$\frac{\alpha : D; \beta : D \to E}{\beta \otimes \alpha : E}$$

where $\beta \otimes \alpha$ is a new list giving details of what is used in the proof of E, constructed from α and β.

Relevance logic will accept a proof of E from a set of assumptions provided all assumptions *were used* at least once. Linear logic insists that each assumption *is used* exactly once.

See also Section 5.2.

2.3 Another look at intuitionistic logic

We saw in Chapter 1 that classical logic is content with regarding $A \to B$ as being true when A and B are both false. Intuitionistic logic takes a more 'dynamic' view that the assertion $A \to B$ must be properly tested. In other words, A must be made true and then we must check for B which must follow from A in all circumstances in which A is true. Thus when we are presented with a candidate for a valid argument of the form $A_1, \ldots, A_m \vdash (A \to B)$, we do not use truth tables to confirm it, i.e. check whether $A_1 \wedge \cdots \wedge A_m \to (A \to B)$ is a tautology, but check instead whether by adding A to the assumptions, $A_1, \ldots, A_m, A \vdash B$ is a valid argument.

2.3. ANOTHER LOOK AT INTUITIONISTIC LOGIC

Intuitionistic logic for the language with $\wedge, \vee, \to$ and $\neg$ can be defined as the minimal logic (by which we mean the logic with the least number of valid arguments) which allows for the equivalence

$$A_1, \ldots, A_m \vdash (A \to B) \text{ iff } A_1, \ldots, A_m, A \vdash B$$

and satisfies reflexivity, monotonicity and cut and also rules 1, 2, 3, 4, 6 and 7 on page 19.

Classical logic allows for a stronger equivalence, namely

$$A_1, \ldots, A_n \vdash (A \to B) \vee C \text{ iff } A_1, \ldots, A_n, A \vdash B \vee C$$

Given a set of assumptions $\{A_1, \ldots, A_n\}$ and a conclusion B, in classical logic we can look, by building truth tables, for all the possibilities (rows of the truth table) in which the assumptions are all true. For each of these possibilities we can check the truth table of B. If B is true in each of these possibilities, then classical logic will regard the argument $A_1, \ldots, A_n \vdash B$ as valid.

Intuitionistic logic would also check whether B is true in all circumstances in which the assumptions are all true. But when the conclusion is an implication, say $C \to D$, the validity of the argument depends on just those possibilities which make the assumptions together with C true. Again, if D is true in all those possibilities then intuitionistic logic will regard the argument $A_1, \ldots, A_n \vdash B$ as valid.

We illustrate the difference with an example. Suppose that there is a single assumption $p \vee q$, and the conclusion is $\neg p \to q$. The truth table for classical implication gives

	p	q	$p \vee q$	$\neg p \to q$
1	T	T	T	T
2	T	$\bot$	T	T
3	$\bot$	T	T	T
4	$\bot$	$\bot$	$\bot$	$\bot$

Rows 1, 2 and 3 all make the assumption true, so the argument is valid in classical logic if and only if these rows make the conclusion true, which they do. Intuitionistic logic requires us to consider what happens when we demand that the antecedent of the implication $\neg p$ is true together with the assumption, which is the case in rows 2 and 3.

More formally we adopt a constructive view of $A \to B$. To assert $A \to B$ we must have a method which transforms any proof of A into a proof of B. For part of the language without disjunction, we are able to motivate intuitionistic logic further as a variation of classical logic. Consider the following formulation of classical logic. We know that $A_1, \ldots, A_n \vdash B$ in classical logic if, according to the truth tables, $\bigwedge A_i \to B$ is a tautology. To check whether the above is a tautology we check each row in the truth table, and ensure that B is true in each row where all of the A_i are true. Each row in the truth table gives values to the atomic propositions. If $q_1, q_2, \ldots, q_m$ are the atoms, a row in the truth table can be taken as giving values to $q_1, q_2, \ldots, q_m$ or adding the assumptions $q_i^\pm$ to the data, where q_i^+ is q_i, and q_i^- is $\neg q_i$. Thus we have $A_1, \ldots, A_n \vdash B$ in classical logic iff for all choices of vectors $x_1, x_2, \ldots, x_m \in \{+, -\}$, we have $A_1, \ldots, A_n, q_1^{x_1}, q_2^{x_2}, \ldots, q_m^{x_m} \vdash B$. It is enough to generate the above choices by the single rule $A_1, \ldots, A_n \vdash B$ if for all formulae C, we have both $A_1, \ldots, A_n, C \vdash B$ and $A_1, \ldots, A_n, \neg C \vdash B$. Intuitionistic logic without disjunction allows for the same equation but only for positive q_i, i.e. $A_1, \ldots, A_n \vdash B$ iff for all subsets $\{p_1, p_2, \ldots, p_k\} \subseteq \{q_1, q_2, \ldots, q_m\}$, we have $A_1, \ldots, A_n, p_1, p_2, \ldots, p_k \vdash B$.

Let us check an example to bring out the difference in the presence of $\vee$. Suppose there is a board meeting of Imperial Petroleum convened to vote on a motion calling for the chairman's resignation. It is expected that the vote will be very close. Information was given anonymously to the chairman that a deal was made between Mr Jones and Mr Smith, two members of the board, that for £100,000 one of them will vote in favour of the motion. It was not clear which of them sold out their vote, but one thing is clear. Either Mr Smith will vote 'yes' if Mr Jones votes 'yes', or Mr Jones will vote 'yes' if Mr Smith votes 'yes'. We can represent this in symbols as

$$(S \to J) \vee (J \to S) \qquad (2.2)$$

using J and S to stand for Mr Jones will vote 'yes' and Mr Smith will vote 'yes' respectively. The chairman was supposed to check this 'deal', but having only been taught classical logic, he did not bother. The classical truth of the allegation is logically evident, as the following truth table shows:

2.3. ANOTHER LOOK AT INTUITIONISTIC LOGIC

S	J	$S{\to}J$	$J{\to}S$	$(S{\to}J)\vee(J{\to}S)$
T	T	T	T	T
T	$\bot$	$\bot$	T	T
$\bot$	T	T	$\bot$	T
$\bot$	$\bot$	T	T	T

The formula (2.2) is true under all circumstances in classical logic. Intuitionistic logic does not follow tables. To establish the truth of $S{\to}J$, we assume S is true and check J. Let us now check the first disjunct:

Database	Query
S is true	Is J true?
(Temporarily assume true)	

The answer is 'no', since we have no further evidence connecting S with J. Let us now check the other alternative:

Database	Query
J is true	Is S true?
(Temporarily assume true)	

Again the answer is 'no', since we have no further evidence connecting J with S. Therefore the allegation is not logically self-evident in intuitionistic logic and needs to be investigated.

Note that our intuitive understanding of intuitionistic implication is hypothetical. This makes intuitionistic logic and intuitionistic implication natural bases for describing database updating and time-dependent reasoning from data.

From the point of view of intuitionistic logic, what component do we add to intuitionistic reasoning to obtain classical (logic) reasoning? We have arrived at intuitionistic logic by questioning and weakening the logic of the truth table. If we use intuitionistic logic, what is it that we can do to make the logic turn into classical logic? The answer is the following: in the course of the computation (or reasoning) in intuitionistic logic, if we fail to delete temporary assumptions when we have finished using them, we get classical logic.

If we check the validity of $(S{\to}J)\vee(J{\to}S)$ again and fail to delete the assumptions, we should succeed. First we check whether $(S{\to}J)$ is true. To do that intuitionistically, we add (or assume) that S is true and ask for J. Since we have no further evidence, we cannot show that J is true and hence the computation (or checking) of whether $S{\to}J$ is true failed. At this point of the computation, S is marked by us as true, because we temporarily assumed S is true to check whether J follows. To pass to the next stage, namely checking whether $J \to S$, we ought to delete the S. Suppose we do not do that and continue the computation by checking whether $J \to S$ is true. To do that we temporarily assume J is true and ask whether S must be true. But S now has to be true because it is still marked true from the previous computation. Thus $(S \to J) \vee (J \to S)$ comes out valid, if we leave old data around.

This happens for all formulae. Suppose we are given a proper computation procedure for asking questions A from a database $\mathcal{P}$, using intuitionistic logic. In the middle of the computation $\mathcal{P}$ may be temporarily increased to $\mathcal{P}'$ (a bigger database). If, at any time in the middle of the computation, we allow the query A to be asked again from $\mathcal{P}'$ (i.e. we *restart* with A, see Section 4.1.2), without having to return to the original $\mathcal{P}$, but continuing with $\mathcal{P}'$, then the answers we get are those of classical logic.

2.4 Worked examples

Example 2.4.1 (Worked examples for Section 2.1)
Exercises

1. Does there exist an L_2-formula $A(p, q)$ with the following truth table?

p	q	$A(p,q)$
1	1	1
1	1/2	0
1	0	1/2
1/2	1	0
1/2	1/2	0
1/2	0	0
0	1	0
0	1/2	0
0	0	1

2. Are the formulae $A \to B$ and $\neg A \vee B$ equivalent in the logic L_2?

3. Show that the connective $\to$ cannot be expressed in terms of $\vee, \wedge, \neg$ in the logic L_n if $n > 1$.

4. Show that a formula without $\to$ cannot be an L_n-tautology if $n > 1$.

2.4. WORKED EXAMPLES

5. *Show that if A is an L_n-tautology for some n, then $\neg A$ is not an L_m-tautology for any m.*

6. *Show that $L_\infty \subseteq L_n$ for any n.*

7. *Show that $L_\infty \neq L_n$ for any finite n.*

8. *Show that $L_\infty = \bigcap \{L_n \mid n \geq 0\}$.*

Solutions

1. No. Notice that every L_2-formula takes a value 0 or 1 when each of its arguments is either 0 or 1. This is easily proved by induction. But this property fails for the truth table in question.

2. No, because $1/2 \to 1/2 = 1$, and $\neg 1/2 \vee 1/2 = 1/2 \vee 1/2 = 1/2$.

3. To show this, it is sufficient to prove that for any L_2-formula $A(p_1, \ldots, p_k)$ built up using the connectives $\neg, \vee, \wedge$ only, we have $A(1/2, \ldots, 1/2) = 1/2$. This is proved by induction.

    (a) If $A = p_i$, the claim is obvious
    (b) If $A = B \vee C$, then $A(1/2, \ldots, 1/2) = B(1/2, \ldots, 1/2) \vee C(1/2, \ldots, 1/2) = 1/2 \vee 1/2$ (by the induction hypothesis) $= 1/2$.
    (c) If $A = B \wedge C$, the proof is similar.
    (d) If $A = \neg B$, then $A(1/2, \ldots, 1/2) = \neg B(1/2, \ldots, 1/2) = \neg 1/2$ (by the induction hypothesis) $= 1/2$.

4. Similarly to the previous exercise, one can prove by induction that for every formula A of the corresponding type, $A(1/n, \ldots, 1/n) \neq 1$.

5. To prove this, observe that $A(1, \ldots, 1) = 1$ in any logic L_m (because this is true in L_n and truth values in L_m for this specific case are computed by the same rules).

6. Note that $A(x_1, \ldots, x_k)$ in $L_n = A(x_1, \ldots, x_k)$ in L_∞ (for any $x_1, \ldots, x_k \in \{0, 1/n, \ldots, 1\}$). This happens because truth tables in L_n and in L_∞ are computed by the same rules.

    Now if A is an L_∞-tautology, it follows that A is an L_n-tautology.

7. Otherwise $L_n \subseteq L_{n+1}$ by the previous exercise, in contradiction with Exercise 2.1.1.4.

8. $L_\infty \subseteq \bigcap\{L_n \mid n \geq 0\}$ by Exercise 7. For the converse, suppose A is not an L_∞-tautology and show that A is not an L_n-tautology for some n. By the assumption, $A(x_1, \ldots, x_k) \neq 1$ (in L_∞) for some values $x_1, \ldots, x_k$ of atoms occurring in A. Since $x_1, \ldots, x_k$ are rational numbers, there exists n, such that $x_1, \ldots, x_k \in \{0, 1/n, \ldots, (n-1)/n, 1\}$. Then $A(x_1, \ldots, x_k)$ (in L_∞) $= A(x_1, \ldots, x_k)$ (in L_n), according to the definitions. Hence A is not an L_n-tautology.

Example 2.4.2 (Worked examples for Section 2.2.1)
Exercises

1. *Check if the following formulae are temporal L_1-tautologies:*

    (a) $B \to (A \to B)$

    (b) $A \wedge B \to B$

    (c) $A \wedge B \twoheadrightarrow B$

    (d) $B \twoheadrightarrow (A \twoheadrightarrow B)$

    (e) $\Box A \twoheadrightarrow A$

    (f) $\Box A \twoheadrightarrow \Diamond A$

    (g) $\Box A \twoheadrightarrow \neg\Diamond\neg A$

    (h) $\Box A \wedge \Box B \twoheadrightarrow \Box(A \wedge B)$

    (i) $\Box(A \wedge B) \twoheadrightarrow \Box A \wedge \Box B$

    (j) $\Diamond(A \vee B) \twoheadrightarrow \Diamond A \vee \Diamond B$

    (k) $\Box\Box A \twoheadrightarrow \Box A$

2. *Prove that a formula without $\twoheadrightarrow$ (and $\Box$) is a temporal L-tautology iff it is an L-tautology.*

3. *Show that (in L_1) the formula $\Box p$ is non-equivalent to any formula built up from the connectives $\vee, \wedge, \neg$.*

4. *Call formulae A, B temporally L-equivalent if $H_t(A) = H_t(B)$ in any model for any t. Show that A, B are temporally L-equivalent iff*

$$\vDash (A \twoheadrightarrow B) \wedge (B \twoheadrightarrow A)$$

5. *Construct infinitely many formulae with a single atom p, which are pairwise temporally L_1-non-equivalent.*

2.4. WORKED EXAMPLES

Solutions

1. (a), (b). These formulae are classical tautologies. Hence they are also temporal tautologies, because truth values at every moment of time are computed according to classical rules.

   (c) We know already that $C = (A \wedge B \to B)$ is a tautology. Now $\Box C$ is a temporal tautology, because $H_t(\Box C) = \top$ iff for all $s \geq t$, $H_s(C) = \top$, and the latter is true.

   (d) This formula is not a temporal tautology. For, let A, B be atoms, and consider a flow of time with two moments, t and s, such that $t < s$. Let $H_t(A) = H_s(A) = 1, H_t(B) = 1, H_s(B) = 0$. Then $H_t(A \twoheadrightarrow B) = 0$, but $H_t(B) = 1$; hence $H_t(B \twoheadrightarrow (A \twoheadrightarrow B)) = 0$.

   (e) This is also a temporal tautology. As in case (c), it suffices to show that $\Box A \to A$ is a temporal tautology. For this, assume that $H_t(\Box A) = 1$, and show that $H_t(A) = 1$. By the assumption, $H_s(A) = 1$ for any $s \geq t$. Hence $H_t(A) = 1$ because $\leq$ is reflexive.

   (f) Again, it is sufficient to check that $H_t(\Box A \to \Diamond A) = 1$ for any t. This follows again from the reflexivity of the time flow. For, assume that $H_t(\Box A) = 1$ then $H_t(A) = 1$, and then $H_t(\Diamond A) = 1$ (since $t \leq t$).

   (g) $\neg \Diamond \neg A = \neg \neg \Box \neg \neg A$. This formula has the same truth value (at any moment of time) as $\Box \neg \neg A$, which is equivalent to $\Box A$ (since $\neg \neg A \equiv A$ in classical logic). Therefore, $\Box A \twoheadrightarrow \neg \Diamond \neg A$ is an L_1-tautology.

   (h, i) Let us check that $H_t(\Box A \wedge \Box B) = H_t(\Box (A \wedge B))$: $H_t(\Box A \wedge \Box B) = 1$ iff $H_t(\Box A) = H_t(\Box B) = 1$ iff $H_s(A) = H_s(B) = 1$ for any $s \geq t$ iff $H_s(A \wedge B) = 1$ for any $s \geq t$ iff $H_t(\Box (A \wedge B)) = 1$.

   (j) Let us show that $H_t(\Diamond (A \vee B) \to \Diamond A \vee \Diamond B) = 1$. For this assume that $H_t(\Diamond (A \vee B)) = 1$. Then $H_s(A \vee B) = 1$ for some $s \geq t$. Hence for this s, $H_s(A) = 1$ or $H_s(B) = 1$. Then $H_t(\Diamond A) = 1$ in the first case, $H_t(\Diamond B) = 1$ in the second.

   (k) This formula is a temporal tautology because $\Box A \twoheadrightarrow A$ is a tautology for any A (see (e) above). So we can replace A by $\Box A$.

2. Take a formula A built by $\vee, \wedge, \neg$. If A is an L-tautology then $H_t(A) = 1$ for any moment t, because truth values at t are found by the same rules as in L.

For the converse, suppose $A(p_1, \ldots, p_k)$ is not an L-tautology. Then there are values $x_1, \ldots, x_k$ of atoms $p_1, \ldots, p_k$ such that $A(x_1, \ldots, x_k) \neq 1$ (in L). Take the flow of time $(T, \leq)$ with a single moment t, and the interpretation h such that

$$H_t(p_1) = x_1, \ldots, H_t(p_k) = x_k$$

Then $H_t(A) = A(x_1, \ldots, x_k)$ in $L \neq 1$. Thus A is not a temporal L-tautology.

3. Suppose the contrary, i.e. that for some classical formula A, $\Box p \Leftrightarrow A$ is a temporal tautology. If such A exists, then we can further assume that A is built using a single variable p. Really, if $\Box p \Leftrightarrow A(p, q_1, \ldots, q_k)$ is a temporal tautology, then $\Box p \Leftrightarrow A(p, \top, \top, \ldots, \top)$ is a temporal tautology (because $\Box p$ does not depend on $q_1, \ldots, q_k$) and we can consider $A' = A(p, \top, \ldots, \top)$ instead of A.

But in classical logic, there are only four pairwise non-equivalent formulae with one variable p, $\neg p$, $\top$, $\bot$, and none of the equivalences (1) $\Box p \Leftrightarrow p$, (2) $\Box p \Leftrightarrow \neg p$, (3) $\Box p \Leftrightarrow \top$, (4) $\Box p \Leftrightarrow \bot$ are temporal tautologies. Corresponding countermodels are:

(1) $T = \{t, s\}, t < s, H_t(p) = 1, H_s(p) = 0$

(2), (4) $T = \{t\}, H_t(p) = 1$

(3) $T = \{t\}, H_t(p) = 0$.

4. Assume that A, B are temporally equivalent, and consider an arbitrary model.

Since for any s, $H_s(A) = H_s(B)$, we have that $H_s(A \to B) = H_s(B \to A) = 1$. Hence $H_t(A \twoheadrightarrow B) = H_t(B \twoheadrightarrow A) = 1$ for any t, and therefore $\vDash_L (A \twoheadrightarrow B) \wedge (B \twoheadrightarrow A)$.

Conversely, suppose $\vDash_L (A \twoheadrightarrow B) \wedge (B \twoheadrightarrow A)$, and take any model. Then for any t,

$$H_t(A \twoheadrightarrow B) = H_t(B \twoheadrightarrow A) = 1$$

which implies that

$$H_t(A \to B) = H_t(B \to A) = 1$$

Then by Łukasiewicz truth tables, we see that $H_t(A) \leq H_t(B)$ and $H_t(B) \leq H_t(A)$, and thus A, B are temporally L-equivalent.

2.4. WORKED EXAMPLES

5. Consider the following formulae:

$$A_1 = \neg p$$
$$A_2 = p \wedge \Diamond \neg p$$
$$A_3 = \neg p \wedge \Diamond(p \wedge \Diamond \neg p)$$
$$\vdots$$
$$A_{2n} = p \wedge \Diamond A_{2n-1}$$
$$A_{2n+1} = \neg p \wedge \Diamond A_{2n}, \ldots$$

Consider the following model: $(\mathsf{T}, \leq)$ is the set $\{-1, -2, -3, \ldots\}$ ordered by the standard relation '$\leq$',

$$H_t(p) = \begin{cases} 1, & \text{if } t \text{ is even} \\ 0, & \text{if } t \text{ is odd} \end{cases}$$

```
···  ─────●─────●─────●─────●─────●
         −5    −4    −3    −2    −1
         ¬p     p    ¬p     p    ¬p
```

Then the following holds:

(a) $H_{-n}(A_n) = 1$;

(b) $H_{-n}(A_m) = 0$ for any $m > n$.

(a) is proved by induction. If $n = 1$ we have

$$H_{-1}(A_1) = H_{-1}(\neg p) = 1 - H_{-1}(p) = 1$$

If (a) is proved for $n = 2k - 1$, we have

$$H_{-2k}(A_{2k}) = 1 \text{ iff } H_{-2k}(p) = H_{-2k}(\Diamond A_{2k-1}) = 1$$

But $H_{-2k}(p) = 1$ by definition, and $H_{-2k}(\Diamond A_{2k-1}) = 1$ since, by the inductive hypothesis, $H_{-(2k-1)}(A_{2k-1}) = 1$. Thus $H_{-2k}(A_{2k}) = 1$.

```
     ─────●─────────●─────────●──────────→
       −(2k+1)    −2k      −(2k−1)
          ¬p        p       A_{2k−1}
```

Similarly,

$$H_{-(2k+1)}(A_{2k+1}) = 1 \text{ iff } H_{-(2k+1)}(\neg p) = H_{-(2k+1)}(\Diamond A_{2k}) = 1$$
$$H_{-(2k+1)}(\neg p) = 1 \text{ since } H_{-(2k+1)}(p) = 0$$
$$\text{and } H_{-(2k+1)}(\Diamond A_{2k}) = 1 \text{ since } H_{-2k}(A_{2k}) = 1$$

This completes the proof of (a).

(b) is also proved by induction on n. Let $n = 1$. Suppose $H_{-1}(A_m) = 1$ for some $m > 1$. Then m must be odd (otherwise $H_{-1}(p) = 1$ which is not true). But if $m = 2k+1$, we have $H_{-1}(\Diamond A_{2k}) = 1$ which implies that $H_{-1}(A_{2k}) = 1$ (because the only future of (-1) is (-1) itself).

But this cannot be true, as we have observed. Now assume that (b) is proved for any pair m', n' such that $m' > n', n' < n$, and consider a pair (m, n) such that $m > n$. Suppose $H_{-n}(A_m) = 1$. Then $H_{-n}(\Diamond A_{m-1}) = 1$ (as it follows from the definition of A_m), and thus

(i) $H_{-n}(A_{m-1}) = 1$

or

(ii) $H_{-n'}(A_{m-1}) = 1$ for some $n' < n$.

The latter contradicts our assumption because $m - 1 > n - 1 \geq n'$.

As for (i), it is incompatible with $H_{-n}(A_m) = 1$ because one of the numbers
$m, (m - 1)$ is even and the other is odd, and $p, \neg p$ cannot both be true at $(-n)$. Therefore $H_{-n}(A_m) = 0$.

Example 2.4.3 (Worked examples for Section 2.2.2)
Exercises

1. *Check, using Kripke models, if the following formulae are intuitionistic tautologies:*

   (a) $\neg A \lor A$
   (b) $\neg\neg A \to A$
   (c) $A \to \neg\neg A$
   (d) $(A \to B) \lor (B \to A)$
   (e) $\neg A \land \neg B \to \neg(A \lor B)$
   (f) $\neg\neg(A \lor \neg A)$
   (g) $A \to (B \to A \land B)$

2. *Describe Kripke structures with the least moment of time, where the formula $p \lor \neg p$ is true at any moment.*

Solutions

1. (a) No. Take the two-element Kripke model (for A atomic):

2.4. WORKED EXAMPLES

Then $h(u, \neg A) = 0$, and so $h(u, A \vee \neg A) = 0$.

(b) No. Take the same model as in the previous case. We have $h(u, \neg\neg A) = 1$ because $h(u, \neg A) = h(v, \neg A) = 0$. Also $h(u, A) = 0$, and thus $h(u, \neg\neg A \to A) = 0$.

(c) Yes. Consider an arbitrary Kripke model, and suppose $h(t, A) = 1$. We have to show that $h(t, \neg\neg A) = 1$. Suppose the contrary. Then $h(s, \neg A) = 1$ for some $s \geq t$, which implies $h(s, A) = 0$. But this contradicts persistence of A (Exercise 2.2.4).

(d) No. Take the three-element Kripke model (for A, B atomic):

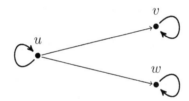

Let $h(v, A) = h(w, B) = 1$, $h(v, B) = h(u, B) = h(u, A) = h(w, A) = 0$.

Then $h(u, A \to B) = 0$ because $v \geq u$, $h(v, A) = 1$, $h(v, B) = 0$; also $h(u, B \to A) = 0$ because $w \geq u$, $h(w, B) = 1$, $h(w, A) = 0$. Hence $h(u, (A \to B) \vee (B \to A)) = 0$.

(e) Yes. In any Kripke model, if $h(t, \neg A \wedge \neg B) = 1$ then $h(t, \neg A) = h(t, \neg B) = 1$, and thus $h(s, A) = h(s, B) = 0$, for any $s \geq t$.

The latter means that $h(s, A \vee B) = 0$. Therefore for every t, $h(t, \neg A \wedge \neg B) = 1$ implies $h(t, \neg(A \vee B)) = 1$.

(f) Yes. Suppose the contrary. Then in some Kripke model, for some t we have $h(t, \neg\neg(A \vee \neg A)) = 0$ and so for some $s \geq t$, $h(s, \neg(A \vee \neg A)) = 1$, and further on, $h(r, A \vee \neg A) = 0$ for any $r \geq s$, i.e. $h(r, A) = h(r, \neg A) = 0$ for any $r \geq s$. Now $h(r, \neg A) = 0$ only if $h(r', A) = 1$ for some $r' \geq r$, and on the other hand, $h(r', A) = 0$ since $r' \geq r \geq s$. This is a contradiction.

(g) Yes. It is sufficient to show that, in any Kripke model, $h(t, A) = 1$ implies $h(t, B \to (A \wedge B)) = 1$.

So suppose $h(t, A) = 1$; take any $s \geq t$, and suppose also that $h(s, B) = 1$. Then $h(s, A \wedge B) = 1$ because $h(s, A) = 1$ due to persistence (Exercise 2.2.4).

2. These are structures where p has the same truth value at all moments of time. Really, assume that $h(t_0, p \vee \neg p) = 1$ and t_0 is the initial moment. Then either $h(t_0, p) = 1$ or $h(t_0, \neg p) = 1$. If $h(t_0, p) = 1$ then $h(t, p) = 1$ for any t (by persistence). If $h(t_0, \neg p) = 1$ then for any t, $h(t, \neg p) = 1$ by persistence, and thus $h(t, p) = 0$.

The converse is clear, because if $h(t, p) = 0$ for any t, then $h(t_0, \neg p) = 1$. Otherwise $h(t_0, p) = 1$ (since $h(t_0, p \vee \neg p) = 1$).

Example 2.4.4 (Three-valued Smetanich logic) *Consider Kripke structures with two moments of time making a chain: $T_2 = \{t_0, t_1\}, t_0 < t_1$. Restrict the definition of an intuitionistic tautology to this case. That is, we define $\vDash_S A$ iff for all Kripke structures $(T_2, \leq, h), h(t, A) = 1$ for any t.*

In general, the possible combinations of truth values of a formula A at the moments t_0, t_1 can be $(0,0), (0,1)$ or $(1,1)$ (because of persistence). Let us consider them as corresponding to the numerical values 0 (for $(0,0)$), $1/2$ (for $(0,1)$) and 1 (for $(1,1)$). Then we can treat Smetanich logic as three-valued. For example, $A \wedge B$ has the following truth table:

A	B	A ∧ B
0	0	0
1/2	0	0
1	0	0
0	1/2	0
1/2	1/2	1/2
1	1/2	1/2
0	1	0
1/2	1	1/2
1	1	1

1. Compute truth tables for all connectives in Smetanich logic and compare them to L_2 truth tables.

2. Check if the following formulae are Smetanich tautologies:

2.4. WORKED EXAMPLES

(a) $\neg A \vee A$

(b) $\neg\neg A \to A$

(c) $(A \to B) \vee (B \to A)$

(d) $\neg A \vee \neg\neg A$

(e) $A \vee (A \to B) \vee \neg B$

3. Let $(T, \leq)$ be a partially ordered set with the first element and at least two other elements. Show that the formula $A \vee (A \to B) \vee \neg B$, with A, B atomic, can be refuted in some Kripke structure $(T, \leq, h)$.

4. Let S be the set of all Smetanich tautologies. Show that $L_2 \not\subseteq S$, and $S \not\subseteq L_2$.

5. Show that L_2-negation is non-expressible in Smetanich logic, but that intuitionistic negation is expressible in L_2.

Solutions

1.

A	B	$\neg A$	$A \wedge B$	$A \vee B$	$A \to B$
1	1	0	1	1	1
1	1/2	0	1/2	1	1/2
1	0	0	0	1	0
1/2	1	0	1/2	1	1
1/2	1/2	0	1/2	1/2	1
1/2	0	0	0	1/2	0
0	1	1	0	1	1
0	1/2	1	0	1/2	1
0	0	1	0	0	1

If we compare this to Table 2.1, we see that $A \wedge B, A \vee B$ are the same functions, whereas $\neg A, A \to B$ are not. In particular, $\neg A$ never gets the value 1/2 in Smetanich logic.

2. (a) $\neg p \vee p$ is not a tautology (for p atomic). We can take the same model as in Exercise 2.4.3.1(a).

   (b) No. Use the model from Exercise 2.4.3.1(b).

   (c) Unlike the general intuitionistic case, $(A \to B) \vee (B \to A)$ is a Smetanich tautology.

Indeed, consider an arbitrary Kripke model $(T_2, \leq, h)$. Owing to persistence, it is sufficient to show that the formula is true at t_0, i.e. that

(c.1) $h(t_0, A \to B) = 1$ or $h(t_0, B \to A) = 1$.

Assume that $h(t_0, A \to B) = 0$, and show that $h(t_0, B \to A) = 1$. $h(t_0, A \to B) = 0$ implies

(c.1.1) $h(t_0, A) = 1, h(t_0, B) = 0$, or

(c.1.2) $h(t_1, A) = 1, h(t_1, B) = 0$.

From (c.1.1) it follows that $h(t_0, B \to A) = 1$ since A is persistent.

From (c.1.2) it also follows that $h(t_0, B \to A) = 1$ since $h(t_0, B) = h(t_1, B) = 0$, owing to the persistence of B.

(d) Similarly to the previous one, this is a tautology. Consider a Kripke model $(T_2, \leq, h)$ and show that $h(t_0, \neg A) = 1$ or $h(t_0, \neg\neg A) = 1$.

Assume that $h(t_0, \neg A) = 0$. Then $h(t_0, A) = 1$ or $h(t_1, A) = 1$. In both cases $h(t_1, A) = 1$, and thus $h(t_0, \neg\neg A) = 1$ (because $\neg A$ is false at t_0 and at t_1).

(e) This is a tautology. Indeed, take any model $(T_2, \leq h)$. Assume that $h(t_0, A) = h(t_0, A \to B) = 0$, and show that $h(t_0, \neg B) = 1$. We have $h(t_0, A) = 0$ and also $h(r, A) = 1, h(r, B) = 0$ for some $r \geq t_0$. Since $h(r, A) \neq h(t_0, A)$, it follows that $r = t_1$. Now $h(t_1, B) = 0$ implies $h(t_0, B) = 0$ by persistence. Therefore, $h(t_0, \neg B) = 1$. Thus (e) is true at t_0 and also at t_1 (by persistence).

3. Let s_0 be the first element of T and let s_1, s_2 be two other elements. We can assume further that $s_2 \not\leq s_1$ (otherwise denote them the other way round). Now take a Kripke model such that

$$h(u, A) = 1 \text{ iff } u \geq s_1$$
$$h(u, B) = 1 \text{ iff } u \geq s_2$$

Then obviously, $h(s_0, A) = 0$. Also $h(s_0, \neg B) = 0$, since $s_0 \leq s_2$ and $h(s_2, B) = 1$. Furthermore, $h(s_1, A) = 1, h(s_1, B) = 0$ (since $s_2 \not\leq s_1$) yield that $h(s_0, A \to B) = 0$. Therefore $A \vee (A \to B) \vee \neg B$ is false at s_0.

2.4. WORKED EXAMPLES

4. $L_2 \not\subseteq S$ for example, because $\neg\neg A \to A \in L_2$, but $\neg\neg A \to A \notin S$ (by 2(b)). $S \not\subseteq L_2$, because $A \vee (A \to B) \vee \neg B \in S$ (by (2(e)), but $A \vee (A \to B) \vee \neg B \notin L_2$. To see this, take the values $A = 1/2, B = 0$. Then $A \vee (A \to B) \vee \neg B = 1/2 \vee (1/2 \to 0) \vee 0 = 1/2 \vee 1/2 = 1/2$.

5. (I) To show that L_2-negation is non-expressible in S, we consider truth tables in S for the following formulae built of a single atom x:

x	$\neg x$	$x \vee \neg x$	$\neg\neg x$	$\top$	$\bot$
0	1	1	0	1	0
1/2	0	1/2	1	1	0
1	0	1	1	1	0

Let us prove that any formula A with a single atom is equivalent to one of these six; this will solve the problem, because L_2-negation has a different truth table (cf. Table 2.1).

The proof goes by induction on the length of A.

- If $A = x, \top$ or $\bot$, there is nothing to prove.
- If $A = B \wedge C$ and $B \equiv C', C \equiv C'$ and B', C' are among the six, then $A \equiv B' \wedge C'$ and there are several cases to be considered.
    (a) $C' = \top$. Then $A \equiv (B' \wedge \top) \equiv B'$ and there is nothing to prove.
    (b) $B' = \top$. Similarly to (a).
    (c) $C' = \bot$ or $B' = \bot$. Then $A \equiv \bot$.
    (d) $A \equiv x \wedge \neg x$. Then $A \equiv \bot$ (this is seen from the truth tables).
    (e) $A \equiv x \wedge (x \vee \neg x)$. From the truth tables one can see that $A \equiv x$.
    (f) $A \equiv x \wedge \neg\neg x$. Then it follows that $A \equiv x$.
    (g) $A \equiv \neg x \wedge (x \vee \neg x)$. Then $A \equiv \neg x$.
    (h) $A \equiv \neg x \wedge \neg\neg x$. Then $A \equiv \bot$.
    (i) $A \equiv (x \vee \neg x) \wedge \neg\neg x$. Then $A \equiv x$.
    (j) $B' = C'$. Then $A \equiv B'$.
- If $A = B \vee C, B \equiv B', C \equiv C'$ and B', C' are among the six, then $A \equiv B' \vee C'$, and again we consider cases:
    (a) $C' = \top$ or $B' = \top$. Then $A \equiv \top$.
    (b) $C' = \bot$. Then $A \equiv B' \vee \bot \equiv B'$.

(c) $B' = \bot$. Then $A \equiv \bot \vee C' \equiv C'$.
(d) $B' = x, C' = \neg x$. Then $A \equiv x \vee \neg x$.
(e) $B' = \neg x, C' = x$. Then $A \equiv x \vee \neg x$.
(f) $A \equiv x \vee (x \vee \neg x)$. Then $A \equiv x \vee \neg x$.
(g) $A \equiv x \vee \neg\neg x$. Then $A \equiv \neg\neg x$.
(h) $A \equiv \neg x \vee (x \vee \neg x)$. Then $A \equiv x \vee \neg x$.
(i) $A \equiv (x \vee \neg x) \vee \neg\neg x$. Then $A \equiv \top$.
(j) $A \equiv \neg x \vee \neg\neg x$. Then $A \equiv \top$.
(k) $B' = C'$. Then $A \equiv B'$.

- If $A = B \to C, B' \equiv B, C' \equiv C$ and B', C' are among the six, then again we have several cases to consider:

(a) $C' = \top$. Then $A \equiv (B' \to \top) \equiv \top$.
(b) $B' = C'$. Then $A \equiv (B' \to B') \equiv \top$.
(c) $A = x \to \bot$. Then $A \equiv \neg x$.
(d) $A \equiv \neg x \to \bot$. Then $A \equiv \neg\neg x$.
(e) $A \equiv (x \vee \neg x) \to \bot$. Then $A \equiv \bot$.
(f) $A \equiv \neg\neg x \to \bot$. Then $A \equiv \neg x$.
(g) $A \equiv \neg x \to x$. Then $A \equiv \neg\neg x$.
(h) $A \equiv \neg x \to x \vee \neg x$. Then $A \equiv \top$.
(i) $A \equiv x \to x \vee \neg x$. Then $A \equiv \top$.
(j) $A \equiv x \to \neg\neg x$. Then $A \equiv \top$.
(k) $A \equiv x \to \neg x$. Then $A \equiv \neg x$.
(l) $A \equiv \neg x \to \neg\neg x$. Then $A \equiv \neg\neg x$.
(m) $A \equiv (x \vee \neg x) \to x$. Then $A \equiv \neg\neg x$.
(n) $A \equiv (x \vee \neg x) \to \neg x$. Then $A \equiv \neg x$.
(o) $A \equiv (x \vee \neg x) \to \neg\neg x$. Then $A \equiv \neg\neg x$.
(p) $A \equiv \neg\neg x \to x$. Then $A \equiv x \vee \neg x$.
(q) $A \equiv \neg\neg x \to \neg x$. Then $A \equiv \neg x$.
(r) $A \equiv \neg\neg x \to (x \vee \neg x)$. Then $A \equiv x \vee \neg x$.
(s) $B' = \top$. Then $A \equiv (\bot \to C') \equiv \top$.

- If $A = \neg B$, then $A \equiv B \to \bot$, and this is reduced to the previous cases.

(II) Now let $\to$ be the L_2-implication, $\neg$ be the L_2-negation.

2.4. WORKED EXAMPLES

Then $\neg(\neg x \to x)$ is the intuitionistic negation. To show this, consider the truth table:

x	$\neg x$	$\neg x \to x$	$\neg(\neg x \to x)$
0	1	0	1
1/2	1/2	1	0
1	0	1	0

3
INTRODUCING FORWARD RULES

In Chapter 1 we presented a method for checking whether a conclusion B follows logically from a set of assumptions $A_1, \ldots, A_n$, namely that of using truth tables to check whether

$$[A_1 \wedge \cdots \wedge A_n] \to B$$

is a tautology. This method, although always effective, is not always the most efficient and immediate. Consider the following example:

$$\frac{\begin{array}{c} p \to q \\ \neg r \vee q \to s \vee p \\ q \end{array}}{s \vee q}$$

If we follow our method, we have to check whether the following formula is a tautology:

$$[(p \to q) \wedge [(\neg r \vee q) \to (s \vee p)] \wedge q] \to s \vee q$$

Because there are four propositions in the formula, we need a table with $2^4 = 16$ rows to do that. Yet we can immediately see that the conclusion follows from the assumption, because we have q as an assumption and $s \vee q$ as a conclusion.

This suggests a new approach. Given a set of assumptions, we would like to manipulate and combine them, step by step, until we can see that the conclusion follows. We are looking for small combinatorial steps

69

which can lead from the assumptions to the conclusion. The approach we shall present relies on the observation that if $A_1 \wedge \cdots \wedge A_n \to B$ is a tautology and $A_1 \wedge \cdots \wedge A_n \wedge B \to C$ is a tautology, then $A_1 \wedge \cdots \wedge A_n \to C$ is a tautology. We can write this in the form of an argument, so that

$$\frac{\begin{array}{c} A_1 \wedge \cdots \wedge A_n \to B \\ A_1 \wedge \cdots \wedge A_n \wedge B \to C \end{array}}{A_1 \wedge \cdots \wedge A_n \to C}$$

The above rule (which is called the *cut* rule, for cutting out the intermediate formula B) allows us to find the intermediate B which help us derive the required C. We are using the B as a lemma, which is a perfectly natural way of thinking for humans. For example,

$$\frac{\begin{array}{c} \text{I have one orange and one apple implies I have two fruit} \\ \text{I have one orange and one apple and I have two fruit implies} \\ \text{I have an even number of fruit} \end{array}}{\text{I have one orange and one apple implies I have an even number of fruit}}$$

Notice that we do not need to use all of the $A_1, \ldots, A_n$ when deriving the C from the intermediate formula B. Certainly we can conclude that I have an even number of fruit from the fact that I have two fruit, without knowing what those fruit are.

Example 3.0.1 *From the assumptions below, we will attempt to derive D by using the* modus ponens *argument from Chapter 1, namely that from P and $P \to Q$ we can obtain Q.*

1. $A \to B$

2. A

3. $B \to C \vee D$

4. $\neg C$

We can argue as follows (the numbered lines give the next steps in the argument):

5. We can get B from assumptions 1 and 2:

$$\frac{A, \; A \to B}{B}$$

6. From intermediate step 5 and assumption 3 we can get $C \vee D$:

$$\frac{B,\ B \to C \vee D}{C \vee D}$$

7. From assumption 4 and intermediate step 6 we get D:

$$\frac{\neg C,\ C \vee D}{D}$$

Here we have only used simple elementary steps, with lines 5 and 6 being the intermediate steps for getting to the conclusion.

We want to formulate a body of rules which can be used to reach conclusions from assumptions and do the same job as the truth tables. Deriving such a method is crucial in the case of predicate logic (the topic of later chapters) because truth tables cannot be drawn up for predicate logic, and the only semi-mechanical way to generate the set of all valid formulae of predicate logic is via rules. In this book we shall present two methods of using rules. In the above two examples, we worked forwards from the assumptions to the conclusion, and thus we were using a *forward method*. In later chapters we shall change to using *backward methods* in which we start with the conclusion, and work backwards to the assumptions.

We use the notation

$$\frac{A_1, \ldots, A_n}{B}$$

to indicate a logic rule which can be used in reasoning from data. The reasoning is performed step by step; in each step we must indicate where and how the $A_1, \ldots, A_n$ are obtained and which rule is used to obtain B. The choice of rules is up to us and there are various standard systems of rules. Section 3.4 will give the formal definition of the reasoning process. Let us denote by

$$A_1, \ldots, A_n \vdash_R B$$

the notion of B being obtainable from $A_1, \ldots, A_n$ using the set of rules R. We must show that R does exactly what the table does, i.e. we must show that R satisfies the following *soundness and completeness* conditions:

Soundness: If $A_1 \wedge \cdots \wedge A_n \vdash_R B$ then $A_1 \wedge \cdots \wedge A_n \to B$ is a tautology.

Completeness: If $A_1 \wedge \cdots \wedge A_n \to B$ is a tautology then $A_1 \wedge \cdots \wedge A_n \vdash_R B$.

In Chapter 5, we will prove that the backward reasoning method to be introduced in Chapter 4 satisfies the soundness and completeness condition.

3.1 Natural deduction rules

We now discuss the rules for reasoning forwards from the assumptions to the conclusion. The rules will be divided into two categories, *introduction* rules and *elimination* rules. The introduction rules will enable us to combine formulae by introducing conjunctions, disjunctions, implications, and so forth. The elimination rules permit us to break up formulae into valid constituent subformulae. By judicious use of the introduction and elimination rules, we are able to combine subformulae of our assumptions to reach the desired conclusion (should it be a logical consequence of the assumptions). To explain further the rationale behind the rules, we will look at the individual rules, starting with those for conjunction.

3.1.1 Rules for conjunction

The truth table for conjunction is

A	B	$A \wedge B$
$\top$	$\top$	$\top$
$\bot$	$\top$	$\bot$
$\top$	$\bot$	$\bot$
$\bot$	$\bot$	$\bot$

We want to write rules which correspond exactly to the truth table. The rules will have two forms, corresponding to the two directions of the truth table:

1. $A = \top$ and $B = \top$ together imply $A \wedge B = \top$.

2. $A \wedge B$ is $\top$ only in the case mentioned in point 1.

The first form is dealt with in the *introduction* rules, because we introduce the conjunction $A \wedge B$ when both A and B are true. So we write

3.1. NATURAL DEDUCTION RULES

$$(\wedge I)\frac{A,\ B}{A \wedge B}$$

which states that $A = \top$ and $B = \top$ together imply that $A \wedge B = \top$. The $(\wedge I)$ stands for 'and introduction'. The *elimination* rule deals with the second rule form. Once we know $A \wedge B = \top$, what can we say about the rows in the truth table which give this result? The rule has two components, and says

$$(\wedge E)\frac{A \wedge B}{A} \text{ and } \frac{A \wedge B}{B}$$

which states that $A \wedge B = \top$ implies $A = \top$ and $A \wedge B = \top$ implies $B = \top$. In other words, when a conjunction is true, both its left and right conjuncts are true. The $\wedge E$ stands for 'and elimination'.

Example 3.1.1 *We will show that from the assumptions $p \wedge q$ and $r \wedge s$ we can derive the conclusion $p \wedge s$. Applying the first component of the $(\wedge E)$ rule to $p \wedge q$, we can derive p, and using the second component on $r \wedge s$ gives us s. Combining these via the $(\wedge I)$ rule on p and s gives us the desired $p \wedge s$.*

3.1.2 Rules for disjunction

The truth table for disjunction is

A	B	$A \vee B$
$\top$	$\top$	$\top$
$\bot$	$\top$	$\top$
$\top$	$\bot$	$\top$
$\bot$	$\bot$	$\bot$

The first three rows in the table, those in which either $A = \top$ or $B = \top$, yield $A \vee B = \top$. Hence the 'or introduction' rule is

$$(\vee I)\frac{A}{A \vee B} \text{ and } \frac{B}{A \vee B}$$

We could have a rule which said that $A \vee B = \top$ when $A = \top$ and $B = \top$, but this is subsumed by the two rules we have presented. No other rows give $A \vee B = \top$, and we must reflect this fact through the 'or elimination' rule. So we must say that $A \vee B$ is true only through A being true or through B being true. There are many ways of doing this. For example, we can write

$$(\vee E1)\frac{A\vee B,\ \neg A}{B} \qquad (\vee E2)\frac{A\vee B,\ \neg B}{A}$$

which will achieve the desired effect. Alternatively we can use the following rule, which is usually to be found in textbooks:

$$(\vee E)\frac{A\to C,\ B\to C,\ A\vee B}{C}$$

($\vee$E) is a better rule to adopt, and we take it as part of our official set of natural deduction rules. The reasons for adopting one rule and not another are mainly because of convenience, and elegance. In this book we will use all three rules for $\vee$: we use ($\vee$E) as a basic rule, and since ($\vee$E1) and ($\vee$E2) can be proved from ($\vee$E) and the rules for $\neg$, we can use them as well.

Example 3.1.2 *From the assumptions $p\wedge\neg q$ and $q\vee r$ we can derive $r\vee s$ as a conclusion. Using the second component of the ($\wedge$E) rule on $p\wedge\neg q$, we can derive $\neg q$, and using ($\vee$E1) on $\neg q$ and $q\vee r$ gives us r. From this, we use the ($\vee$I) rule to introduce the new disjunct s, giving the conclusion $r\vee s$.*

3.1.3 Rules for implication

The table for $\to$ is

A	B	$A\to B$
T	T	T
⊥	T	T
T	⊥	⊥
⊥	⊥	T

1. If $A = \bot$ or $B = \top$ then $A\to B = \top$.

2. There are no other cases than point 1 where $A\to B = \top$.

We can write these as

$$(\to I1)\ \frac{\neg A}{A\to B} \qquad (\to I2)\ \frac{B}{A\to B} \qquad (\to E1)\ \frac{A\to B}{\neg A\vee B}$$

but these rules involve $\vee$ and $\neg$ and are thus not rules which purely deal with implication. Further on in this book, we shall show how conjunction, disjunction and negation may be expressed in terms of implication, which we shall take as a 'fundamental' connective. It is better, therefore, for us to use pure rules for $\to$. These are

3.1. NATURAL DEDUCTION RULES

$$(\to E) \; \frac{A, \; A \to B}{B}$$

and

$(\to I)$ If $\dfrac{\{\text{assumptions}\}, A}{B}$ is valid then $\dfrac{\{\text{assumptions}\}}{A \to B}$ is also valid

We will use $(\to I)$ and $(\to E)$ as our rules for $\to$.

Example 3.1.3 *To show s from the assumptions*

1. $p \to q$
2. p
3. $q \vee r \to s$

we proceed as follows.

4. *From assumptions 1 and 2 using $(\to E)$ we get q.*
5. *From intermediate result 4 using $(\vee I)$ we get $q \vee r$.*
6. *From intermediate result 5 and assumption 3 using $(\to E)$ we get s.*

We can show this as a structured argument:

$$\dfrac{\dfrac{\dfrac{p, \; p \to q}{q}}{q \vee r} \quad q \vee r \to s}{s}$$

3.1.4 Rules for negation

The truth table for $\neg$ is

A	$\neg A$
$\top$	$\bot$
$\bot$	$\top$

1. If $A = \bot$ then $\neg A = \top$.
2. $\neg A = \top$ only if $A = \bot$.

If we write the rules in the usual way we get

$$\frac{\neg A}{\neg A} \text{ and } \frac{\neg A}{\neg A}$$

which are not particularly useful, so let us write other rules which are more effective:

$$(\neg\text{E1}) \ \frac{\neg\neg A}{A} \text{ and } (\neg\text{I1}) \ \frac{A}{\neg\neg A}$$

These are not enough. We must also say that any A has two possible truth values, $\top$ and $\bot$, and no more. Thus we take

$$(\neg 2) \ \frac{A\to B \quad \neg A\to B}{B}$$

The usual rules for $\neg$ are

$$(\neg\text{E}) \ \frac{\neg A\to B, \ \neg A\to \neg B}{A} \text{ and } (\neg\text{I}) \ \frac{A\to B, \ A\to \neg B}{\neg A}$$

We will adopt ($\neg$E) and ($\neg$I) as our basic rules, although we will find ($\neg$E1), ($\neg$I1) and ($\neg$2) most useful. They can be proved from the basic rules.

Example 3.1.4 *We must show r from these two assumptions:*

1. $p\to(q\to r)$

2. $p\wedge q$.

We do this by the following steps:

3. *From assumption 2 using ($\wedge$E) we derive p.*

4. *From intermediate result 3 and assumption 1 using ($\to$E) we have $q\to r$.*

5. *From assumption 2 using ($\wedge$E) we derive q.*

6. *From intermediate results 4 and 5 using ($\to$E) we have r.*

In argument form this is

$$\cfrac{\cfrac{p\wedge q}{p} \quad p\to(q\to r)}{q\to r} \quad \cfrac{p\wedge q}{q}$$
$$r$$

3.2. USING SUBCOMPUTATIONS

Exercise 3.1.5 *Reasoning with these rules is tricky, because there is very little guidance about which rule to apply in which situation. In the following exercise, the aim is to explore the possibilities that the introduction and elimination rules permit, rather than actually to prove, the conclusion from the assumptions. Using the rules is a little like playing chess—there are many ways in which the pieces may be moved, but not all moves are good ones.*

Attempt to show that the following arguments are valid (they all are):

1. $\dfrac{p \wedge q}{p \vee q}$

2. $\dfrac{p \rightarrow q,\ \neg q}{\neg p}$

3. $\dfrac{p \vee r,\ \neg p,\ r \rightarrow q}{q}$.

3.2 Using subcomputations

So far we have not altered the set of assumptions during the derivation of the conclusion. The following example requires us to do just that.

Example 3.2.1 *This example involves showing that the implication $p \rightarrow r$ follows from these assumptions:*

1. $p \rightarrow q$

2. $q \rightarrow r$

Recall that the $(\rightarrow\text{I})$ rule is

$$\text{If } \dfrac{\{assumptions\},\ A}{B} \text{ is valid then } \dfrac{\{assumptions\}}{A \rightarrow B} \text{ is also valid}$$

So that by $(\rightarrow\text{I})$ it is sufficient to show that r follows from the assumptions 1, 2 and p to show $p \rightarrow r$. Hence we make a third assumption:

3. p

before we derive r.

4. Now assumptions 1 and 3 give q by $(\rightarrow\text{E})$.

5. Intermediate result 4 and assumption 2 give r by (→E).

Hence we have shown that p→r follows from the assumptions.

The preceding example was a simple illustration of a technique which we shall often employ in this book, namely reducing the problem of proving one formula to a subproblem of proving a reduced formula from an increased set of assumptions. When making use of this technique, it is important to be aware of the boundaries of the subcomputation. If we had been trying to prove that (A→C)∧ A followed from the assumptions in Example 3.2.1, it would have been no good suddenly to add a new step 6, in which we took the A that we had assumed for the subproblem and used (∧I) to derive (A→C) ∧ A. The A must only be used within the bounds of the subproblem (A→C). To illustrate this, we can represent the argument in Example 3.2.1 in the following diagram:

(1)   A→B   data
(2)   B→C   data
(3)   A→C   from subcomputation

$$\underline{C}$$

(3.1)   A   assumption
(3.2)   B   from (1) and (3.1)
(3.3)   C   from (2) and (3.2)

The underlined formula in the upper right-hand corner of the box is what we wish to prove. If we manage to prove it (as in this case) it will also be the last formula in the box. The first formulae in the box are the assumptions. Further formulae can be added after the assumptions by using the natural deduction rules as in the previous examples. We need to mark each line whether it is an assumption or indicate from what previous lines it is derived and by which rules. If we need to use a subcomputation to prove any of the new formulae, including the 'goal' formula, we start a new box after the formula we wish to prove. This new box has its 'subgoal' formula (in this case C) in the upper right-hand corner, and as the last formula in the box. The added assumptions are put in at the start of the box (in this case A). Again, the natural deduction rules are used to derive any intermediate formulae needed. *The added assumptions may only be used within their own box (or any box included in their box)*, although subcomputations can access assumptions in their outer boxes. If we attempt to draw a diagram for the attempted derivation of (A→C)∧A given above, we run into problems.

3.2. USING SUBCOMPUTATIONS

$$\begin{array}{l} A \to B \\ B \to C \\ A \to C \\ A \\ C \\ \times \quad (A \to C) \wedge A \end{array}$$

The last line in the outer box cannot be derived, because the A is not available outside the box in which it was introduced. A more formal description of the box representation of proofs will be given later.

To prove that a formula is a tautology, we must show that it can be derived from the empty set of assumptions. Recall that an argument with assumptions $A_1, \ldots, A_n$ and conclusion B is logically valid if and only if $[A_1 \wedge \cdots \wedge A_n] \to B$ is a tautology. Now any formula B is equivalent to the formula $\top \to B$, where $\top$ is truth. So to show that a formula B is a tautology is the same as showing that $\top \to B$ is a tautology, and thus that the argument with assumption $\top$ and conclusion B is valid. Since $\top$ is always given, we should show that the argument with no assumptions and the conclusion B is valid.

Example 3.2.2 *$p \vee \neg p$ is a tautology provided*

$$\frac{nothing}{p \vee \neg p}$$

is valid. By the $(\to I)$ rule, since

$$\frac{p}{p \vee \neg p}$$

is valid (by the $(\vee I)$ rule), so is

$$\frac{nothing}{p \to (p \vee \neg p)}$$

Similarly, since

$$\frac{\neg p}{p \vee \neg p}$$

is valid (again by the $(\vee I)$ rule), so is

$$\frac{nothing}{\neg p \to (p \vee \neg p)}$$

By using $(\neg 2)$ we have

$$\frac{p \to p \vee \neg p \quad \neg p \to p \vee \neg p}{p \vee \neg p}$$

as required.

Exercise 3.2.3 *Prove the following tautologies from the earlier Exercise 1.2.2, this time using the forward rules instead of the truth tables used then.*

1. $(p \vee q) \wedge \neg p \to q$

2. $(\neg q \to \neg p) \to (p \to q)$

Consider the tautology $((p \to q) \to p) \to p$ which was also shown to be a tautology in Exercise 1.2.2. To show this, we must demonstrate that the argument

$$\frac{nothing}{((p \to q) \to p) \to p}$$

is valid. By $(\to I)$ we have to show

$$\frac{(p \to q) \to p}{p}$$

How can we show p from the data $(p \to q) \to p$? The only way we have is to show $(p \to q)$ and then use $(\to E)$. We thus show

$$\frac{(p \to q) \to p}{p \to q}$$

which by $(\to I)$ is shown by proving

$$\frac{p, (p \to q) \to p}{q}$$

At this point we grind to a halt, as none of the rules seem to help to prove this argument. In fact this argument is invalid; we can let $p = \top$ and $q = \bot$. In Chapter 1, however, we showed our original argument to be a tautology via truth tables, so our method of reasoning with rules should be powerful enough to prove it as well.

We will leave this problem until the next chapter, since the solution lies in a different way of using rules. In the meantime remember the following: truth tables are mechanical—when you use truth tables you are assured of finding an answer. When you use natural deduction rules you need ingenuity; you may not find an answer at all. There are certain problems in the logic literature involving a few simple assumptions and a simple conclusion, which have required hundreds of reasoning steps to prove. Do not, therefore, be surprised if several intermediate steps (though hopefully not hundreds) are needed to solve Exercise 3.2.4 below.

Exercise 3.2.4

1. Derive the rules ($\to$I1), ($\to$I2) and ($\to$E1) from ($\to$I) and ($\to$E) and the negation rules.

2. Show the following using the natural deduction rules:

   (a) $\dfrac{p\to(q\to r)}{p\wedge q\to r}$

   (b) $\dfrac{p\vee q\to r}{(p\to r)\wedge(q\to r)}$

   (c) $\dfrac{nothing}{q\to(p\to q)}$

   (d) $\dfrac{nothing}{p\to p}$

   (e) $\dfrac{\neg(p\to q)}{p\wedge\neg q}$

   (f) $\dfrac{\neg p\vee q}{p\to q}$

3.3 Example proofs

In this section, we present a series of proofs using the box notation. Together they illustrate how the natural deduction rules should be used and, in conjunction with the exercises included in the section, give proofs of some of the common equivalences of propositional logic. We begin with some proofs which use the introduction and elimination rules for ∧, ∨ and →, before progressing to the rules for negation, which often involve more convoluted proofs.

Given a problem to prove of the form $\mathcal{P} \vdash A$ (sometimes we write $\mathcal{P} \vdash ?A$ to indicate that we want to show this) we refer to the wffs in $\mathcal{P}$ as *data* and to A as a goal.[1] In the course of the proof (computation), we may need to show $C \to D$; we start a new subcomputation box with C as the *assumption* (actually additional data) and D as the goal within that subcomputation.

3.3.1 Proofs using ∧, ∨ and → rules

We start by proving that ∧ is both commutative and associative.

1. $P \wedge Q \vdash Q \wedge P$

(1)	$P \wedge Q$	data
(2)	Q	(∧E) on (1)
(3)	P	(∧E) on (1)
(4)	$Q \wedge P$	(∧I) on (3),(2)

2. $P \wedge (Q \wedge R) \vdash (P \wedge Q) \wedge R$

(1)	$P \wedge (Q \wedge R)$	data
(2)	P	(∧E) on (1)
(3)	$Q \wedge R$	(∧E) on (1)
(4)	Q	(∧E) on (3)
(5)	$P \wedge Q$	(∧I) on (4),(2)
(6)	R	(∧E) on (3)
(7)	$(P \wedge Q) \wedge R$	(∧I) on (5),(6)

Next we prove some properties of implication, showing that anything implies something which is true, that implication distributes over conjunctions on the right, and that nested implications have the effect of conjunctive conditions.

---

[1] In Section 10.1 $\mathcal{P}$ is referred to as the *database* and A as the *query*. The reasoning process in Chapter 10 is non-monotonic.

3.3. EXAMPLE PROOFS

3. $Q \vdash P \to Q$

   (1) Q data
   (2) $P \to Q$ subcomputation

		Q
(2.1)	P	assumption
(2.2)	Q	from (1)

4. $P \to Q \land R \vdash (P \to Q) \land (P \to R)$

   (1) $P \to Q \land R$ data
   (2) $P \to Q$ subcomputation

		Q
(2.1)	P	assumption
(2.2)	$Q \land R$	($\to$E) on (1),(2.1)
(2.3)	Q	($\land$E) on (2.2)

   (3) $P \to R$ subcomputation

		R
(3.1)	P	assumption
(3.2)	$Q \land R$	($\to$E) on (1),(3.1)
(3.3)	R	($\land$E) on (3.2)

   (4) $(P \to Q) \land (P \to R)$ ($\land$I) on (2),(3)

5. $P \to (Q \to R) \vdash P \land Q \to R$

   (1) $P \to (Q \to R)$ data
   (2) $P \land Q \to R$ subcomputation

		R
(2.1)	$P \land Q$	assumption
(2.2)	P	($\land$E) on (2.1)
(2.3)	$Q \to R$	($\to$E) on (1),(2.2)
(2.4)	Q	($\land$E) on (2.1)
(2.5)	R	($\to$E) on (2.3),(2.4)

Finally in this group, we illustrate the use of ($\lor$E).

6. $B, R{\vee}S{\rightarrow}A, R{\vee}S, A{\wedge}R{\rightarrow}C, B{\wedge}S{\rightarrow}C \vdash C$

   (1) B data
   (2) $R{\vee}S{\rightarrow}A$ data
   (3) $R{\vee}S$ data
   (4) $A{\wedge}R{\rightarrow}C$ data
   (5) $B{\wedge}S{\rightarrow}C$ data
   (6) A ($\rightarrow$E) on (2),(3)
   (7) $R{\rightarrow}C$ subcomputation

> $\underline{C}$
> (7.1) R assumption
> (7.2) $A{\wedge}R$ ($\wedge$I) on (6),(7.1)
> (7.3) C ($\rightarrow$E) on (4),(7.2)

   (8) $S{\rightarrow}C$ subcomputation

> $\underline{C}$
> (8.1) S assumption
> (8.2) $B{\wedge}S$ ($\wedge$I) on (1),(8.1)
> (8.3) C ($\rightarrow$E) on (5),(8.2)

   (9) C ($\vee$E) on (3),(7),(8)

Exercise 3.3.1 *Show that the following arguments are valid, using the rules ($\wedge$I), ($\wedge$E), ($\vee$I), ($\vee$E), ($\rightarrow$I) and ($\rightarrow$E).*

1. $P{\wedge}Q{\rightarrow}R \vdash P{\rightarrow}(Q{\rightarrow}R)$

2. $(P{\rightarrow}Q){\wedge}(Q{\rightarrow}R) \vdash P{\rightarrow}R$

3. $P{\rightarrow}R, Q{\rightarrow}S \vdash P{\wedge}Q{\rightarrow}R{\wedge}S$

4. $(P \vee Q){\rightarrow}R \vdash (P{\rightarrow}R){\wedge}(Q{\rightarrow}R)$

3.3.2 Proofs involving negation rules

We start with a few relatively simple proofs using the negation rules. We use the $\neg$I rule $A{\rightarrow}B, A{\rightarrow}\neg B \vdash \neg A$ in both proofs, in the following manner. We select a formula to play the role of B, and use two subcomputations to show that we have both $A{\rightarrow}B$ and $A{\rightarrow}\neg B$ and thus we must have $\neg A$.

3.3. EXAMPLE PROOFS

7. $P \to Q \vdash \neg(P \wedge \neg Q)$
   - (1) $P \to Q$ data
   - (2) $(P \wedge \neg Q) \to Q$ subcomputation
     > $\underline{Q}$
     > (2.1) $P \wedge \neg Q$ assumption
     > (2.2) P ($\wedge$E) on (2.1.1)
     > (2.3) Q ($\to$E) on (1),(2.1.2)
   - (3) $(P \wedge \neg Q) \to \neg Q$ subcomputation
     > $\underline{\neg Q}$
     > (3.1) $P \wedge \neg Q$ assumption
     > (3.2) $\neg Q$ ($\wedge$E) on (2.2.1)
   - (4) $\neg(P \wedge \neg Q)$ ($\neg$I) on (2),(3)

8. $\vdash \neg(P \wedge \neg P)$
   - (1) $(P \wedge \neg P) \to P$ subcomputation
     > $\underline{P}$
     > (1.1.1) $P \wedge \neg P$ assumption
     > (1.1.2) P ($\wedge$E)on (1.1.1)
   - (2) $(P \wedge \neg P) \to \neg P$ subcomputation
     > $\underline{\neg P}$
     > (1.2.1) $P \wedge \neg P$ assumption
     > (1.2.2) $\neg P$ ($\wedge$E)on (1.2.1)
   - (3) $\neg(P \wedge \neg Q)$ ($\neg$I) on (1),(2)

9. $P \to Q, \neg Q \vdash \neg P$
   - (1) $P \to Q$ data
   - (2) $\neg Q$ data
   - (3) $P \to Q$ subcomputation
     > $\underline{Q}$
     > (3.1) P assumption
     > (3.2) Q ($\to$E) on (1),(3.1)

  (4) $P{\to}\neg Q$ subcomputation

> $\underline{\neg Q}$
>
> (4.1) P assumption
> (4.2) $\neg Q$ from (2)

  (5) $\neg P$ ($\neg$I) on (3),(4)

10. $(P{\to}\neg P){\to}(\neg P{\to}P) \vdash P$

 (1) $(P{\to}\neg P){\to}(\neg P{\to}P)$ data
 (2) $\neg P{\to}P$ subcomputation

> $\underline{P}$
>
> (2.1) $\neg P$ assumption
> (2.2) $P{\to}\neg P$ subcomputation
>
> > $\underline{\neg P}$
> >
> > (2.2.1) P assumption
> > (2.2.2) $\neg P$ from (2.1)
>
> (2.3) $\neg P{\to}P$ ($\to$E) on (1),(2.2)
> (2.4) P ($\to$E) on (2.1),(2.3)

 (3) $\neg P{\to}\neg P$ subcomputation

> $\underline{\neg P}$
>
> (3.1) $\neg P$ assumption
> (3.2) $\neg P$ from (3.1)

 (4) P ($\neg$E) on (2),(3)

We complete this section with a convoluted example to show how involved these proofs can become, and that there may be more than one way of proving a formula.

3.3. EXAMPLE PROOFS

11. $(P{\to}Q){\to}Q,\ Q{\to}P \vdash P$

   (1) $(P{\to}Q){\to}Q$ data
   (2) $Q{\to}P$ data
   (3) $\neg P{\to}(P{\to}Q)$ subcomputation

                                                                                       $\underline{P{\to}Q}$

      (3.1) $\neg P$ assumption
      (3.2) $P{\to}Q$ subcomputation

                                                                               $\underline{Q}$

         (3.2.1) P assumption
         (3.2.2) $\neg Q{\to}P$ assumption

                                                                                             $\underline{P}$

            (3.2.2.1) $\neg Q$ assumption
            (3.2.2.2) P from (3.2.1)

         (3.2.3) $\neg Q{\to}\neg P$ assumption

                                                                              $\underline{\neg P}$

            (3.2.3.1) $\neg Q$ assumption
            (3.2.3.2) $\neg P$ from (3.1)

         (3.2.4) Q ($\neg$E) on (3.2.2),(3.2.3)

   (4) $\neg P{\to}\neg(P{\to}Q)$ subcomputation

                                                                      $\underline{\neg(P{\to}Q)}$

      (4.1) $\neg P$ assumption
      (4.2) $Q{\to}\neg P$ subcomputation

                                                           $\underline{\neg P}$

         (4.2.1) Q assumption
         (4.2.2) $\neg P$ from (4.1)

      (4.3) $\neg Q$ ($\neg$I) on (1),(4.2)
      (4.4) $(P{\to}Q){\to}\neg Q$ subcomputation

                                                                      $\underline{Q}$

         (4.4.1) $P{\to}Q$ assumption
         (4.4.2) $\neg Q$ (4.3)

      (4.5) $\neg(P{\to}Q)$ ($\neg$I)(1),(4.4)

   (5) P ($\neg$E) on (3),(4)

A shorter box proof is the following:

(1) $(P \to Q) \to Q$ data
(2) $Q \to P$ data
(3) $\neg P \to P$ subcomputation

> $\underline{P}$
>
> (3.1) $\neg P$ assumption
> (3.2) $P \to Q$ subcomputation
>
> > $\underline{Q}$
> >
> > (3.2.1) P assumption
> > (3.2.2) $\neg Q \to P$ subcomputation
> >
> > > $\underline{P}$
> > >
> > > (3.2.2.1) $\neg Q$ assumption
> > > (3.2.2.2) P from (3.2.1)
> >
> > (3.2.3) $\neg Q \to \neg P$ subcomputation
> >
> > > $\underline{\neg P}$
> > >
> > > (3.2.3.1) $\neg Q$ assumption
> > > (3.2.3.2) $\neg P$ from (3.1)
> >
> > (3.2.4) Q ($\neg$E) on (3.2.2),(3.2.3)
>
> (3.3) Q ($\to$ E) on (1), (3.2)
> (3.4) P ($\to$ E) on (2), (3.3)

(4) $\neg P \to \neg P$ subcomputation

> $\underline{\neg P}$
>
> (4.1) $\neg P$ assumption
> (4.2) $\neg P$ from (4.1)

(5) P ($\neg$E) on (3),(4)

Exercise 3.3.2 *Show, using negation rules, that the following arguments are valid.*

1. $\neg(P \wedge \neg Q) \vdash P \to Q$

2. $\vdash ((P \to Q) \to P) \to P$

3. $\vdash \neg(P \to Q) \to P$

3.4 Formal descriptions of proofs

Definition 3.4.1 (Indices) *We now give a series of precise definitions for the notion of a correct forward proof.*

1. *Let $\mathbb{N}$ be the set of natural numbers and let $\mathbb{N}^*$ be the set of non-empty sequences of natural numbers. For $\alpha, \beta \in \mathbb{N}^*$, let $\alpha S \beta$ mean that α is an initial segment of β.*

2. *Let $\alpha < \beta$ be defined as follows:[2]*

$$(a_1 \ldots a_n) < (b_1 \ldots b_m) \quad \textit{iff} \quad (m \geq n) \wedge \left(\bigwedge_{i=1}^{n-1} a_i = b_i \wedge a_n < b_n \right).$$

3. *Let $\alpha \oplus \beta$ mean the concatenation of the sequences α and β, and let Λ be the empty sequence.*

4. *Let $\alpha R \beta$ mean that α comes before β in the lexicographic ordering. It can be defined as*

   $\alpha R \beta$ *iff for some γ we have $[\gamma < \beta \wedge \gamma S \alpha]$, γ is possibly empty*

Example 3.4.2 *We need the above indices for the description of computations and subcomputations. Suppose we want to prove that from the data $A \rightarrow (B \rightarrow C)$ it follows that $B \rightarrow (A \rightarrow C)$. The first line of the proof would be*

$$(1) \quad A \rightarrow (B \rightarrow C) \quad \text{data}$$

The second line of the proof is a subcomputation showing the goal $B \rightarrow (A \rightarrow C)$, by assuming B and showing $A \rightarrow C$. We write

(2)   Show $B \rightarrow (A \rightarrow C)$
(2.1) B assumption for subcomputation 2
(2.2)   Show $A \rightarrow C$

We show $A \rightarrow C$ by going to another subcomputation, namely we assume A and show C. We write

---
[2] We want to say that $\alpha = (a_1, \ldots, a_{n-1}, a_n)$ and $\beta = (a_1, \ldots, a_{n-1}, b_n, \ldots, b_m)$ and $a_n < b_n$. This means that at the point $(a_1, \ldots, a_{n-1}, b_n)$ we started a new box subcomputation and β is somewhere inside the box.

(2.2.1) A   assumption for subcomputation (2.2)
(2.2.2) B→C   (2.2.1) and from (1) and (→E)
(2.2.3) C   from (2.1) and (2.2.2) and (→E)

The indices tell us what is a subcomputation of what and what is a box. Given an index α then all indices (α, a) are elements of the box α. $\alpha < \beta$ means that α is an assumption or result that β can use (e.g. in the rule (→E)). $\alpha R \beta$ simply orders the lines linearly. If we were to draw the boxes we would get

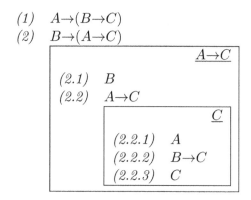

Definition 3.4.3 (Correct annotated computation) Let $\mathcal{P} = \{A_1, \ldots, A_n\}$ be a database and let G be a goal. Consider a set of indexed wffs $\pi = \{(\alpha_1, B_1, a_1), (\alpha_2, B_2, a_2), \ldots\}$. π is said to be a forward proof of the goal from the data iff the following conditions are satisfied:

1. π contains $((1), A_1, \text{data}), \ldots, ((n), A_n, \text{data})$ and $((m), G, a)$, for some m and a.

2. Any element $(\alpha, B, a) \in \pi$ satisfies exactly one of the following conditions.

    (a) $a = \text{'data'}$ and $B \in \mathcal{P}$ or for some b and for some $\beta < \alpha$, $(\beta, B, b) \in \pi$ and $a = \text{'reiteration from } \beta\text{'}$.

    (b) For some a_1, a_2 and for some $\gamma_1, \gamma_2 < \alpha$, and some B_1, B_2, $(\gamma_1, B_1, a_1) \in \pi$ and $(\gamma_2, B_2, a_2) \in \pi$ and $B_2 = B_1 \to B$ and $a = \text{'from } \gamma_1, \gamma_2 \text{ using rule } (\to E)\text{'}$.

    (c) For some b and for some $\beta < \alpha$ and some B_2, $B = B_1 \vee B_2$ and $(\beta, B_1, b) \in \pi$ and $a = \text{'from } \beta \text{ using rule } (\vee I)\text{'}$.

3.4. FORMAL DESCRIPTIONS OF PROOFS

(d) For some b and for some $\beta < \alpha$, and some B_2, $B = B_2 \vee B_1$ and $(\beta, B_1, b) \in \pi$ and $a = $ 'from β using rule ($\vee$I)'.

(e) For some b and for some $\beta < \alpha$ and some B_1, $(\beta, B \wedge B_1, b) \in \pi$ and $a = $ 'from β using rule ($\wedge$E)'.

(f) For some b and for some $\beta < \alpha$ and some B_1, $(\beta, B_1 \wedge B, b) \in \pi$ and $a = $ 'from β using rule ($\wedge$E)'.

(g) B is $B_1 \rightarrow B_2$ and $(\alpha \oplus (1), B_2, \text{assumption}) \in \pi$ and for some b and for some k, $(\alpha \oplus (k), B_2, b)$ is in π, and $a = $ 'subcomputation'.

(h) B is $\neg B_1$ and for some b_1, b_2, x_1, x_2 and for some wff X, $(\alpha \oplus (1,1), B_1, b_1) \in \pi$ and $(\alpha \oplus (2,1), B_1, b_2) \in \pi$, and for some k_1 and k_2, $(\alpha \oplus (1, k_1), X, x_1) \in \pi$ and $(\alpha \oplus (2, k_2), \neg X, x_2) \in \pi$, and $a = $ 'subcomputation', and $b_1 = b_2 = $ 'assumption'.

(i) For some b_1, b_2, x_1, x_2 and for some wff X, $(\alpha \oplus (1,1), \neg B, b_1) \in \pi$, and $(\alpha \oplus (2,1), \neg B, b_2) \in \pi$, and for some k_1 and k_2, $(\alpha \oplus (1, k_1), X, x_1) \in \pi$ and $(\alpha \oplus (2, k_2), \neg X, x_2) \in \pi$, and $a = $ 'subcomputation', and $b_1 = b_2 = $ 'assumption'.

(j) For some b, b_1, b_2, x_1, x_2 and for some $\beta < \alpha$, $(\beta, C \vee D, b) \in \pi$, and $(\alpha \oplus (1,1), C, b_1) \in \pi$ and $(\alpha \oplus (2,1), D, b_2) \in \pi$, and for some k_1, k_2, $(\alpha \oplus (1, k_1), B, x_2) \in \pi$ and $(\alpha \oplus (2, k_2), B, x_2) \in \pi$, and $a = $ 'subcomputation', and $b_1 = b_2 = $ 'assumption'.

(k) $B = B_1 \wedge B_2$ and for some b_1, b_2 and for some $\beta_1, \beta_2 < \alpha$, $(\beta_1, B_1, b_1) \in \pi$ and $(\beta_2, B_2, b_2) \in \pi$, and $a = $ 'from β_1, β_2 and rule ($\wedge$I)'.

(l) α has the form $\alpha_0 \oplus (1)$ or $\alpha_0 \oplus (1,1)$ or $\alpha_0 \oplus (2,1)$ and $a = $ 'assumption'.

Definition 3.4.4 (Box display) *Let π be a proof. For each $(\alpha, B_1, a) \in \pi$ let $Box(\alpha) = \{(\alpha_1, C, a_1) | \alpha S \alpha_1 \text{ and } \alpha \neq \alpha_1\}$. $Box(\alpha)$ may be empty. The following is a definition of box display. We present π linearly according to the lexicographic order of the indices. If (α, B, a) is a line we write $\alpha : \beta$; annotation.*

To be a box display, each line must satisfy one of the conditions of the previous definition. The annotation of line α describes the condition which the line satisfies. If the line $\alpha : \beta$ has a box associated with it ($Box(\alpha)$ is non-empty) we indent the line inwards and put in a graphical box. At the top right-hand corner of a box we write the goal (last line of the box).

Definition 3.4.5 (Box consequence) Let $\mathcal{P} \vdash_{\text{Box}} G$ be defined to mean that there exists a forward proof π from $\mathcal{P}$ as data to G as goal, in the sense of Definition 3.4.3.

Theorem 3.4.6 (Soundness of box consequence) If $\mathcal{P} \vdash_{\text{Box}} G$ then $\bigwedge \mathcal{P} \to G$ is a tautology.

Proof. Assume $\{A_1, \ldots, A_n\} \vdash_{\text{Box}} G$, with a box proof π. Then by definition, the following items are in π: $((i), A_i, \text{data}) \in \pi, i = 1, \ldots, n$, and $((m), G, a) \in \pi$ for some m. We now show that $E = \bigwedge_i A_i \to G$ is a tautology, by induction on the size of π (complexity of the proof) and the number m.

Case 1. $m \leq n$
In this case $b = $ 'data' and $G = A_m$ and clearly E is a tautology.

Case 2. $m > n$
This case means that some proof steps were taken to get G. These steps are according to item 2 of Definition 3.4.3, conditions (a)–(k). We examine each condition in turn.

(a)   This case follows from the induction hypothesis (case (1)).

(b)   By the induction hypothesis $\bigwedge A_i \to G_1$ and $\bigwedge A_i \to (G_1 \to G)$ are tautologies and therefore so is E.

(c), (d)   By the induction hypothesis, since $G = G_1 \vee G_2$, we get that $\bigwedge A_i \to G$ is a tautology since $\bigwedge A_i \to G_1$ is a tautology.

(e), (f)   These cases are similar to (d) and (e).

(g)   In this case G is $G_1 \to G_2$ and we enter a subcomputation whose assumption is $((m, 1), G_1, \text{assumption})$ and conclusion (goal) is $((m, k), G_2, b)$.

Let $\pi_m = \{(\alpha^*, X, x) \mid ((m) \oplus \alpha, X, x) \in \pi\} \cup \{((i), B_i, \text{data}) \mid i = 1, \ldots m - 1 \text{ and } ((i), B_i, b_i) \in \pi\}$, where α^* is defined as follows:

$$\alpha^* = ((n + y), X, x), \text{ whenever } \alpha = (y, X, x)$$

Then π_m is a smaller proof of G_2 from $\{A_1, \ldots, A_n, B_{n+1}, B_{m-1}\}$. By the induction hypothesis $\bigwedge A_j \wedge \bigwedge B_j \wedge G_1 \to G_2$ and $\bigwedge A_i \to B_j, j = 1$, are all tautologies. Hence E is a tautology.

3.4. FORMAL DESCRIPTIONS OF PROOFS

(h) In this case $G = \neg G_1$. We create $\pi_{(m,1)}$ and $\pi_{(m,2)}$ in a similar manner to the previous case, letting

$$\pi_{(m,i)} = \{(\alpha^*, X, x) \mid ((m,i) \oplus \alpha, X, x) \in \pi\} \cup \\ \{((i), B_i, \text{data}) \mid i = 1, \ldots, m-1, ((i), B_i, b_i) \in \pi\}$$

We get $\bigwedge A_i \wedge \bigwedge_j B_j \wedge G_1 \to X$ and $\bigwedge A_i \wedge \bigwedge_j B_j \wedge G_1 \to \neg X$ are tautologies. Hence so is E.

(i) This case is similar to the previous case.

(j) This case is a combination of $C \vee D$ and $C \to G$ and $D \to G$. We get that $\bigwedge A_i \to (C \to G)$, $\bigwedge A_i \to (D \to G)$ and $\bigwedge A_i \to C \vee D$ are all tautologies and hence E is a tautology.

(k) In this case $G = G_1 \wedge G_2$ and we get by the induction hypothesis that $\bigwedge A_i \to G_1$ and $\bigwedge A_i \to G_2$ are tautologies and therefore so is E.

(l) This case does not arise.

This completes the proof of the theorem. ∎

Summary

Let us conclude this chapter by presenting in Figure 3.1 a summary of the natural deduction rules.

$$(\wedge\text{I})\frac{A, B}{A \wedge B} \qquad (\wedge\text{E})\frac{A \wedge B}{A} \text{ and } \frac{A \wedge B}{B}$$

$$(\vee\text{I})\frac{A}{A \vee B} \text{ and } \frac{B}{A \vee B} \qquad (\vee\text{E1})\frac{A \vee B, \neg A}{B} \qquad (\vee\text{E2})\frac{A \vee B, \neg B}{A}$$

$$(\vee\text{E})\frac{A \to C, B \to C, A \vee B}{C}$$

$$(\to\text{I1})\frac{\neg A}{A \to B} \qquad (\to\text{I2})\frac{B}{A \to B} \qquad (\to\text{E1})\frac{A \to B}{\neg A \vee B}$$

$$(\to\text{E})\frac{A, A \to B}{B}$$

$(\to\text{I})$: If $\dfrac{P, A}{B}$ is shown to be valid then $\dfrac{P}{A \to B}$ is also valid

$$(\neg\text{E1})\frac{\neg\neg A}{A} \qquad (\neg\text{I1})\frac{A}{\neg\neg A} \qquad (\neg 2)\frac{\neg A \to B, A \to B}{B}$$

$$(\neg\text{E})\frac{\neg A \to B, \neg A \to \neg B}{A} \qquad (\neg\text{I})\frac{A \to B, A \to \neg B}{\neg A}$$

$$\text{Cut rule: } \frac{P}{A} \text{ and } \frac{P, A}{B} \text{ imply } \frac{P}{B}$$

The rules ($\vee$E1), ($\vee$E2), ($\neg$E1), ($\neg$2), ($\neg$I1) and ($\neg$I2) can be proved from the other rules. The cut rule can also be proved.

Figure 3.1: Summary of natural deduction rules

4

FROM FORWARD TO BACKWARD RULES

Let us summarize the knowledge we have so far. The basic problem is that we have a database (a set of assumptions) and a query (a conclusion) and we want to know whether logically the answer to the query is yes (i.e. whether the conclusion follows logically from the assumptions). We have learnt two methods for checking that this is so. The first method, introduced in Chapter 1, constructs a truth table to check whether an implication from the assumptions to conclusion is a tautology. The second method, presented in Chapter 3, uses forward rules to reason from the assumptions, searching for the conclusion. The truth table method is mechanical and always gives an answer. The forward rules method is not completely mechanical and requires some intuition and ingenuity.

Our task in this chapter is to make the method of using rules as mechanical as possible. This attempt will lead us to a third method—the method of backward-going rules and rewrites of the database. We will develop this method from the forward rules, and progress slowly through examples.

Suppose we are given the following typical situation:

$$\text{Assumptions} \quad \text{Conclusion}$$
$$\mathcal{P} \quad \vdash \quad C$$

Our question is whether C follows from the assumptions. We would like to give advice on how to use the rules of the previous chapter so that their application becomes as automatic and mechanical as possible.

4.1 Introductory heuristics

Some *heuristics* might be

H1 When the argument is of the form $\mathcal{P} \vdash A \wedge B$ then first show $\mathcal{P} \vdash A$ and then show $\mathcal{P} \vdash B$.

H2 When the argument is of the form $\mathcal{P} \vdash A \rightarrow B$ then add A to the assumptions and have B as the conclusion, i.e. show $\mathcal{P} \cup \{A\} \vdash B$.

H3 When the argument is of the form $\mathcal{P} \vdash \neg A$ then it is advisable to use the $\neg$I rule; that is, to show A guess a 'good' x and show $\mathcal{P} \vDash A \rightarrow x$ and $\mathcal{P} \vDash A \rightarrow \neg x$.

The problem is how to guess such an x. We do this by going through all the assumptions and choosing x to be a formula whose negation appears as one of the assumptions. If no negated formula can be found in the assumptions we resort to choosing x to be any formula in the assumptions. This may not always work—although if we try all the x which are subformulae of the data and the goal then it will work. However, the guessing, although systematic, is not efficient. We now give some worked examples using H1–H3 as heuristics.

Example 4.1.1 *We will attempt to show that* $\neg A \vee B \vdash A \rightarrow B$.

Step 1 *We will apply H2. Since the conclusion has the form of an implication we ask instead whether* $\neg A \vee B, A \vdash B$. *Looking at the rules of Chapter 3, we see that the nearest rule to use is* $(\vee E1)$:$x \vee y, \neg x \vdash y$. *Its form is not exactly the same as that of our problem, but if we take $x = \neg A$ and $y = B$ we get:* $\neg A \vee B, \neg \neg A \vdash B$. *If we could replace A by $\neg \neg A$ we could use the above rule. But we can do this because we have another rule, namely* $(\neg I1)$.

Step 2 *Since* $(\neg I1)$:$A \vdash \neg \neg A$ *we can add $\neg \neg A$ and we have to show:* $\neg A \vee B, \neg \neg A \vdash B$.

Step 3 *We can now get B using the $(\vee E1)$ rule.*

Example 4.1.2 *To show* $A \rightarrow (B \rightarrow C) \vdash \neg C \wedge B \rightarrow \neg A$ *we use H2 and try to show*

4.1. INTRODUCTORY HEURISTICS

$$\frac{A\to(B\to C), \neg C\wedge B}{\neg A}$$

Since we want to show a negation $\neg A$, we follow the advice given in H3, i.e. try to guess an x with which we can show

$$\frac{A\to(B\to C), \neg C\wedge B}{A\to x} \qquad \frac{A\to(B\to C), \neg C\wedge B}{A\to \neg x}$$

Which x should we guess? The advice says look for negated assumptions. We have $\neg C\wedge B$ in the data which is really $\neg C, B$; thus let us try $x = C$. It may not work, in which case we shall have to try something else. We now have to show

$$\frac{A\to(B\to C), \neg C\wedge B}{A\to C} \qquad \frac{A\to(B\to C), \neg C\wedge B}{A\to \neg C}$$

Putting the A into the assumptions, in accordance with H2 we have to show

$$\frac{A\to(B\to C), \neg C\wedge B, A}{C} \qquad \frac{A\to(B\to C), \neg C\wedge B, A}{\neg C}$$

The right-hand argument is obvious (x was chosen especially for that purpose). We are left now with the left-hand argument. Since we have $A, A\to(B\to C)$ we get $B\to C$ and since we also have B we get C. Thus we have shown what was necessary, and the exercise is solved successfully.

Exercise 4.1.3 *Show $(A\to B), \neg B \vdash \neg A$.*

4.1.1 Adding positive and negative assumptions

We now introduce a further, very useful, heuristic rule:

H4 To show $\mathcal{P} \vdash C$ find a 'good' x and show $\mathcal{P}, x \vdash C$ and $\mathcal{P}, \neg x \vdash C$.

This rule is often useful. As for H3, the question is how to choose x? There are two criteria to help us. One is to choose x such that many conclusions can be obtained from the assumptions $\mathcal{P}, x$. For example, if $\mathcal{P}$ happens to be

$$\{A\to B,\quad A\to E,\quad A\to D\}$$

then it may be worthwhile to take $x = A$. Another criterion is to choose an x which will give the conclusion C. For example, if C is $A\vee B$ we may want to take $x = B$. Let us look at an example.

Example 4.1.4 *To show $A \to B \vdash \neg A \lor B$ we follow H4 and show*

$$\frac{A \to B, x}{\neg A \lor B} \quad \text{and} \quad \frac{A \to B, \neg x}{\neg A \lor B}$$

$x = A$ will yield B for the first case, and directly give $\neg A$ for the second, and make both cases valid. We could have tried $x = B$ but that would have turned out to be fairly complicated. For the choice $x = A$ we have to show

$$\frac{A \to B, A}{\neg A \lor B} \quad \text{and} \quad \frac{A \to B, \neg A}{\neg A \lor B}$$

*This is valid since   valid by direct
we can get B rule $(\lor I)$
and use $(\lor I)$*

For the choice $x = B$ we must show

$$\frac{A \to B, B}{\neg A \lor B} \quad \text{and} \quad \frac{A \to B, \neg B}{\neg A \lor B}$$

*valid by direct rule   valid since $A \to B, \neg B \vdash \neg A$
by Exercise 4.1.3*

Our procedures are more mechanical now than before but still they are not completely automatic. Rules H1 and H2 seem automatic enough but H3 and H4 involve choices. Consider

$$\frac{A \to (B \land C), A}{B \land C} \quad \text{and} \quad \frac{A \to (B \to C), A}{B \to C}$$

In these arguments we immediately see that the conclusion follows from the assumption because we can use *modus ponens* immediately. If we follow heuristics H1 and H2 mechanically, for the first of the examples we have to show both the following:

$$\frac{A \to (B \land C), A}{B} \quad \text{and} \quad \frac{A \to (B \land C), A}{C}$$

and for the second example we must show

$$\frac{A \to (B \to C), A, B}{C}$$

4.1. INTRODUCTORY HEURISTICS

which are both more complicated. We want to make our rules very strict so that there will not be much choice left. Notice that the heuristic rules H1 and H2 reduce the complexity of the conclusion C:

H1 To check $\mathcal{P} \vdash A \wedge B$ we check $\mathcal{P} \vdash A$ and $\mathcal{P} \vdash B$.

H2 To check $\mathcal{P} \vdash A \rightarrow B$ we check $\mathcal{P}, A \vdash B$

so the query we have to check becomes simpler and simpler. For negation, we have

H3 To check $\mathcal{P} \vdash \neg A$ we choose an x and check $\mathcal{P}, A \vdash x$ and $\mathcal{P}, A \vdash \neg x$.

Our guidelines for x were to look through the assumptions $\mathcal{P}$ for some negated formula $\neg D$ in $\mathcal{P}$ and take $x = D$. D may be, however, much more complicated than A. We encounter a similar problem in the case of $\mathcal{P} \vdash A \vee B$ where we use H4, namely:

H4 To check $\mathcal{P} \vdash A \vee B$, look for an x such that $\mathcal{P}, x \vdash A \vee B$ and $\mathcal{P}, \neg x \vdash A \vee B$ are both valid.

Again we may be making matters more complicated, unless the choice for x is a 'good' one, in which case we cannot expect a mechanical choice. We thus want to get rid of cases such as $\neg A$ and $A \vee B$, so let us agree on a new heuristic rule:

H5 To show $\mathcal{P} \vdash A \vee B$ show instead $\mathcal{P} \vdash \neg A \rightarrow B$ which leads to showing $\mathcal{P}, \neg A \vdash B$.

Here we are using a rewrite equivalence (see Section 1.2.1) of $A \vee B$ in terms of $\neg$ and $\rightarrow$. Thus we are rid of $\vee$ at the expense of gaining a negation. So using H5, instead of showing $A \rightarrow B \vdash \neg A \vee B$ we show $A \rightarrow B, \neg \neg A \vdash B$ which is very easy. The next example is more complicated.

Example 4.1.5 *To show $\neg(\neg A \wedge \neg B) \vdash A \vee B$, according to H5 we show $\neg(\neg A \wedge \neg B), \neg A \vdash B$. The next step is not so obvious. It is much more convenient when all our assumptions have the form x or the form $y \rightarrow z$, without $\neg$ in front. Now we know that*

$$\neg(\neg A \wedge \neg B) \equiv \neg A \rightarrow B$$

(you can check by truth tables), and so we have to show in this case that $\neg A \rightarrow B, \neg A \vdash B$, which is much easier. However, we just made a lucky guess. Is there a policy we can follow?

Note that since we are dealing with an essentially mechanical procedure we have first to transform the assumptions to the 'ready for computation' form and then apply our rules. The fact that we know (or guessed) the two sides are equivalent is not relevant.

This gives us the idea that we should look into the structure of A when we are asked to show $\mathcal{P} \vdash \neg A$. Since we have the following equivalences,

$$\begin{aligned}
\neg(A \wedge B) &\equiv A \to \neg B \\
\neg(A \vee B) &\equiv \neg A \wedge \neg B \\
\neg(\neg A) &\equiv A \\
\neg(A \to B) &\equiv A \wedge \neg B
\end{aligned}$$

we are led to yet another heuristic:

H6 We can replace $\neg$ by these rules:

1. Replace $\mathcal{P} \vdash \neg(A \wedge B)$ by $\mathcal{P}, A \vdash \neg B$
2. Replace $\mathcal{P} \vdash \neg(A \vee B)$ by $\mathcal{P} \vdash \neg A$ and $\mathcal{P} \vdash \neg B$
3. Replace $\mathcal{P} \vdash \neg\neg A$ by $\mathcal{P} \vdash A$
4. Replace $\mathcal{P} \vdash \neg(A \to B)$ by $\mathcal{P} \vdash A$ and $\mathcal{P} \vdash \neg B$.

In each case the negative conclusion to show is reduced to a simpler conclusion, or several such simpler conclusions.

Our situation now is this: given rules H1, H2, H5 and H6, we can replace any problem of the form $\mathcal{P} \vdash C$. where C is complex, by several auxiliary problems $\mathcal{P}_i \vdash x_i$ where the x_i are either atoms q_i, or negations of atoms $\neg q_i$. The reduction can be done mechanically.

Example 4.1.6 To show $\neg A \vdash A \to B$ using the official rules, i.e. without using rules $\to$I1 and $\to$I2. Using H1 we have to show $\neg A, A \vdash B$. This requires some ingenuity. Since we have the rule $\neg\neg B \vdash B$ we show $\neg A, A \vdash \neg\neg B$. (This is a non-mechanical 'guess'.) Using recommendation H3 we show for some x (which we also guess)

$$\frac{\neg A, A}{\neg B \to x} \quad \text{and} \quad \frac{\neg A, A}{\neg B \to \neg x}$$

i.e. we show

4.1. INTRODUCTORY HEURISTICS

$$\frac{\neg A, A, \neg B}{x} \quad \text{and} \quad \frac{\neg A, A, \neg B}{\neg x}$$

The obvious 'guess' is $x = A$. The step of choosing to show $\neg A, A \vdash \neg\neg B$ instead of $\neg A, A \vdash B$ is an 'inspiration'. A machine would not have known what to do, given the rules and recommendations which we have developed so far.

Example 4.1.7 Here is another example of this sort: $\neg q \to q \vdash q$ is a valid argument. How can it be shown using the rules? Since $\neg q \to q$ is given in the assumptions, we might attempt to show $\neg q \to q \vdash \neg q$ and then use $\neg q \to q, \neg q \vdash q$. However, $\neg q \to q \vdash \neg q$ is not valid (take $q = \top$). We need a non-mechanical inspiration here. We take $\neg\neg q$, instead of q, and show $\neg q \to q \vdash \neg\neg q$ and since $\neg\neg q \vdash q$ is a rule, we will get the desired conclusion. We show $\neg q \to q \vdash \neg\neg q$. Since this is a negation we show instead (by H3)

$$\frac{\neg q \to q}{\neg q \to x} \quad \text{and} \quad \frac{\neg q \to q}{\neg q \to \neg x}$$

We have to guess a good x. The x to guess is $x = q$. We thus show

$$\frac{\neg q \to q}{\neg q \to q} \quad \text{and} \quad \frac{\neg q \to q}{\neg q \to \neg q}$$

which can be easily done.

We see that we still need to be creative. Any hope of automatic use of the rules requires better heuristics.

Example 4.1.8 To show $\neg(A \to B) \vdash A \land \neg B$, following H1 we show both the following:

$$\text{(a)} \quad \frac{\neg(A \to B)}{A} \quad \text{and} \quad \text{(b)} \quad \frac{\neg(A \to B)}{\neg B}$$

We start with (b). We choose an x and show

$$\frac{\neg(A \to B)}{B \to x} \quad \text{and} \quad \frac{\neg(A \to B)}{B \to \neg x}$$

What do we take for x? The recommendation is to look for some negative information among the assumptions. In this case we have $\neg(A \to B)$. We thus take $x = (A \to B)$. We thus have to show

$$(b1) \quad \frac{\neg(A \to B)}{B \to (A \to B)} \quad \text{and} \quad (b2) \quad \frac{\neg(A \to B)}{B \to \neg(A \to B)}$$

Using H2 we show

$$(b1^*) \quad \frac{\neg(A \to B), B}{A \to B} \quad \text{and} \quad (b2^*) \quad \frac{\neg(A \to B), B}{\neg(A \to B)}$$

The second case $(b2^*)$ is valid and so we are left with the first case $(b1^*)$ which can be written as

$$\frac{\neg(A \to B), A, B}{B}$$

which is also valid. This concludes case (b) of the example. Let us return to case (a), namely $\neg(A \to B) \vdash A$. How about our new 'inspiration' of showing $\neg\neg A$ instead of A? Let us try to show $\neg(A \to B) \vdash \neg\neg A$. We choose an x and show

$$\frac{\neg(A \to B)}{\neg A \to x} \quad \text{and} \quad \frac{\neg(A \to B)}{\neg A \to \neg x}$$

Following our heuristics, we take $x = (A \to B)$. We must thus show

$$\frac{\neg(A \to B)}{\neg A \to (A \to B)} \quad \text{and} \quad \frac{\neg(A \to B)}{\neg A \to \neg(A \to B)}$$

Applying H2, we have to show

$$\frac{\neg(A \to B), \neg A, A}{B} \quad \text{and} \quad \frac{\neg(A \to B), \neg A}{\neg(A \to B)}$$

The first is valid because $\neg A, A \vdash B$ was shown in Example 4.1.6, and the second is obviously valid.

In the last two examples we had

$$\frac{\neg(A \to B), \text{something}}{\neg(A \to B)}$$

and we stopped checking and said that it was valid. What would happen if we were to use an automatic method of computation? Let us go back to Example 4.1.8, to the point where we have to show

$$\frac{\neg(A \to B), B}{\neg(A \to B)}$$

4.1. INTRODUCTORY HEURISTICS

If we were to continue mechanically using our rules and use heuristic H6 we would continue removing the negation, and trying to show

$$\frac{\neg(A \to B), B}{A \land \neg B}$$

which brings us back to showing $\neg B$ and we may enter an infinite loop.

4.1.2 Restart rule

Let us collect here the mechanical principles we have so far. Given a problem of the form $\mathcal{P} \vdash C$ we can mechanically use rules H1, H2, H5 and H6 to reduce the original problem to problems of the form $\mathcal{P}_i \vdash x_i$ where x_i are atoms or their negations (i.e. $x_i = q$ or $x_i = \neg q$). We also saw, through examples such as $\neg q, q \vdash p \neg q \to q \vdash q$ and $(p \to q) \to p \vdash p$, that we still need rules like H3 and H4 involving ingenuity, and we have just seen that we may get stuck in a loop. Can things be made more mechanical?

Let us look at H4. This rule says that when we are stuck in showing $\mathcal{P}^* \vdash C$ we choose an x and show instead $\mathcal{P}^*, x \vdash C$ and $\mathcal{P}^*, \neg x \vdash C$. Let us try to mechanize this rule and see what happens. Suppose we start our problem with $\mathcal{P} \vdash A$. We might feel that later in our mechanical attempts to get A we are going to be stuck with some $\mathcal{P}^* \vdash C$. Now $\mathcal{P}$ contains less data than $\mathcal{P}^*$ (because our computation always increases the data) so before we embark on our computation of $\mathcal{P} \vdash A$ we might use rule H4 with $x = A$ which is a good guess. We thus replace $\mathcal{P} \vdash A$ by both $\mathcal{P}, A \vdash A$ and $\mathcal{P}, \neg A \vdash A$. Of course, $\mathcal{P}, A \vdash A$ is valid, so we have to show $\mathcal{P}, \neg A \vdash A$. What have we gained? We now have $\neg A$ as an extra assumption. Imagine that we continue the computation and get stuck at the point $\mathcal{P}^* \vdash C$. Now we also have $\neg A$, so really we are stuck at $\mathcal{P}^*, \neg A \vdash C$. Here is an ingenious trick. Suppose instead of asking $\mathcal{P}^*, \neg A \vdash C$ we ask $\mathcal{P}^*, \neg A \vdash A$ and suppose we succeed in showing A. Now we are going to have

$$\frac{\mathcal{P}^*, \neg A}{A} \text{ and } \frac{\mathcal{P}^*, \neg A}{\neg A}$$

both valid. We can certainly show C because $A \land \neg A \vdash C$ is a correct rule, and we have certainly shown that $\mathcal{P}^*, \neg A \vdash A \land \neg A$. But look what we have done. We started with $\mathcal{P} \vdash A$ and then we got to $\mathcal{P}^* \vdash C$ and were stuck; we asked $\mathcal{P}^* \vdash A$ instead and we knew that to succeed with that is the same as to succeed with $\mathcal{P}^* \vdash C$. We have thus discovered the *restart rule*:

Restart If when starting with $\mathcal{P} \vdash A$ we get stuck at $\mathcal{P}^* \vdash C$ we can restart A and ask $\mathcal{P}^* \vdash A$ instead.[1]

You may want to clarify one point about our computation: we had $\mathcal{P}^*, \neg A \vdash A$ before, while the restart rule allows for $\mathcal{P}^* \vdash A$. This is not a problem, because we saw above that we can always add $\neg A$ to the assumptions. The restart rule is mechanical. You do not have to guess an x. All you have to do is, whenever you are stuck, try to prove the original A again.

Example 4.1.9 *To show $(p{\to}q){\to}p \vdash p$ we try to get $p{\to}q$ while remembering that the original aim was to get p. Thus we ask $(p{\to}q){\to}p \vdash p{\to}q$ which by use of H1 becomes $(p{\to}q){\to}p, p \vdash q$. In Section 3.2 we saw that as an original question the above argument is invalid (because we can have, we recall, $p = \top$ and $q = \bot$ and we will get $\top \vdash \bot$). However, this is not an original question but part of the computation of the original argument $(p{\to}q){\to}p, p \vdash q$. We are thus allowed to use the restart rule and ask p again. So instead of $(p{\to}q){\to}p, p \vdash q$ we ask $(p{\to}q){\to}p, p \vdash p$ which succeeds.*

Another way to justify the restart rule is to notice that $((q{\to}c){\to}q){\to}q$ is a tautology. Thus to ask $\mathcal{P} \vdash q$ is the *same* as to ask $\mathcal{P}, ((q{\to}c){\to}q){\to}q \vdash q$. By the rule ${\to}$E for atoms we can try and ask $\mathcal{P}, ((q{\to}c){\to}q){\to}q \vdash (q{\to}c){\to}q$ and by the rule for $\to$ we ask $\mathcal{P}, ((q{\to}c){\to}q){\to}q, q{\to}c \vdash q$. We are thus back to our original query q but we have $q{\to}c$ in the data. Since c can be any proposition, we can repeat this process over and over. Therefore we can show $\mathcal{P} \vdash q$ by showing $\mathcal{P}, q{\to}c_i \vdash q$ for $i = 1, 2, \ldots$. Thus if we have the query c_i in the middle of a computation we can always use *modus ponens* backwards on $q{\to}c_i$ and ask for q, i.e. we can restart q.

---

[1] In fact we can restart with any previous goal (to the current goal C) and not necessarily with the original goal A. Thus we can limit restart to be activated at atomic goals C only. In general, however, there are logics where restart cannot be restricted to be activated at atomic goals only. We need to prove formally that for any $\mathcal{P}$, and A we have: $\mathcal{P} \vdash A$ with restart iff $\mathcal{P}, \neg A \vdash A$ without restart. This will be done later in Lemma 6.6.6 in the context of the completeness theorem.

4.1.3 A heuristic for negation

So far we have mechanized the computation for the cases of $\wedge$, $\rightarrow$ and $\vee$. We have not dealt with negation—we have not replaced rule H3, which still requires intuition. H3 says that to show $\mathcal{P} \vdash \neg A$ we choose an x wisely and show both

$$\frac{\mathcal{P}}{A \rightarrow x} \text{ and } \frac{\mathcal{P}}{A \rightarrow \neg x}$$

If we show $A \rightarrow x$ and we show $A \rightarrow \neg x$ then we have shown $A \rightarrow (\neg x \wedge x)$. This is a valid intuitive rule. $\neg x \wedge x$ is always $\bot$. So $\neg A$ and $A \rightarrow \bot$ have the same truth table. Thus we can always write $A \rightarrow \bot$ instead of $\neg A$ and use the rules for $\rightarrow$. Let us check.

To show $\mathcal{P} \vdash \neg A$ we choose x and show

$$\frac{\mathcal{P}}{A \rightarrow x} \text{ and } \frac{\mathcal{P}}{A \rightarrow \neg x}$$

so that we must show

$$\frac{\mathcal{P}, A}{x} \text{ and } \frac{\mathcal{P}, A}{\neg x}$$

which is the same as showing

$$\frac{\mathcal{P}, A}{x \wedge \neg x}$$

(since $x, \neg x \vdash x \wedge \neg x$), but $x \wedge \neg x \equiv \bot$, so we therefore show $\mathcal{P}, A \vdash \bot$. If, on the other hand, right from the start, we write $A \rightarrow \bot$ instead of $\neg A$ and use the rule for $\rightarrow$ then to show $\mathcal{P} \vdash \neg A$ we show $\mathcal{P} \vdash A \rightarrow \bot$ and hence show $\mathcal{P}, A \vdash \bot$, which is the same as before.

We have seen that $A, \neg A \vdash B$ is a valid rule. In terms of $\bot$ this rule will be $\bot \vdash B$ for any B.

We are thus led to a heuristic for negation: rewrite $\neg A$ as $A \rightarrow \bot$ and use $\rightarrow$ rules, pretending $\bot$ is just another atom. The only additional rule needed is $\bot \vdash B$ for any B. (Check this via truth tables.)

Example 4.1.10 *Let us try $A, \neg A \vdash B$ which required ingenuity before we had the negation heuristic. Rewrite the argument as $A, A \rightarrow \bot \vdash B$. Now $A, A \rightarrow \bot \vdash \bot$ is successful since by $\rightarrow$E we get $\bot$ and the new negation rule yields B. This was both easy and mechanical.*

Example 4.1.11 *Showing $\neg q \to q \vdash q$ also required ingenuity. The argument can be rewritten as $(q \to \bot) \to q \vdash q$. To get q (our original query) we try to ask for $q \to \bot$ because $(q \to \bot) \to q$ is in the data. Thus we must show $(q \to \bot) \to q \vdash q \to \bot$. By using H1 we may ask $(q \to \bot) \to q, q \vdash \bot$. None of the normal rules now apply, so we are stuck. In this situation we are able to use restart and ask instead for the original q: $(q \to \bot) \to q, q \vdash q$ which succeeds. Thus the original argument is valid.*

Compare with Example 4.1.9, where we can replace p by q and q by $\bot$.

Suppose we have the query $\mathcal{P} \vdash A$. We know by truth tables that $A \equiv (\neg A \to A)$, so we can replace the above argument by $\mathcal{P} \vdash \neg A \to A$ and hence use H1 to reduce this to $\mathcal{P}, \neg A \vdash A$, and after translation to $\mathcal{P}, A \to \bot \vdash A$. If later in the computation of A we get stuck at $\mathcal{P}^*, A \to \bot \vdash C$, we can always continue by using the negation rule and asking for A because if A succeeds then we get $\bot$ from $A, A \to \bot$, and when we have $\bot$ we can get anything we want, including C.

4.1.4 Rewriting assumptions

We are close to having a fully mechanical reasoning method. Let us summarize where we stand now. Given $\mathcal{P} \vdash A$ to show, we can mechanize the proof as follows:

1. All negations $\neg D$ are rewritten as $D \to \bot$, where $\bot$ is falsity.

2. $A \vee B$ is rewritten as $\neg A \to B$, which by 1. above, can be further rewritten to $(A \to \bot) \to B$.

3. Rules H1 and H2 deal with $\wedge$ and $\to$ and these are mechanical rules.

4. Rule H4 was replaced by the restart rule.

5. Rule H3 for $\neg$ was replaced by the rule $\bot \vdash B$.

We now have to formulate the exact mechanical use of the rules. We know that any argument $\mathcal{P} \vdash A$ can be reduced to a list of several arguments $\mathcal{P}_i \vdash x_i$ where x_i are atoms (we do not have $\neg$ any more and so x_i can be $\bot$, which is also considered to be an atom). It would be useful to have all the data in $\mathcal{P}_i$ in the form $[A \wedge B \wedge C] \to x$, i.e.

['body' of formulae] →[atomic 'head']

4.1. INTRODUCTORY HEURISTICS

so that we can, say, to show $\mathcal{P} \vdash x$ just show $\mathcal{P} \vdash A{\wedge}B{\wedge}C$.

The next step is to deal with $\mathcal{P} \vdash x$ where x is of the above form, namely an atom or $\bot$. We saw in an earlier example that sometimes we make the computation more complicated by our reduction. For example, $A{\rightarrow}(B{\wedge}C), A \vdash B{\wedge}C$ becomes $A{\rightarrow}(B{\wedge}C), A \vdash B$ and $A{\rightarrow}(B{\wedge}C), A \vdash C$. If we reduce B and C further we get something like

$$\frac{A{\rightarrow}B{\wedge}C \\ A \\ \text{extra added in the process of simplification}}{\text{atom } x}$$

where x is embedded somewhere in A, B, C. It is therefore to our advantage to make sure that the assumptions themselves have the convenient form $A{\rightarrow}x$, where x is atomic or $A{\rightarrow}\bot$, so that when we simplify the conclusion we get things like $A_i{\rightarrow}x_i \vdash y$ and we just look whether we have y appearing among the x_i and whether we can show the relevant A_i.

We use two more equations to reduce anything of the form $A{\rightarrow}B$, where B is an arbitrarily complex formula, to the form $C{\rightarrow}x$, where x is atomic. These are

H7 $A{\rightarrow}(x{\wedge}y) \equiv (A{\rightarrow}x){\wedge}(A{\rightarrow}y)$

H8 $A{\rightarrow}(x{\rightarrow}y) \equiv (A{\wedge}x){\rightarrow}y$

Thus if $(A{\rightarrow}x{\wedge}y)$ is an assumption we replace it by the two assumptions $A{\rightarrow}x$ and $A{\rightarrow}y$, and if $A{\rightarrow}(x{\rightarrow}y)$ is an assumption we replace it by $(A{\wedge}x){\rightarrow}y$. In both cases the 'head' becomes less complicated. We continue to rewrite all the assumptions until all 'heads' become atomic.

Example 4.1.12 *Showing* $\neg(\neg A{\wedge}\neg B) \vdash A{\vee}B$ *involves first rewriting wffs of the form* $\neg x$ *as* $x{\rightarrow}\bot$ *and* $x{\vee}y$ *as* $(x{\rightarrow}\bot){\rightarrow}y$. *So we show instead:*

$$\frac{[(A{\rightarrow}\bot){\wedge}(B{\rightarrow}\bot)]{\rightarrow}\bot}{(A{\rightarrow}\bot){\rightarrow}B}$$

Using H2 we show

$$\frac{[(A{\rightarrow}\bot){\wedge}(B{\rightarrow}\bot)]{\rightarrow}\bot, A{\rightarrow}\bot}{B}$$

To continue we ask: does anything in the assumptions give B? The answer is 'no' because the heads of the assumptions are $\bot$ not B. But remember that if we can prove $\bot$ then we can prove any formula. Thus the correct question to ask is: do we have an assumption giving B or $\bot$? The answer is 'yes', two of them. We proceed by mechanically checking the first of them and, if it does not work, trying the second and so forth.

The first assumption with $\bot$ as head is $[(A\to\bot)\wedge(B\to\bot)]\to\bot$. If we can show $(A\to\bot)\wedge(B\to\bot)$ we will get $\bot$ which will give $\neg B$. To show a conjunction we have to show each conjunct; thus we have to show

$$\frac{((A\to\bot)\wedge(B\to\bot))\equiv\bot, A\to\bot}{A\to\bot} \quad \text{and} \quad \frac{(A\to\bot)\wedge(B\to\bot)\to\bot, A\to\bot}{B\to\bot}$$

We must remember the original query to show, which was B, as we may want to restart. Clearly the first argument is valid and can be proved mechanically. To show the second we use H2:

$$\frac{[(A\to\bot)\wedge(B\to\bot)]\to\bot, A\to\bot, B}{\bot}$$

To show $\bot$ we can use the first assumption but that will put us into a loop. Using the second assumption will lead us to try to show A which does not seem to help either. So we restart with the original B which succeeds. Thus the entire example succeeds.

Example 4.1.13 To show $\neg(A\vee B) \vdash \neg A\wedge\neg B$ first we rewrite the formula $A\vee B$ as $(A\to\bot)\to B$, $\neg(A\vee B)$ as $((A\to\bot)\to B)\to\bot$, and $\neg A$ as $A\to\bot$ and $\neg B$ as $B\to\bot$. We have to show

$$\frac{((A\to\bot)\to B)\to\bot}{(A\to\bot)\wedge(B\to\bot)}$$

so by H1 we must show both $(A\to\bot)$ and $(B\to\bot)$. To show $A\to\bot$ we use H2 and show

$$\frac{((A\to\bot)\to B)\to\bot, A}{\bot}$$

which leads to

$$\frac{((A\to\bot)\to B)\to\bot, A}{(A\to\bot)\to B}$$

because $((A\to\bot)\to B)\to\bot$ is an assumption with head $\bot$. Now using H2 we reach

$$\frac{((A{\to}\bot){\to}B){\to}\bot, A, A{\to}\bot}{B}$$

To show that we match B with the $\bot$ of $A{\to}\bot$ *(if we used the first assumption we would loop)* and show

$$\frac{((A{\to}\bot){\to}B){\to}\bot, A, A{\to}\bot}{A}$$

which succeeds.

To show $B{\to}\bot$ we use H2 and show

$$\frac{((A{\to}\bot){\to}B){\to}\bot, B}{\bot}$$

which leads to

$$\frac{((A{\to}\bot){\to}B){\to}\bot, B}{(A{\to}\bot){\to}B}$$

and by H2 we must show

$$\frac{((A{\to}\bot){\to}B){\to}\bot, B, A{\to}\bot}{B}$$

which succeeds.

4.1.5 Summary of computation rules

H1 To show $\mathcal{P} \vdash A{\wedge}B$ show $\mathcal{P} \vdash A$ and $\mathcal{P} \vdash B$.

H2 To show $\mathcal{P} \vdash A{\to}B$ show $\mathcal{P}, A \vdash B$.

H3 $\bot \vdash A$ for any A.

H4 If you start with $\mathcal{P} \vdash A$ then at any time later in the computation you may start showing A again.

H5 Rewrite $A{\vee}B$ as $(A{\to}\bot){\to}B$.

H7 Rewrite $A{\to}(B{\wedge}C)$ as $(A{\to}B){\wedge}(A{\to}C)$.

H8 Rewrite $A{\to}(B{\to}C)$ as $A{\wedge}B{\to}C$.

H9 Rewrite $\neg A$ as $A \to \bot$.

Note that heuristic H6 no longer applies.

The outline computation is that to show $\mathcal{P} \vdash C$ use the rules to get to $\mathcal{P}_i \vdash x_i$ where x_i are atoms or $\bot$. Rewrite any assumption in $\mathcal{P}_i$ into the form $A \to y$ where y is an atom, or $\bot$. If $x_i = y$ or $y = \bot$ then to show $\mathcal{P}_i \vdash x_i$ show $\mathcal{P}_i \vdash A$. If you need to restart with C you can show $\mathcal{P}_i \vdash x_i$ by showing $\mathcal{P}_i \vdash C$.

Exercise 4.1.14 *Show*

1. $(p \to q) \wedge (q \to r) \wedge \neg r \vdash \neg p$

2. $\neg\neg p \vdash p$

3. $\neg p \to p \vdash p$

4. $p \vee q, \neg p \vdash q$

5. $q \wedge r \to s, p \vee r \to q, \neg p \vdash r \to s$

6. $p \vee q, p \to r, q \to r \vdash r$

4.2 Worked examples

We now give a summary of the official backward rules for dealing with the computational problem of whether a conclusion follows from some assumptions. There are two principles to follow for representing both the assumptions and the conclusion:

use only $\wedge$, $\to$ and $\bot$ We rewrite everything, assumptions and conclusion, with the connectives $\wedge$, $\to$ and $\bot$. Thus we translate $\neg x$ as $x \to \bot$ for any x, and we translate $A \vee B$ as $(A \to \bot) \to B$. Note that $A \vee B$ also has the same truth table as $(A \to B) \to B$ and the same table as $(B \to A) \to A$, and the same table as $(B \to \bot) \to A$, so any of the above can serve as a translation. We officially choose $A \vee B$ to be translated as $(A \to \bot) \to B$. Here is a truth table showing all the equivalences:

A	B	$A \vee B$	$A \to \bot$	$(A \to \bot) \to B$	$(A \to B)$	$(A \to B) \to B$
T	T	T	$\bot$	T	T	T
$\bot$	T	T	T	T	T	T
T	$\bot$	T	$\bot$	T	$\bot$	T
$\bot$	$\bot$	$\bot$	T	$\bot$	T	$\bot$

4.2. WORKED EXAMPLES

$(A{\to}B){\to}B$ has a symmetrical table in A and B and hence equals $(B{\to}A){\to}A$. Notice that $(A{\to}\bot){\to}B$ has a symmetrical table in A and B and hence equals $(B{\to}\bot){\to}A$.

use only clausal form Even in the case of data (assumptions) and query (conclusion) written using $\to$, $\wedge$ and $\bot$ only, we want to write the data in a special form, namely as a set of implications of the form $B{\to}q$ where q is atomic, or $\bot$. Each $B{\to}q$ is called a *clause*, with B as *body* and q as *head*.

Given a set of assumptions, can we always write the assumptions in the above form? The answer is yes. Given an assumption E, we can apply the following rules:

E has the form:	Rewrite E to:
atomic	no change
$A{\to}\bot$	no change
$A{\to}q$, q atomic	no change
$A{\wedge}B$	A and B
$A{\to}(B_1{\wedge}B_2)$	$A{\to}B_1$ and $A{\to}B_2$
$A{\to}(B_1{\to}B_2)$	$A{\wedge}B_1{\to}B_2$

Where E is rewritten, the above rules must be applied again to the products of the rewrite of E, until there is no change. This is to ensure that assumptions such as $p{\wedge}q{\to}(r{\wedge}p{\to}s)$ are fully rewritten to clausal form. We thus see that any database can be rewritten and replaced by an equivalent database, whose elements have the desired form.

There are several computational steps for clausal form databases, given by the following:

For atoms q To show $\mathcal{P} \vdash q$ succeeds, we distinguish several cases:

- $q \in \mathcal{P}$, thus we succeed.
- Some $B{\to}q \in \mathcal{P}$, then we must show that $\mathcal{P} \vdash B$ succeeds.
- $\bot \in \mathcal{P}$, thus we succeed.
- For some $B{\to}\bot \in \mathcal{P}$, then we must show that $\mathcal{P} \vdash B$ succeeds.

For conjunction To show $\mathcal{P} \vdash A{\wedge}B$ we must show $\mathcal{P} \vdash A$ and $\mathcal{P} \vdash B$.

For implication To show $\mathcal{P} \vdash A \to B$ show $\mathcal{P}, A \vdash B$. Remember that A may need to be rewritten when being added to the database. This is known as the *deduction theorem*.

Restart rule Assume our original problem is to show $\mathcal{P} \vdash A$. In the middle of the computation we have to show $\mathcal{P}^* \vdash C$. Usually C will be atomic. We can, at any time, choose to continue with $\mathcal{P}^* \vdash A$ instead (even if we do not have to).

We can now give some worked examples. After the examples we will discuss ways of automating the computation. Remember—loops are possible but if A logically follows from $\mathcal{P}$ then there is always a way to use the rules to get from $\mathcal{P}$ to A, without loops.

Example 4.2.1
$(p \to q) \to p \vdash p$
$\quad\quad\Big|\ \textit{rule for atoms}$
$(p \to q) \to p \vdash p \to q$
$\quad\quad\Big|\ \textit{rule for} \to$
$p, (p \to q) \to p \vdash q$
$\quad\quad\Big|\ \textit{restart}$
$p, (p \to q) \to p \vdash p$
$\quad\quad\Big|$
$\quad\quad \textit{success}$

Example 4.2.2
$(a \to b) \to b \vdash c$
$\quad a \to c$
$\quad b \to c$
$\quad\quad\Big|\ \textit{rule for atoms used with } b \to c$
$(a \to b) \to b \vdash b$
$\quad a \to c$
$\quad b \to c$
$\quad\quad\Big|\ \textit{rule for atoms used with } (a \to b) \to b$

4.2. WORKED EXAMPLES

$(a \to b) \to b \vdash a \to b$
  $a \to c$
  $b \to c$

  | rule for $\to$

$(a \to b) \to b \vdash b$
  $a \to c$
  $b \to c$
  a

  | *Notice here that we can use the rule for atoms for b used with $(a \to b) \to b$ but we will loop because we will be back where we are. Let us use restart instead.*
  | restart

$(a \to b) \to b \vdash c$
  $a \to c$
  $b \to c$
  a

  | *Notice now that if we use the rule for atoms with $b \to c$ we will ask for b and loop again. It is wise (almost mechanical though) to use $a \to c$.*
  | rule for atoms with $a \to c$

$(a \to b) \to b \vdash a$
  $a \to c$
  $b \to c$
  a

  |
  success

Example 4.2.3

$(a \to b) \to b \vdash (b \to a) \to a$

  | rule for $\to$

$(a \to b) \to b \vdash a$
  $b \to a$

  | rule for atoms for $b \to a$

$(a{\to}b){\to}b \;\vdash\; b$
$\quad b{\to}a$

$\qquad \Big|\; \text{rule for atoms using } (a{\to}b){\to}b$

$(a{\to}b){\to}b \;\vdash\; a{\to}b$
$\quad b{\to}a$

$\qquad \Big|$

$(a{\to}b){\to}b \;\vdash\; b$
$\quad b{\to}a$
$\quad a$

$\qquad \Big|\; \text{restart}$

$(a{\to}b){\to}b \;\vdash\; (b{\to}a){\to}a$
$\quad b{\to}a$
$\quad a$

$\qquad \Big|\; \text{rule for } \to$

$(a{\to}b){\to}b \;\vdash\; a$
$\quad b{\to}a$
$\quad a$
$\quad b{\to}a$

$\qquad \Big|$

$\qquad success$

Example 4.2.4

$(a{\to}b){\to}b \;\vdash\; a \vee b$

$\qquad \Big|\; rewrite$

$(a{\to}b){\to}b \;\vdash\; (a{\to}\bot){\to}b$

$\qquad \Big|\; \text{rule for } \to$

$(a{\to}b){\to}b \;\vdash\; b$
$\quad (a{\to}\bot)$

4.2. WORKED EXAMPLES

> *We have two choices here. One is to use the rule for atoms with $a \to \bot$ and the second to use the rule for atoms with $(a \to b) \to b$. You can guess that the first case loops so we use the second case. A computer will use each case in the order of writing and if it loops (we can use loop detection techniques) will try the next. So we continue*
>
> *rule for atoms with $(a \to b) \to b$*

$(a \to b) \to b \vdash a \to b$
$(a \to \bot)$

> *rule for* $\to$

$(a \to b) \to b \vdash b$
$(a \to \bot)$
a

> *rule for atoms with $a \to \bot$*

$(a \to b) \to b \vdash a$
$(a \to \bot)$
a

success

Example 4.2.5
$(q \to q) \vdash q$

> *(This should not succeed!)*
> *rule for atoms*

$(q \to q) \vdash q$

This loops with no way out. Even the restart rule is of no use to us.

Example 4.2.6
$\neg q \to q \vdash q$

This should succeed; check by truth tables.

$\quad$ | rewrite

$((q{\to}\bot){\to}q) \ \vdash q$

$\quad$ | rule for atoms

$((q{\to}\bot){\to}q) \ \vdash q{\to}\bot$

$\quad$ | rule for $\to$

$((q{\to}\bot){\to}q) \ \vdash \bot$
$\quad q$

Note that the rule for atoms involving $\bot$ applies only when $\bot$ is in the database, i.e. if we had $(q{\to}\bot) \vdash a$, we could ask for q but the rule does not apply in the case above. Therefore we use the restart rule.

$\quad$ | restart

$((q{\to}\bot){\to}q) \ \vdash q$
$\quad q$

$\quad$ |

$\quad$ success

Example 4.2.7

$(a{\to}b) \ \vdash (x{\vee}a){\to}(x{\vee}b)$

$\quad$ | rewrite

$(a{\to}b) \ \vdash ((x{\to}\bot){\to}a){\to}((x{\to}\bot){\to}b)$

$\quad$ | rule for $\to$

$(a{\to}b) \qquad \vdash ((x{\to}\bot){\to}b)$
$((x{\to}\bot){\to}a)$

$\quad$ | rule for $\to$

$(a{\to}b) \qquad \vdash b$
$((x{\to}\bot){\to}a)$
$\quad (x{\to}\bot)$

$\quad$ | rule for atoms using $a{\to}b$

4.2. WORKED EXAMPLES

$(a \to b) \quad \vdash a$
$((x \to \bot) \to a)$
$(x \to \bot)$

| rule for atoms using $((x \to \bot) \to a)$

$(a \to b) \quad \vdash x \to \bot$
$((x \to \bot) \to a)$
$(x \to \bot)$

| *Note that you cannot say we have $x \to \bot$ in the database and hence success, since the rules are mechanical. Success can be obtained only in the rule for atoms. $x \to \bot$ is not an atom. So:*
| *rule for* $\to$

$(a \to b) \quad \vdash \bot$
$((x \to \bot) \to a)$
$(x \to \bot)$
x

| *rule for atoms with* $x \to \bot$

$(a \to b) \quad \vdash x$
$((x \to \bot) \to a)$
$(x \to \bot)$
x

|
success

Example 4.2.8
$a \quad \vdash a \lor b$

| *rewrite*

$a \quad \vdash (a \to \bot) \to b$

| *rule for* $\to$

$a \qquad \vdash b$
$(a \to \bot)$

| *rule for atoms using* $a \to \bot$

$$a \quad \vdash a$$
$$(a{\to}\bot)$$
$$\big|$$
$$success$$

Example 4.2.9
$$(a{\to}b) \quad \vdash \neg a$$
$$\neg b$$
$$\Big| \; rewrite$$
$$(a{\to}b) \quad \vdash a{\to}\bot$$
$$b{\to}\bot$$
$$\Big| \; rule\ for\ {\to}$$
$$(a{\to}b) \quad \vdash \bot$$
$$b{\to}\bot$$
$$a$$
$$\Big| \; rule\ for\ atoms\ using\ b{\to}\bot$$
$$(a{\to}b) \quad \vdash b$$
$$b{\to}\bot$$
$$a$$
$$\Big| \; rule\ for\ atoms\ using\ a{\to}b$$
$$(a{\to}b) \quad \vdash a$$
$$b{\to}\bot$$
$$a$$
$$\big|$$
$$success$$

Example 4.2.10
$$\vdash a \lor (a{\to}b)$$
$$\Big| \; rewrite$$
$$\vdash (a{\to}\bot){\to}(a{\to}b)$$

4.2. WORKED EXAMPLES

$$a{\to}\bot \ \vdash \ a{\to}b$$
$\quad\quad\quad$ | *rule for* $\to$
$$a{\to}\bot \ \vdash \ b$$
$\quad a$
$\quad\quad\quad$ | *rule for* $\to$
$$a{\to}\bot \ \vdash \ a$$
$\quad a$
$\quad\quad\quad$ | *rule for atoms using* $a{\to}\bot$

$\quad\quad$ *success*

Example 4.2.11
$(a{\vee}b){\vee}c \ \vdash \ a{\vee}(b{\vee}c)$
$\quad\quad$ | *rewrite*
$((a{\to}\bot){\to}b){\vee}c \ \vdash \ a{\vee}((b{\to}\bot){\to}c)$
$\quad\quad$ | *rewrite*
$(((a{\to}\bot){\to}b){\to}\bot){\to}c \ \vdash \ (a{\to}\bot){\to}((b{\to}\bot){\to}c)$
$\quad\quad$ | *rule for* $\to$
$(((a{\to}\bot){\to}b){\to}\bot){\to}c \ \vdash \ (b{\to}\bot){\to}c$
$\quad (a{\to}\bot)$
$\quad\quad$ | *rule for* $\to$
$(((a{\to}\bot){\to}b){\to}\bot){\to}c \ \vdash \ c$
$\quad (a{\to}\bot)$
$\quad b{\to}\bot$

The rule for atoms can be used with each of the three assumptions. If you want to use the system automatically try all three. Otherwise try the first one because it will make you add to the data and thus increase the chances of success.

| rule for atoms using $(((a{\rightarrow}\bot){\rightarrow}b){\rightarrow}\bot){\rightarrow}c$

$(((a{\rightarrow}\bot){\rightarrow}b){\rightarrow}\bot){\rightarrow}c \;\vdash\; ((a{\rightarrow}\bot){\rightarrow}b){\rightarrow}\bot$
    $(a{\rightarrow}\bot)$
    $b{\rightarrow}\bot$

  | rule for $\rightarrow$

$(((a{\rightarrow}\bot){\rightarrow}b){\rightarrow}\bot){\rightarrow}c \;\vdash\; \bot$
    $(a{\rightarrow}\bot)$
    $b{\rightarrow}\bot$
    $(a{\rightarrow}\bot){\rightarrow}b$

  | rule for atoms using $b{\rightarrow}\bot$

$(((a{\rightarrow}\bot){\rightarrow}b){\rightarrow}\bot){\rightarrow}c \;\vdash\; b$
    $(a{\rightarrow}\bot)$
    $b{\rightarrow}\bot$
    $(a{\rightarrow}\bot){\rightarrow}b$

  | rule for atoms using $(a{\rightarrow}\bot){\rightarrow}b$

$(((a{\rightarrow}\bot){\rightarrow}b){\rightarrow}\bot){\rightarrow}c \;\vdash\; a{\rightarrow}\bot$
    $(a{\rightarrow}\bot)$
    $b{\rightarrow}\bot$
    $(a{\rightarrow}\bot){\rightarrow}b$

  | rule for $\rightarrow$

$(((a{\rightarrow}\bot){\rightarrow}b){\rightarrow}\bot){\rightarrow}c \;\vdash\; \bot$
    $(a{\rightarrow}\bot)$
    $b{\rightarrow}\bot$
    $(a{\rightarrow}\bot){\rightarrow}b$
    a

  | rule for atoms using $a{\rightarrow}\bot$

4.2. WORKED EXAMPLES

$(((a\to\bot)\to b)\to\bot)\to c \vdash a$
$\quad\quad (a\to\bot)$
$\quad\quad\quad b\to\bot$
$\quad\quad (a\to\bot)\to b$
$\quad\quad\quad\quad a$
$\quad\quad\quad\quad |$
$\quad\quad\quad success$

Exercise 4.2.12 *Use backward rules to prove the following:*

1. *that $\neg p$ follows from $p\to q$ and $\neg q$*
2. *that $p\vee q$ follows from $\neg p\to q$*
3. *that $((a\to b)\to a)\to a$ is a tautology*
4. *that p follows from $(p\to q)\to q$ and $q\to p$*
5. *that $\neg(p\wedge\neg p)$ is a tautology*
6. *that $\neg(p\to q)\to p$ is a tautology*
7. *that $\neg(p\to q)\to\neg q$ is a tautology*
8. *that $b\vee(b\to c)$ is a tautology*
9. *that $(p\to\neg p)\to\neg p$ is a tautology*
10. *that $(\neg p\to p)\to p$ is a tautology*
11. *that $(p\wedge q)\to\neg(\neg p\vee\neg q)$ is a tautology*
12. *that $a\to(\neg b\to\neg(a\to b))$ is a tautology*
13. *that $\neg p\to((p\vee q)\to q)$ is a tautology*
14. *that b follows from $a \Leftrightarrow ((a\wedge b)\vee(\neg a\wedge\neg b))$.*

4.3 A deterministic goal-directed algorithm

We need to write a precise algorithm of how the backward computation works. We propose the definition below. The database $\mathcal{P}$ is a sequence of assumptions of the form $B \to q$ or $B \to \bot$, for q atomic. When we add data to $\mathcal{P}$ we append the data to the end of the list. When we have an atomic goal q and we look for data items in $\mathcal{P}$ of the form $B \to q$ or $B \to \bot$, we scan the list from beginning to end and take the first such clause. If the computation fails for this first choice, then we go to the next one in the list.

Thus, for example, when we try

1. $p \to q$

2. $q \to q \vdash q$

3. q

we first use the $\to$ rule with clause 1. We try $\vdash p$; this fails because we have no clause with head p. So we go back to the list and use the next clause in line with head q. This is clause 2 and so we try $\vdash q$. We now scan the list and take the first clause with head q, which is clause 1, and so we ask $\vdash p$, and we are in a loop. If we have a historical loop checker we will know that we have already tried $\vdash p$ so we stop and pass to clause 3 and succeed.

Another way of avoiding looping is to throw out clauses that have been used. This is called the policy of *diminishing resource*. If we do that then we do not need a historical loop checker. However, if we adopt the policy of diminishing resource we need to prove that there are no cases where some clauses need to be used twice, or if there are such cases, we need to compensate by additional rules.

As we shall see later, the restart rule can compensate for the policy of diminishing resource and we can manage with using clauses at most once.

Definition 4.3.1 (Translation into ready for computation form)

1. *A ready for computation clause can be defined inductively as follows:*

    - *An atom q or $\bot$ is a ready for computation clause. $\bot, q$ are called the heads. This clause has no body.*

4.3. A DETERMINISTIC GOAL-DIRECTED ALGORITHM

- Let $A_i = (B_i \to x_i)$, $i = 1, \ldots, n$, be ready for computation clauses with bodies B_i and heads x_i (x_i is either atomic or $\bot$). Let x be either atomic or $\bot$; then $\bigwedge_i A_i \to x$ is a ready for computation clause with body $\bigwedge_i A_i$ and head x.

2. The following is a translation $*$ of an aribtrary wff into a conjunction of ready for computation clauses (which is classically equivalent to it).[2]

   A is rewritten as A^* using the following steps:

   $$(A \to B \wedge C)^* = (A \to B)^* \wedge (A \to C)^*$$
   $$(A \to (B \to C))^* = (A \wedge B \to C)^*$$
   $$(A \to q)^* = (A^* \to q), q \text{ atomic}$$
   $$(A \to \bot)^* = (A^* \to \bot)$$
   $$(\neg A)^* = (A \to \bot)^*$$
   $$(A \vee B)^* = ((A \to \bot) \to B)^*$$
   $$(A \wedge B)^* = A^* \wedge B^*$$

Definition 4.3.2 (Backward computation with diminishing resource)

1. A database is a list of clauses annotated with numbers 0 and 1; 1 means available for computation and 0 means not available.

   We present the list as

   $$\mathcal{P} = ((\alpha_1, A_1), \ldots, (\alpha_n, A_n))$$

   where A_i is a clause and $\alpha_i \in \{0, 1\}$.

2. Let $\mathcal{P}$ be an annotated database as above and A a conjunction of clauses of the form[3]

   $$A = \bigwedge_{i=1}^{k} (B_i \to x_i)$$

---

[2] The translation A^* of A is classically equivalent to A. It is *not* intuitionistically equivalent to A, because of the way we translate disjunction. Thus for example in Exercise 4.2.18 item 8, $B \vee (B \to C)$ is not an intuitionistic theorem but its translation $(B \to \bot) \to (B \to C)$ is an intuitionistic theorem and indeed in the solution at the end of the book, restart is not used.

[3] We can view the conjunction as a list $((B_1 \to x_1), \ldots, (B_k \to x_k))$. Although in classical logic the order of the conjuncts is not important, the computation can take account of the order if necessary.

where x_i is either atomic or $\bot$. We let $\mathcal{P} + A$ mean the annotated database

$$((\alpha_1, A_1), \ldots, (\alpha_n, A_n), (1, B_1 \to x_1), \ldots, (1, B_k \to x_k))$$

3. A goal list is a list of the form $H = (x_1, \ldots, x_n)$ where x_i are atoms or $\bot$.

4. A computation *state* has the form $\langle \mathcal{P}, H, G \rangle$, where $\mathcal{P}$ is an annotated database, H is a goal list and G a clause called the goal or query.

Definition 4.3.3 (Diminishing resource computation procedures) *The following are recursive definitions of the notion of success or finite failure of a computation state.*

1. G atomic or $\bot$, immediate success

   - $\langle \mathcal{P}, H, x \rangle$ *succeeds immediately* without restart *for x atomic or $\bot$ if for some $i, (\alpha_i, y) \in \mathcal{P}$ and $\alpha_i = 1$ and either $y = x$ or $y = \bot$.*
   - *It succeeds immediately* with bounded restart *if for some $a_i \in H$, $a_i = x$ and for some $a_j \in H, j \geq i$ and $(1, a_j) \in \mathcal{P}$.*
   - *It succeeds immediately* with restart *if for some $a \in H, (1, a) \in \mathcal{P}$.*
   - *It succeeds immediately if it succeeds according to one of the above.*[4]

2. Case $G =$ atom or $\bot$, immediate failure
   $\langle \mathcal{P}, H, x \rangle$ *fails immediately* without restart *if for all the clauses $B \to y$ such that $(1, B \to y) \in \mathcal{P}$, we have $y \neq \bot$ and $y \neq x$.*

   - *It fails immediately* even with bounded restart *if for any $a_i = x, a_i \in H$ and any $a_j, i \leq j, a_j \in H$, there is no clause B with head a_j such that $(1, B) \in \mathcal{P}$.*
   - *It fails immediately* even with restart *if for any $a \in H$, there is no clause B with head a such that $(1, B) \in \mathcal{P}$.*

---

[4]If the current query is x and the history of previous queries is $(a_1, \ldots, a_k = x)$ then restart allows to be asked any previous query a_j. Bounded restart allows to be asked any a_j, which is 'trapped' between 'now' (i.e. $a_k = x$) and a previous occurrence of x (e.g. $a_i = x$ and $i < j < k$).

4.3. A DETERMINISTIC GOAL-DIRECTED ALGORITHM 125

3. Case of success of G atom or $\bot$, using a clause in the data
   $\langle \mathcal{P}, H, y \rangle$, $y =$ atom or $\bot$, succeeds (succeeds with bounded restart, succeeds with restart) if for some $(\alpha_i, A_i) \in \mathcal{P}$ we have $\alpha_i = 1$, and $A_i = (\bigwedge_j (B_j \to x_j)) \to x$ and either $x = \bot$ or $x = y$ and for each j, the following succeeds (succeeds with bounded restart, succeeds with restart, respectively)

$$\langle \mathcal{P}'_j, H + y, x_j \rangle$$

   where $\mathcal{P}'_j$ is obtained from $\mathcal{P} + B_j$ by changing $\alpha_i = 1$ to $\alpha'_i = 0$, and where $H + y$ is the result of appending y to the end of the list H.

4. Case of failure of G atom or $\bot$ using a clause in the data
   - We say $\langle \mathcal{P}, H, y \rangle$ above fails for the choice of (α_i, A_i) if at least one of $\langle \mathcal{P}'_j, H + y, x_j \rangle$ fails (respectively with or without restart or bounded restart).
   - We say $\langle \mathcal{P}, H, y \rangle$ fails if for all (α, A) in $\mathcal{P}$ such that $\alpha = 1$ and A has head $x = \bot$ or $x = y$, $\langle \mathcal{P}, H, y \rangle$ fails for the choice of (α, A).

5. $\langle \mathcal{P}, H, A \to y \rangle$ succeeds (resp. fails) iff $\langle \mathcal{P} + A, H, y \rangle$ succeeds (resp. fails).

6. $\langle \mathcal{P}, H, A_1 \wedge A_2 \rangle$ succeeds (resp. fails) iff both (resp. at least one) of $\langle \mathcal{P}, H, A_i \rangle$ succeed (resp. fail).

Remark 4.3.4 *It is possible to show that (compare with Exercise 4.3.8)*

- $\mathcal{P} \vdash A$ *in classical logic iff* $\langle \mathcal{P}, \varnothing, A \rangle$ *succeeds with restart.*

- $\mathcal{P} \vdash A$ *in intuitionistic logic iff* $\langle \mathcal{P}, \varnothing, A \rangle$ *succeeds with bounded restart.*

Let us redo some of the examples using restart and diminishing resource. What we shall do is take some of the examples in the previous section and add 0 or 1 annotations and remember the history H of previous goals. We are not writing the history H as a sequence since it can be seen from the record of the computation on the page. The initial list is of course empty since obviously there are no previous goals.

Example 4.3.5

$(1, (p{\to}q){\to}p) \vdash p$

   | *rule for atoms*

$(0, (p{\to}q){\to}p) \vdash p{\to}q$

   | *rule for* $\to$

$(1, p), (0, (p{\to}q){\to}p) \vdash q$

   | *restart*

$(1, p), (0, (p{\to}q){\to}p) \vdash p$

   success

Example 4.3.6

$(1, (a{\to}b){\to}b) \vdash c$
  $(1, a{\to}c)$
  $(1, b{\to}c)$

   | *rule for atoms used with* $b{\to}c$

$(1, (a{\to}b){\to}b) \vdash b$
  $(1, a{\to}c)$
  $(0, b{\to}c)$

   | *rule for atoms used with* $(a{\to}b){\to}b$

$(0, (a{\to}b){\to}b) \vdash a{\to}b$
  $(1, a{\to}c)$
  $(0, b{\to}c)$

   | *rule for* $\to$

$(0, (a{\to}b){\to}b) \vdash b$
  $(1, a{\to}c)$
  $(0, b{\to}c)$
  $(1, a)$

4.3. A DETERMINISTIC GOAL-DIRECTED ALGORITHM

> *Notice here that we cannot use the rule for atoms for b with $(a \to b) \to b$ because it is annotated 0. Thus we not only get that the databases decrease, but also cannot loop! Let us use restart.*
>
> restart

$(0, (a \to b) \to b) \vdash c$
  $(1, a \to c)$
  $(0, b \to c)$
    $(1, a)$

> *Notice now that we can only use $a \to c$.*
> rule for atoms with $a \to c$

$(0, (a \to b) \to b) \vdash a$
  $(0, a \to c)$
  $(0, b \to c)$
    $(1, a)$

success

Example 4.3.7

$(1, (a \to b) \to b) \vdash (b \to a) \to a$

> rule for $\to$

$(1, (a \to b) \to b) \vdash a$
  $(1, b \to a)$

> rule for atoms for $b \to a$

$(1, (a \to b) \to b) \vdash b$
  $(0, b \to a)$

> rule for atoms using $(a \to b) \to b$

$(0, (a \to b) \to b) \vdash a \to b$
  $(0, b \to a)$

> *Note that at this stage the database is empty—no active (annotated with 1) data. However, we do have the history which is $H = (a, b)$.*
>
> *We now use rule for $\to$.*

$(0, (a \to b) \to b) \vdash b$
  $(0, b \to a)$
  $(1, a)$

  | restart

$(0, (a \to b) \to b) \vdash a$
  $(0, b \to a)$
  $(1, a)$

  |

  success

Exercise 4.3.8 (Soundness of the diminishing resource computation procedure) *Show that if $\langle \mathcal{P}, H, B \rangle$ succeeds with restart then $\bigwedge \mathcal{P} \to B \vee \bigvee H$ is a classical tautology, where $\bigwedge \mathcal{P}$ means $\bigwedge_{(1, A) \in \mathcal{P}} A$.*

5
METHODOLOGY AND METATHEOREMS

In previous chapters we encountered several kinds of logics defined in a variety of ways. It is time for a methodological discussion about logics and their properties in general, and some metatheorems about classical logic in particular.

In Chapter 2 we said that to present a logical system we need three components:

- A formal language **L** defining the well-formed formulae of the logic.

- A semantic interpretation for the language.

- A family of reasoning rules.

We proceeded to define classical logic in this way. The formulae of the language were built up from atomic propositions and the connectives $\{\neg, \wedge, \vee, \rightarrow, \top, \bot\}$. The semantic interpretation was given through the classical truth tables. We had several options for reasoning rules: forward rules was one and backward rules was another. We also saw how to define our logics by changing the semantic interpretation (defining many-valued logics), or the language (adding modality $\Box$), or modifying the reasoning rules (adopting the policy of diminishing resource). All of these different definitions actually define different consequence relations for the language of the logics. We now need to study this notion in the abstract.

5.1 Consequence relations and Hilbert systems

Let **L** be a language. This section will briefly outline several syntactical methods for defining logical systems for **L**.

Definition 5.1.1 *We define a (transitive) logic (or consequence relation) as any relation $\mathrel{|\!\sim}$ between well-formed formulae satisfying the following:*

Identity:
$$A \mathrel{|\!\sim} A$$

Transitivity:
$$\frac{A \mathrel{|\!\sim} B; B \mathrel{|\!\sim} C}{A \mathrel{|\!\sim} C}$$

Equivalence:[1]
$$\frac{A \mathrel{|\!\sim} B; B \mathrel{|\!\sim} A; C \mathrel{|\!\sim} A}{C \mathrel{|\!\sim} B}$$

$$\frac{A \mathrel{|\!\sim} B; B \mathrel{|\!\sim} A; A \mathrel{|\!\sim} C}{B \mathrel{|\!\sim} C}$$

Such systems can be extended to relations between sets $\mathcal{P}$ of wffs (including the empty set) and single wff, satisfying:

Reflexivity:
$$\mathcal{P} \mathrel{|\!\sim} A, \text{ if } A \in \mathcal{P}$$

Restricted monotonicity:
$$\frac{\mathcal{P} \mathrel{|\!\sim} A; \mathcal{P} \mathrel{|\!\sim} B}{\mathcal{P}, A \mathrel{|\!\sim} B}$$

Cut:
$$\frac{\mathcal{P}, A \mathrel{|\!\sim} B; \mathcal{P} \mathrel{|\!\sim} A}{\mathcal{P} \mathrel{|\!\sim} B}$$

---

[1] *Equivalence* is derivable from *transitivity* and so can be eliminated. However, equivalence is more basic than transitivity. Some notions of a (non-monotonic) logical system do not include transitivity.

5.1. CONSEQUENCE RELATIONS AND HILBERT SYSTEMS

A rule ρ for $\mathrel{\vert\!\sim}$ has the form

$$\rho: \frac{\mathcal{P}_i \mathrel{\vert\!\sim} A_i, i = 1, \ldots, k}{\mathcal{P} \mathrel{\vert\!\sim} A}$$

A consequence relation $\mathrel{\vert\!\sim}$ is said to satisfy (or be closed under) the rule ρ if for any substitution instance $\mathcal{P}', A'_i, A'$ of $\mathcal{P}, A_i$ and A respectively obtained by substituting B_j for q_j, $j = 1, \ldots$, we have that if $\mathcal{P}'_i \mathrel{\vert\!\sim} A'_i$ holds for $i = 1, \ldots, k$ then $\mathcal{P}' \mathrel{\vert\!\sim} A'$ holds.

A logic is defined by a set of rules $\{\rho_i\}$ if it can be presented as the smallest consequence relation closed under these rules.

Example 5.1.2 *Consider a language with the connectives $\{\wedge, \to, \bot\}$. Consider the smallest consequence relation $\mathrel{\vert\!\sim}$ for this language satisfying the following.*

- $A \wedge B \mathrel{\vert\!\sim} A$
- $A \wedge B \mathrel{\vert\!\sim} B$
- $C \wedge A \mathrel{\vert\!\sim} B$ iff $C \mathrel{\vert\!\sim} A \to B$
- $\bot \mathrel{\vert\!\sim} A$
- $A \wedge B \mathrel{\vert\!\sim} B \wedge A$
- $\dfrac{A \mathrel{\vert\!\sim} B_1, A \mathrel{\vert\!\sim} B_2}{A \mathrel{\vert\!\sim} B_1 \wedge B_2}$
- $\dfrac{A \mathrel{\vert\!\sim} B; C \wedge B \mathrel{\vert\!\sim} D}{C \wedge A \mathrel{\vert\!\sim} D}$

We now show that this consequence relation gives intuitionistic logic for this fragment of the language. We prove

Exercise 5.1.3 *Let $\vDash_I$ be the consequence relation of Definition 2.2.3. Show that $\vDash_I$ is the same as $\mathrel{\vert\!\sim}$ of the previous example.*

Hints

1. First show that $\vDash_I$ satisfies the conditions on $\mathrel{\vert\!\sim}$. This will show that $\mathrel{\vert\!\sim} \subseteq \vDash_I$.

2. To show that $\vDash_I \,\subseteq\, \mid\!\sim$, construct a Kripke model as follows.
   Let $A \equiv B$ mean $A \mid\!\sim B$ and $B \mid\!\sim A$.

   - Let $T =$ the set of all equivalence classes of wffs over $\equiv$, except the class of $\bot$. We denote such a class by $A/\equiv$.
   - Let $A/\equiv \,\leq\, B/\equiv$ be defined as $B \mid\!\sim A$. Show that this relation is well defined; in other words, show that $A \equiv A_1, B \equiv B_1$ imply $B_1 \mid\!\sim A_1$ iff $B \mid\!\sim A$.
   - Let $h(A/\equiv, q)$ be defined as holding iff $A \mid\!\sim q$.

   Show that in the model $(T, \leq, h)$ we have

   $$h(A/\equiv, B) = 1 \text{ iff } A \mid\!\sim B$$

In particular the definition of h does not depend on the representative $A \in A/\equiv$. The above shows that if $A \not\mid\!\sim B$ then $A \nvDash_I B$, as we have $h(A/\equiv, B) = 0$ in the above model. For a solution see the proof of Theorem 5.7.1.

Some logics can be completely characterized by the set of theorems, i.e. by $\{A \mid \varnothing \mid\!\sim A\}$. This leads us to a Hilbert formulation of a logic, and to the next section.

5.2 Hilbert formulation of a logic

We saw in previous sections that different logics for the same language, say for $\twoheadrightarrow$, have different sets of tautologies. There is a way of characterizing a logic by generating its tautologies. We need some definitions.

Definition 5.2.1 *Consider a propositional language with atomic propositions and the single binary connective $\twoheadrightarrow$.*

1. *We define the set of wffs in the usual way, namely:*

   - *Let q be atomic; then 'q' is a wff based on $\{q\}$.*
   - *If $A(q_1, \ldots, q_n), B(p_1, \ldots, p_k)$ are wffs based on $\{q_1, \ldots, q_n\}$ and $\{p_1, \ldots, p_k\}$ respectively then $(A \twoheadrightarrow B)$ is a wff based on $\{p_1, \ldots, p_k, q_1, \ldots, q_n\}$.*

5.2. HILBERT FORMULATION OF A LOGIC

2. Assume C is based on $q_1, \ldots, q_n$, then C' is said to be a substitution instance of C if for some $A_i, i = 1, \ldots, n$, we have $C' = C(q_i/A_i), i = 1, \ldots, n$. Recall Definition 1.1.2.

Definition 5.2.2 *A set $\mathcal{P}$ of wffs in a language with a binary connective $\twoheadrightarrow$ is said to be* generated *as a Hilbert system from the formulae (called Hilbert axioms) $C_1, \ldots, C_k$ if $\mathcal{P}$ is the smallest set of wffs, such that*

- $C_1, \ldots, C_k \in \mathcal{P}$.

- *If $C \in \mathcal{P}$ and C' is a substitution instance of C then $C' \in \mathcal{P}$.*

- *If $A \in \mathcal{P}$ and $(A \twoheadrightarrow B) \in \mathcal{P}$ then also $B \in \mathcal{P}$.*

It is customary to compare logics by generating their set of tautologies. The stronger logics will have more axioms.

Example 5.2.3 *It is possible to generate all intuitionistic implicational tautologies by substitution, and the rule of* modus ponens, *namely*

$$\frac{A, A \to B}{B}$$

and the Hilbert axioms

$$A \to (B \to A)$$

and

$$(A \to (B \to C)) \to ((A \to B) \to (A \to C))$$

The following additional tautology, taken as an additional axiom and known as Peirce's rule, helps generate all classical implicational tautologies, i.e. turns $\to$ into classical implication

$$((A \to B) \to A) \to A$$

Exercise 5.2.4 Glivenko theorem

1. Show that A is a classical tautology, if and only if $\neg\neg A$ is an intuitionistic tautology.

2. Show that if $A \equiv B$ in classical logic, then $\neg\neg A \equiv \neg\neg B$ in intuitionistic logic.

Exercise 5.2.5

1. Prove (generate) $A \to A$ from the two Hilbert axioms of intuitionistic $\to$, and the rule of modus ponens.

2. Show that Peirce's rule cannot be generated from the intuitionistic axioms, and modus ponens.

Example 5.2.6 (A Hilbert formulation for classical logic) *The following is a sample of a Hilbert formulation of classical logic, taken from [Nidditch, 1962].*

The axioms are (all substitution instances of)

1. $p \to (q \to p)$

2. $(p \to (q \to r)) \to ((p \to q) \to (p \to r))$

3. $p \to p \vee q$

4. $q \to p \vee q$

5. $(p \to r) \to ((q \to r) \to (p \vee q \to r))$

6. $p \wedge q \to p$

7. $p \wedge q \to q$

8. $(r \to p) \to ((r \to q) \to (r \to p \wedge q))$

9. $(p \to q) \to (\neg q \to \neg p)$

10. $p \to \neg\neg p$

11. $\neg\neg p \to p$

The above axioms together with modus ponens generate all classical tautologies. In fact, if we drop axiom 11, the remaining axioms together with modus ponens constitute a Hilbert formulation for intuitionistic logic and generate all intuitionistic theorems.[2]

A consequence relation $\mathrel{\vdash}$ *can be defined from the Hilbert system by letting* $A_1, \ldots, A_n \mathrel{\vdash} B$ *iff the formula* $A_1 \to (A_2 \to \cdots \to (A_n \to B) \cdots)$ *is generated.*

---

[2]This axiom system is *separated*; that is, any formula A built up from $\to$ and other connectives is provable only from the axioms about $\to$ and these other connectives.

5.2. HILBERT FORMULATION OF A LOGIC

Example 5.2.7 (A Hilbert formulation of three-valued logic)
For comparison with the classical two-valued case, the following is a Hilbert axiom system for Łukasiewicz three-valued logic, for the language with $\{\neg, \wedge, \vee, \rightarrow\}$.

1. $A \rightarrow (B \rightarrow A)$
2. $(A \rightarrow B) \rightarrow ((B \rightarrow C) \rightarrow (A \rightarrow C))$
3. $(A \rightarrow (B \rightarrow C)) \rightarrow (B \rightarrow (A \rightarrow C))$
4. $((A \rightarrow B) \rightarrow B) \rightarrow ((B \rightarrow A) \rightarrow A)$
5. $((((A \rightarrow B) \rightarrow A) \rightarrow A) \rightarrow (B \rightarrow C)) \rightarrow (B \rightarrow C)$
6. $A \wedge B \rightarrow A$
7. $A \wedge B \rightarrow B$
8. $(A \rightarrow B) \rightarrow ((A \rightarrow C) \rightarrow (A \rightarrow (B \wedge C)))$
9. $A \rightarrow A \vee B$
10. $B \rightarrow A \vee B$
11. $(A \rightarrow C) \rightarrow ((B \rightarrow C) \rightarrow (A \vee B \rightarrow C))$
12. $(\neg B \rightarrow \neg A) \rightarrow (A \rightarrow B)$

The inference rule is modus ponens.
See [Avron, 1988] for more details.

Definition 5.2.8 (Consequence relations associated with Hilbert systems) *Let H be a Hilbert system, as defined in Definition 5.2.2. We define a consequence relation $\vdash_H$ as follows: $A_1, \ldots, A_n \vdash_H B$ iff there exists a (proof) sequence of formulae $(D_1, \ldots, D_k)$ such that the following hold.*

1. $D_k = B$.

2. *For each element D_m in the sequence, one of the following cases holds:*

   (a) $D_m = A_i$, *for some $1 \leq i \leq n$.*

   (b) D_m *is a theorem of (i.e. is generated by) the Hilbert system H.*

(c) For some $m_1, m_2 < m$ we have that $D_{m_2} = (D_{m_1} \twoheadrightarrow D_m)$.[3]

5.3 Metatheorems about classical logic consequence

At the start of Chapter 3, we encountered the cut rule, which we restate as a theorem below:

Theorem 5.3.1 (Cut theorem) *If* $\mathcal{P} \models \varphi$ *and* $\mathcal{P}, \varphi \models \psi$ *then* $\mathcal{P} \models \psi$.

Proof. Suppose that $\mathcal{P} \models \varphi$ and $\mathcal{P}, \varphi \models \psi$, and draw a truth table for $\mathcal{P}$, φ and ψ. Each row in the table which has all the formulae in $\mathcal{P}$ true, and also has φ true, will have ψ true (because $\mathcal{P}, \varphi \models \psi$). But every row with all of the formulae in $\mathcal{P}$ true will have φ true (because $\mathcal{P} \models \varphi$). Therefore every row with all of the formulae in $\mathcal{P}$ true will also have ψ true. Hence $\mathcal{P} \models \psi$. ∎

We can see the cut theorem in action as follows. Let $\mathcal{P} = \{a \to b, a \wedge c\}$, and let $\varphi = a$ and $\psi = b$. Now $\mathcal{P} \models a$ because whenever all of $\mathcal{P}$ is true, $a \wedge c$ must be true, and thus a must be true. Moreover, $\mathcal{P}, a \models b$ because whenever $a \to b$ and a are true together, b must be true. Hence by the cut theorem, we must have $\mathcal{P} \models b$. This is indeed the case, as can be seen from the truth table below. Each row which makes all of $\mathcal{P}$ true has b true.

a	b	c	$a \to b$	$a \wedge c$	$\mathcal{P}$	$\mathcal{P} \to a$	$\mathcal{P}, a \to b$
T	T	T	T	T	T	T	T
T	T	⊥	T	⊥	⊥	T	T
T	⊥	T	⊥	T	T	T	T
T	⊥	⊥	⊥	⊥	⊥	T	T
⊥	T	T	T	⊥	⊥	T	T
⊥	T	⊥	T	⊥	⊥	T	T
⊥	⊥	T	T	⊥	⊥	T	T
⊥	⊥	⊥	T	⊥	⊥	T	T

---

[3] Note that for the Hilbert systems of Example 5.2.6, the natural rules for the other connectives can be derived; for example, using instances of axiom 8 we can get that $D_1, D_2 \vdash D_1 \wedge D_2$.

5.3. METATHEOREMS ABOUT CLASSICAL LOGIC CONSEQUENCE

Towards the end of Chapter 1 we stated that logical consequence possessed the property of monotonicity, namely that if a conclusion from a set of assumptions is valid, then the same conclusion is valid from the set of assumptions with some additional formulae added.

Theorem 5.3.2 (Monotonicity) *Let $\mathcal{P}$ and $\mathcal{Q}$ be sets of formulae, and φ be a formula. If $\mathcal{P} \vDash \varphi$ then $\mathcal{P} \cup \mathcal{Q} \vDash \varphi$ for any $\mathcal{Q}$.*

Proof. If $\mathcal{P} \cup \mathcal{Q} \nvDash \varphi$ then we would have a row in the truth table in which all of $\mathcal{P} \cup \mathcal{Q}$ was true, but φ was false. But for every such row, it is the case that $\mathcal{P}$ would be all true, and φ would be false, and thus $\mathcal{P} \nvDash \varphi$. ∎

We can now use these two theorems to prove the transitivity theorem presented earlier.

Theorem 5.3.3 (Transitivity) *For any three propositional formulae α, β and γ, if $\alpha \vDash \beta$ and $\beta \vDash \gamma$ then $\alpha \vDash \gamma$.*

Proof. Suppose that $\alpha \vDash \beta$ and $\beta \vDash \gamma$. From the latter, by using the monotonicity theorem we can show that $\alpha, \beta \vDash \gamma$. The cut theorem on this and $\alpha \vDash \beta$ then gives us $\alpha \vDash \gamma$ as required. ∎

Theorem 5.3.4 (Substitution) *If φ is a tautology which contains a particular proposition p, and we simultaneously replace p with some formula ψ (which does not contain p) throughout φ to get a new formula φ', then the new formula φ' is also a tautology.*

Proof. See Example 1.3.3. ∎

A somewhat surprising theorem about propositional logic shows that if one formula is a logical consequence of another, and there is a subset of propositions common to both formulae, then there is a third formula involving just the common propositions which is a consequence of the first formula, and which in turn has the second formula as a consequence. This is perhaps easier to present symbolically:

Theorem 5.3.5 (Interpolation) *For any two formulae of propositional logic φ and ψ such that $\varphi \vDash \psi$, with φ and ψ having some common atomic propositions, then there exists a formula θ built up solely of the common atomic propositions, such that $\varphi \vDash \theta$ and $\theta \vDash \psi$.*

Proof. Draw up a truth table for the propositions common to both φ and ψ. For each row in the truth table which can be expanded to make ψ false (i.e. by adding the propositions private to ψ to values that make ψ false), let the row have the truth value $\bot$, otherwise let the row have the value $\top$. A formula θ can then be constructed for this table (see Section 1.1.2). It is not possible for a row to be expanded to make φ true and ψ false (since $\varphi \models \psi$); hence there is no conflict as to which value to give to θ. Now all the rows in which θ is true are rows in which ψ cannot be made false; hence $\theta \models \psi$. All the rows which can be expanded to make φ true have θ true; hence $\varphi \models \theta$. ∎

See also Remark 6.7.3.

5.4 Propositional soundness and completeness

It is important that any procedure purporting to be a theorem prover for a logical system should possess the property of *soundness*; that is, any formula which the procedure indicates is a theorem of the logical system actually is a theorem of the system. In addition, our theorem prover should be *complete*. This means that if a formula is a theorem of the logical system, then our procedure will indicate that. In Figure 5.1 we illustrate the four possible interesting relationships between the set of formulae that our procedure indicates is a theorem ($\vdash$) and the set of theorems of the system (taut). As you can see from Figure 5.1, the possibilities are:

(a) there is no subset relationship between $\vdash$ and taut and so some of the formulae that the theorem prover claims are valid are theorems, and some are not. Hence the theorem prover is neither sound nor complete. (The intersection of the two sets may be empty—in which case every answer produced by the theorem prover is wrong.)

(b) the formulae that the theorem prover claims are valid are a proper subset of the theorems of the system, and so the theorem prover is sound (although because there are theorems which lie outside $\vdash$, it is not complete).

(c) the theorems of the system are a proper subset of the formulae that the
theorem prover claims are valid, so the theorem prover is complete

5.4. PROPOSITIONAL SOUNDNESS AND COMPLETENESS

(it proves all valid tautologies) but not sound (there are formulae it claims are tautologies, but which are not).

(d) the set of formulae that the theorem prover claims are valid is equal to the set of theorems of the system, and thus the theorem prover is both sound and complete.

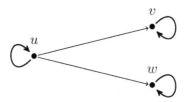

Figure 5.1: Combinations of soundness and completeness

For some logical systems it is not possible to build theorem provers which have the completeness property. Fortunately, for the propositional logic we have been dealing with, it is possible to have theorem-proving procedures which are both sound and complete—the use of truth tables is one such procedure. In this chapter we will show that the backward rules procedure is also sound and complete. From now on every definition and so forth in this chapter should be taken as being with respect to propositional logic.

Definition 5.4.1 *We shall write $\vdash_\Gamma A$ to mean that the procedure Γ indicates that A is a theorem. We say that the Γ-computation of A succeeds. A procedure Γ is said to have the* soundness *property iff*

$$\vdash_\Gamma A \text{ implies that } A \text{ is a tautology}$$

for every wff A. A procedure Γ is said to have the completeness *property iff*

$$A \text{ is a tautology implies } \vdash_\Gamma A$$

for every wff A. These can alternatively be written as

$$\vdash_\Gamma A \text{ implies that } \vDash A \qquad \text{and} \qquad \vDash A \text{ implies } \vdash_\Gamma A$$

If we denote our backward rules procedure by the symbol $\mathcal{B}$, then we need to show that for any wff A,

$$\vdash_{\mathcal{B}} A \quad \text{iff} \quad A \text{ is a tautology}$$

in order to prove that the backward rules procedure is both sound and complete. To carry out the proof, we shall need a formal definition of the backward rules procedure for propositional logic.

We agreed in Chapter 4 that all our formulae shall be written using $\wedge$, $\rightarrow$ and $\bot$ only, and that all the rules will have atomic heads so that each rule has the form $A \rightarrow q$, with q being atomic. We saw that this can be done using the rules

$$\begin{aligned} \neg A &= A \rightarrow \bot \\ A \vee B &= (A \rightarrow \bot) \rightarrow B \\ A \rightarrow x \wedge y &= (A \rightarrow x) \wedge (A \rightarrow y) \\ \text{and} \quad A \rightarrow (x \rightarrow y) &= A \wedge x \rightarrow y. \end{aligned}$$

The following is an inductive definition of the notion of a $\mathcal{B}$-clause (i.e. the type of clause we are dealing with).

Definition 5.4.2 (Propositional $\mathcal{B}$-clauses)

1. *An atom q is a propositional $\mathcal{B}$-clause, as is $\bot$.*

2. *If $A_1, \ldots, A_n$ are $\mathcal{B}$-clauses and q is an atom then*

$$A_1 \wedge \cdots \wedge A_n \rightarrow q \text{ and } A_1 \wedge \cdots \wedge A_n \rightarrow \bot$$

are propositional $\mathcal{B}$-clauses.

3. *A query is a conjunction of propositional $\mathcal{B}$-clauses, e.g.*

$$A_1 \wedge \cdots \wedge A_n$$

where $A_1, \ldots, A_n$ are propositional $\mathcal{B}$-clauses.

The discussion in Chapter 4 showed that any formula of propositional logic can be rewritten as a conjunction of propositional $\mathcal{B}$-clauses. Let us state this formally.

Lemma 5.4.3 *Any formula A of propositional classical logic can be equivalently written as a conjunction of $\mathcal{B}$-clauses.*

5.4. PROPOSITIONAL SOUNDNESS AND COMPLETENESS

The general form of asking a query from a set of data using backward computation is $\mathcal{P} \vdash_\mathcal{B} A$ where $\mathcal{P}$ is a set of $\mathcal{B}$-clauses and A is a conjunction of $\mathcal{B}$-clauses. For example,

$$\left\{ \begin{array}{l} (a \to b) \to b \\ (c \wedge b) \to \bot \\ b \to \bot \end{array} \right\} \vdash_\mathcal{B} (a \to \bot) \to c$$

We extend the notion of a computation succeeding and write $\mathcal{P} \vdash_\mathcal{B} A$ to mean A succeeds from $\mathcal{P}$. Recall that for the purposes of the restart rule, the backward computation is performed relative to the original goal, so we should write $\langle \mathcal{P}, G \rangle \vdash_\mathcal{B} A$ where $\langle \mathcal{P}, G \rangle$ means $\mathcal{P}$ is the data and G is the original goal. The following definition is a summary of the computation rules. Remember, the clauses in $\mathcal{P}$ are always of the form $B \to q$. Thus when we write $\mathcal{P} \cup \{D\}$ and D is $(B_1 \to q_1) \wedge (B_2 \to q_2)$ we mean $\mathcal{P} \cup \{B_1 \to q_1, B_2 \to q_2\}$.

Definition 5.4.4 (Backward computation rules) *We define $\langle \mathcal{P}, G \rangle \vdash_\mathcal{B} A$ by induction over the structure of the $\mathcal{B}$-clause A, and the restart rule.*

1. *When A is an atom q we have $\langle \mathcal{P}, G \rangle \vdash_\mathcal{B} q$ iff either*

   (a) *$q \in \mathcal{P}$ or $\bot \in \mathcal{P}$, or*

   (b) *for some $B \to q \in \mathcal{P}$ or $B \to \bot \in \mathcal{P}$ we have $\langle \mathcal{P}, G \rangle \vdash_\mathcal{B} B$.*

2. *When A is a conjunction $B \wedge C$, we have $\langle \mathcal{P}, G \rangle \vdash_\mathcal{B} B \wedge C$ iff $\langle \mathcal{P}, G \rangle \vdash_\mathcal{B} B$ and $\langle \mathcal{P}, G \rangle \vdash_\mathcal{B} C$.*

3. *When A is an implication $B \to q$ we have $\langle \mathcal{P}, G \rangle \vdash_\mathcal{B} B \to q$ iff $\langle \mathcal{P} \cup \{B\}, G \rangle \vdash_\mathcal{B} q$.*

4. *The restart rule states that $\langle \mathcal{P}, G \rangle \vdash_\mathcal{B} q$ for an atom q if $\langle \mathcal{P}, G \rangle \vdash_\mathcal{B} G$, where G is the original goal of the computation.*

5. *We sometimes abbreviate $\langle \mathcal{P}, Q \rangle \vdash_\mathcal{B} Q$ by $\mathcal{P} \vdash_\mathcal{B} Q$.*

6. *Let us write $\mathcal{P} \vdash_\mathcal{I} Q$ if $\mathcal{P} \vdash_\mathcal{B} Q$ without use of the restart rule (clause 4 above).*

Our use of the backward rules procedure has generally been to show that a formula Q is a tautological consequence of a set of formulae $\mathcal{P}$.

Thus the tautology that we are interested in is not Q, but $\bigwedge \mathcal{P} \to Q$, where $\bigwedge \mathcal{P}$ is the conjunction of all the formulae in $\mathcal{P}$ (see Definition 1.2.6). Hence in showing the soundness and completeness of the backward rules procedure, we are concerned with showing that

$$\langle \mathcal{P}, Q \rangle \vdash_B Q \text{ iff } \bigwedge \mathcal{P} \to Q \text{ is a tautology}$$

5.5 Proof of soundness

The proof of the soundness of the backward computation rules is fairly simple. We show that for each type of query Q that the procedure can claim to follow from a given set of data $\mathcal{P}$, the query is a tautological consequence of the data.

Theorem 5.5.1 (Soundness of the backward computation rules) *If $\langle \mathcal{P}, G \rangle \vdash_B Q$ then $\bigwedge \mathcal{P} \to Q \vee G$ is a tautology.*

Proof. We shall deal with each computation rule in turn. Assume that $\langle \mathcal{P}, G \rangle \vdash_B Q$.

1. If $Q = q$ where q is an atom, then we have one of the following cases:

    (a) $q \in \mathcal{P}$, in which case q cannot be false without $\bigwedge \mathcal{P}$ being false; hence the only situation which could falsify $\bigwedge \mathcal{P} \to q$ cannot arise, so that $\bigwedge \mathcal{P} \to q \vee G$ is a tautology.

    (b) $\bot \in \mathcal{P}$. This means that $\bigwedge \mathcal{P}$ must be false and therefore $\bigwedge \mathcal{P} \to q \vee G$ must be a tautology.

    (c) For some $A \to q \in \mathcal{P}$ we have $\langle \mathcal{P}, G \rangle \vdash_B A$. Now if the subcomputation of A succeeds, then we have $\bigwedge \mathcal{P} \to A \vee G$ as a tautology. Hence whenever $\bigwedge \mathcal{P}$ is true, $A \vee G$ must be true, and by the clause $A \to q$ we must have $q \vee G$ true. Therefore $\bigwedge \mathcal{P} \to q \vee G$ is true.

    (d) For some $A \to \bot \in \mathcal{P}$ we have $\langle \mathcal{P}, G \rangle \vdash_B A \vee G$. If the subcomputation of A succeeds, then we have $\bigwedge \mathcal{P} \to A \vee G$ as a tautology. Assume that $\bigwedge \mathcal{P}$ is true, then $A \vee G$ is true. But $A \to \bot \in \mathcal{P}$ and thus $A \to \bot$ is true, i.e. A is false. Then G must be true, and hence $q \vee G$ is true.

    Therefore $\bigwedge \mathcal{P} \to q \vee G$ is true.

5.6. PROOF OF COMPLETENESS

2. If $Q = A \wedge B$ then we know that $\langle \mathcal{P}, G \rangle \vdash_{\mathcal{B}} A$ and $\langle \mathcal{P}, G \rangle \vdash_{\mathcal{B}} B$ and thus we have $\bigwedge \mathcal{P} \to A \vee G$ and $\bigwedge \mathcal{P} \to B \vee G$ as tautologies.

    Assume that $\bigwedge \mathcal{P}$ is true; then $A \vee G, B \vee G$ are true. Now there are two cases:

    - If G is true then $(A \wedge B) \vee G$ is true.
    - If G is false then A, B must be true; hence $(A \wedge B)$ is true, and again $(A \wedge B) \vee G$ is true.

    So we must have $\bigwedge \mathcal{P} \to ((A \wedge B) \vee G)$ as a tautology.

3. If $Q = A \to B$ then we know that $\langle \mathcal{P} \cup \{A\}, G \rangle \vdash_{\mathcal{B}} B$, and thus we know that $(\bigwedge \mathcal{P}) \wedge A \to B \vee G$ is a tautology. Assume $\bigwedge \mathcal{P}$ is true; hence we know that $A \to B \vee G$ is also true; therefore $\bigwedge \mathcal{P} \to (A \to B) \vee G$ is also true, since $((A \to B) \vee G) \equiv (A \to B \vee G)$.

    Thus $\bigwedge \mathcal{P} \to (A \to B) \vee G$ is a tautology.

4. The soundness of the restart rule follows from the following: G is equivalent to $\neg G \to G$, which is equivalent to $(G \to \bot) \to G$ (which can be checked via truth tables). Thus when we start our computation with $\langle \mathcal{P}, G \rangle \vdash_{\mathcal{B}} G$ we can write $\langle \mathcal{P}, G \rangle \vdash_{\mathcal{B}} (G \to \bot) \to G$ instead, since $[(G \to \bot) \to G] \equiv G$. Using the rule for $\to$, we add the antecedent to the data, and ask $\langle \mathcal{P} \cup \{G \to \bot\}, G \rangle \vdash_{\mathcal{B}} G$. If in the course of the computation we want to restart and ask G again, we can use the rule for $\bot$, since $G \to \bot$ is available, so we have $\langle \mathcal{P} \cup \{G \to \bot\}, G \rangle \vdash_{\mathcal{B}} q$ whenever $\langle \mathcal{P} \cup \{G \to \bot\}, G \rangle \vdash_{\mathcal{B}} G$.

    Thus $\langle \mathcal{P}, G \rangle \vdash_{\mathcal{B}} G$ succeeds with the use of restart iff $\mathcal{P} \cup \{G \to \bot\} \vdash_{\mathcal{B}} G$ succeeds, even without restart. If the latter succeeds then $[(\bigwedge \mathcal{P}) \wedge (G \to \bot)] \to G$ is a tautology, and thus $\bigwedge \mathcal{P} \to G \vee Q$ is a tautology. ∎

5.6 Proof of completeness

Proving completeness is much more tricky than proving soundness. When we prove soundness, we have to show that each 'positive' usage of the computation rules maintains the property of tautological consequence. The computation rules are a small set of ideas to work with. In contrast, when we prove completeness, we have to show that each tautological consequence is demonstrated by the computation rules. In order to be able to work from the rules for the proof procedure, instead of showing that

$$\bigwedge \mathcal{P} \to \varphi \text{ is a tautology implies } \mathcal{P} \vdash_B \varphi$$

we will show the equivalent

$$\mathcal{P} \nvdash_B \varphi \text{ implies } \bigwedge \mathcal{P} \to \varphi \text{ is not a tautology}$$

This means that we have to work with the cases in which our proof procedure fails to show φ from $\mathcal{P}$, and prove that φ is then not a tautological consequence of $\mathcal{P}$. The drawback with this approach is due to the difficulty of identifying the cases when the proof procedure fails to show φ from $\mathcal{P}$. The proof of Theorem 5.6.1 is therefore rather drawn out, and so we sketch the proof in broad outline, before filling in the details.

Theorem 5.6.1 (Completeness of the backward computation rules) *If $\mathcal{P} \nvdash_B \varphi$, then $\bigwedge \mathcal{P} \to \varphi$ is not a tautology.*

Proof sketch: Assume that the proof procedure fails, i.e. $\mathcal{P} \nvdash_B \varphi$. We will show from this that there are truth values which can be assigned to the atoms which make $\bigwedge \mathcal{P}$ true and φ false, and hence that $\bigwedge \mathcal{P} \to \varphi$ is not a tautology.

Proving this is much more complicated than proving soundness, and so we shall proceed in comprehensible stages. We begin by defining the notion of a computation tree which records the progress of the proof of a query, and what it is for such a tree to be 'successful'. We illustrate some of the properties of the trees which we shall use, including replacing branches of the tree with other, equivalent, branches.

Next we show that a version of the cut theorem holds for computation trees, i.e. if we have trees for $\mathcal{P} \vdash_B \varphi$ and $\mathcal{P} \cup \{\varphi\} \vdash_B q$ then we have a tree for $\mathcal{P} \vdash_B q$, where q is an atomic proposition. Then we show how we include restart into the trees. As the final part of the preparation, we prove that if we have a tree for $\mathcal{P} \cup \{R\} \vdash_B \varphi$ and $\mathcal{P} \cup \{\neg R\} \vdash_B \varphi$ then we have a tree for $\mathcal{P} \vdash_B \varphi$.

Definition 5.6.2 (Computation trees without restart) *We define the notion of a successful computation tree of a proof of a formula φ from a set of assumptions $\mathcal{P}$. The tree is for a proof without the restart rule. A basic tree structure is described by three components: $\mathcal{T} = \langle T, \prec, 0 \rangle$. T is a set of nodes which make up the tree, $\prec$ is a binary relation which describes the ordering of nodes in the tree, and 0 is the node which is the root of the tree. The following hold:*[4]

---
[4] The tree is imagined to be 'growing' downwards, with the root 0 at the top.

5.6. PROOF OF COMPLETENESS

- $\prec$ is an irreflexive and transitive relation.

- Let $x \prec^0 y$ be $x = y$ and $x \prec^{n+1} y$ iff for some z, $x \prec z$ and $z \prec^n y$.

  Let $T_t = \{y \mid \text{for some } n, t \prec^n y\}$. Then we require that $T = T_0$ and for all $t \in T$, $t \prec x$ and $t \prec y$ and $x \neq y$ imply $T_x \cap T_y = \emptyset$.

- Let $\mathcal{T}_t$ be the tree $\langle T_t, \prec, t, g \rangle$, where g is defined below.

To turn the basic tree structure into a computation tree, we need a labelling function, which associates information with each node in the tree. Let g be this function. For each $t \in T$, $g(t) = \langle \mathcal{P}(t), G(t) \rangle$, i.e. g labels each node with a set of assumptions $\mathcal{P}(t)$ and a goal formula $G(t)$. The system $\langle T, \prec, 0, g \rangle$ must satisfy the following conditions:

1. $g(0) = \langle \mathcal{P}, \varphi \rangle$ where $\mathcal{P}$ is the initial set of assumptions, and φ is the initial goal formula.

2. If $t \in T$ and $g(t) = \langle \mathcal{P}(t), G(t) \rangle$ and if $G(t) = A_1 \wedge \cdots \wedge A_n$, then the node t has exactly n immediate successors in the tree $(s_1, \ldots, s_n,$ say) with $g(s_i) = \langle \mathcal{P}(t), A_i \rangle$.

3. If $t \in T$ and $g(t) = \langle \mathcal{P}(t), G(t) \rangle$ and if $G(t) = B \to C$, then the node t has one immediate successor s in the tree with $g(s) = \langle \mathcal{P}(t) \cup \{B\}, C \rangle$.

4. If $t \in T$ and $g(t) = \langle \mathcal{P}(t), G(t) \rangle$ and if $G(t) = q$ (an atom), then there are several subcases:

    (a) $q \in \mathcal{P}(t)$ or $\bot \in \mathcal{P}(t)$ and t has no successors;

    (b) t has one immediate successor s, and for some formula $A \to x \in \mathcal{P}(t)$ (called the transition formula of node t) we have $x = q$ or $x = \bot$, and that $g(s) = \langle \mathcal{P}(t), A \rangle$. Note that in this case we may still have either $q \in \mathcal{P}(t)$ or $\bot \in \mathcal{P}(t)$.

Lemma 5.6.3 (Tree properties) *A computation tree has the following properties:*

1. *It computes the proof* $\mathcal{P}(0) \vdash_\mathcal{I} G(0)$.

2. *Any point in the tree, with the label* $\langle \mathcal{P}(t), G(t) \rangle$, *is the root of a computation tree* $\mathcal{T}_t$ *for the proof* $\mathcal{P}(t) \vdash_\mathcal{I} G(t)$,

3. If $t \prec s$, i.e. s is a successor (not necessarily immediate) to t, then $\mathcal{P}(t) \subseteq \mathcal{P}(s)$ (formulae are never removed from the assumptions as we move down the tree).

4. Suppose that $\mathcal{P}'$ is an additional set of assumptions, and that $(T, \prec, 0, g)$ is a successful computation tree for the proof $\mathcal{P}(0) \vdash_{\mathcal{I}} G(0)$, with $g(t) = \langle \mathcal{P}(t), G(t) \rangle$ for each node t. Suppose that we define g' such that $g'(t) = \langle \mathcal{P}(t) \cup \mathcal{P}', G(t) \rangle$; in other words, we add the assumptions $\mathcal{P}'$ to all the nodes in the tree, even though we do not need them, and do not use them. Then $\mathcal{T} = \langle T, \prec, 0, g' \rangle$ is a successful computation tree for the proof $\mathcal{P}(0) \cup \mathcal{P}' \vdash_{\mathcal{I}} G(0)$.

5. We call the tree $\mathcal{T}'$ of (4) above the tree of $\mathcal{T}' + \mathcal{P}'$ obtained by adding $\mathcal{P}'$ to the data of $\mathcal{T}$.

Proof. Straightforward from the definitions. ∎

Notice that a successful tree must be finite. An important operation is the grafting of one tree onto another. We will use grafts to prove a lemma below, by replacing part of a computation tree with another with different properties.

Definition 5.6.4 (Grafting) Let $\mathcal{T} = \langle T, \prec_T, 0, g \rangle$ and $\mathcal{S} = \langle S, \prec_S, 1, h \rangle$ be two successful computation trees with only one point 1 in common, and for some B we have $g(1) = \langle \mathcal{P}(1), B \rangle$ and $h(1) = \langle \mathcal{Q}(1), B \rangle$. Note that 1 is the root of $\mathcal{S}$. We construct a new tree which is the result of grafting $\mathcal{S}$ onto $\mathcal{T}$ at the point 1.

1. The set of points in the new tree is $T - \{t \in T \mid 1 \prec_T t\} \cup S$ so that we have all of S, plus all of T except those points below the common point.

2. The ordering relation over the new tree is given by

$$x \prec y \quad \text{if} \quad \begin{array}{l} x \in T \text{ and } y \in T \text{ and } x \prec_T y \\ \text{or} \quad x \in S \text{ and } y \in S \text{ and } x \prec_S y \\ \text{or} \quad x \in T \text{ and } x \prec_T 1 \text{ and } y \in S \text{ and } 1 \prec_S y \end{array}$$

3. The labelling function f for the new tree is defined for $x \in T$ by

$$f(x) = \langle \mathcal{P}(x) \cup \mathcal{Q}(1), G(x) \rangle$$

5.6. PROOF OF COMPLETENESS

and for $x \in S$ by

$$f(x) = \langle \mathcal{P}(1) \cup \mathcal{Q}(x), H(x) \rangle$$

where $g(x) = \langle \mathcal{P}(x), G(x) \rangle$ in $\mathcal{T}$, and $h(x) = \langle \mathcal{Q}(x), H(x) \rangle$ in $\mathcal{S}$. Notice that when $x = 1$, the two definitions agree.

The lemma we shall now prove is needed for a step in the proof of completeness. The proof of the lemma involves a three-way induction, and the simplicity of the lemma itself belies the awkwardness of its proof. Readers who wish to may omit the proof without hindering their reading of the rest of this book.

Lemma 5.6.5 (Tree cut) *If there exist computation trees for $\mathcal{P} \vdash_{\mathcal{I}} A_i$ for $i = 1, \ldots, n$ and there exists a computation tree for $\mathcal{P} \cup \{A_1\} \cup \cdots \cup \{A_n\} \vdash_{\mathcal{I}} q$, then there exists a computation tree for $\mathcal{P} \vdash_{\mathcal{I}} q$.*

Proof: The proof is a complicated induction on the complexity of the A_i and on the number n and on the maximum number of nested uses of the A_i in the computation of q (which number we shall denote by m).

Base case: All the A_i are atomic propositions; n and m take arbitrary values. Assume that we have a computation tree for $\mathcal{P} \cup \{A_1\} \cup \cdots \cup \{A_n\} \vdash_{\mathcal{I}} q$ in which $t_1, \ldots, t_k$ are the highest points in the tree which involves the A_i in the computation. Since the A_i are atomic, the only way in which they can be used is to solve the goal $G(t_j) = A_i$, so that the tree rooted at t_j is for $\mathcal{P}(t_j) \vdash_{\mathcal{I}} A_i$ for some A_i. Now we are given that $\mathcal{P} \vdash_{\mathcal{I}} A_i$, and from Lemma 5.6.3 we know that $\mathcal{P} \subseteq \mathcal{P}(t_j)$ and further that we have computation trees for $\mathcal{P}(t_j) - \{A_1\} - \cdots - \{A_n\} \vdash_{\mathcal{I}} A_i$.

Hence we can replace the subtrees rooted at each of the $t_1, \ldots, t_k$ with the appropriate subtrees which do not have $A_1, \ldots, A_n$ as assumptions by means of grafting. Thus we have $\mathcal{P}(t_j) - \{A_1\} - \cdots - \{A_n\} \vdash_{\mathcal{I}} G(t_j)$, and therefore $\mathcal{P} \vdash_{\mathcal{I}} q$ (since $A_1, \ldots, A_n$ are not used in the reconstructed tree for $\mathcal{P} \cup \{A_1, \ldots, A_n\} \vdash_{\mathcal{I}} q$).

Induction step: Assume that we have computation trees for $\mathcal{P} \vdash_{\mathcal{I}} A_i \rightarrow x_i$ for $i = 1, \ldots, n$ (so that we have trees for $\mathcal{P} \cup \{A_i\} \vdash_{\mathcal{I}} x_i$) and a computation tree for $\mathcal{P} \cup \{A_1 \rightarrow x_1\} \cup \cdots \cup \{A_n \rightarrow x_n\} \vdash_{\mathcal{I}} q$. Assume that m is the maximum number of nested uses of any $A_i \rightarrow x_i$ in the tree.

Let $\mathcal{T} = \langle T, \prec, 0, g \rangle$ be a computation tree for $\mathcal{P} \cup \{A_1 \to x_1\} \cup \cdots \cup \{A_n \to x_n\} \vdash_\mathcal{I} q$ and let $t_1, \ldots, t_k$ be the lowest points in the tree in which any of the $A_i \to x_i$ are used in the computation, so that below each t_j none of the $A_i \to x_i$ are used. Therefore each of the $t_1, \ldots, t_k$ is labelled such that $\mathcal{P}(t_j) \supseteq \mathcal{P}$ and intersects $\{A_1 \to x_1\} \cup \cdots \cup \{A_n \to x_n\}$ and $G(t_j) = y$ for some proposition y. The single immediate successor to t_j, say t'_j, is derived by the use of one of the $A_i \to x_i$, say $A \to x$, and must therefore compute the goal $G(t'_j) = A$ from $\mathcal{P}(t'_j) = \mathcal{P}(t_j)$. x might be equal to y, or alternatively be $\bot$.

Now we know that the subtree rooted at t'_j does not use any of the $A_i \to x_i$ (by definition) so we know that we have a computation tree for

$$\mathcal{P}(t_j) - \{A_1 \to x_1\} - \cdots - \{A_n \to x_n\} \vdash_\mathcal{I} A$$

Now we are given that $\mathcal{P} \cup \{A\} \vdash_\mathcal{I} x$ and since $\mathcal{P}(t_j) \supseteq \mathcal{P}$, by Lemma 5.6.3 we have $\mathcal{P}(t_j) \cup \{A\} - \{A_1 \to x_1\} - \cdots - \{A_n \to x_n\} \vdash_\mathcal{I} x$.

By the induction hypothesis, from the existence of computation trees for $\mathcal{P}(t_j) - \{A_1 \to x_1\} - \cdots - \{A_n \to x_n\} \vdash_\mathcal{I} A$ and for $\mathcal{P}(t_j) \cup \{A\} - \{A_1 \to x_1\} - \cdots - \{A_n \to x_n\} \vdash_\mathcal{I} x$ we get a computation tree for $\mathcal{P}(t_j) - \{A_1 \to x_1\} - \cdots - \{A_n \to x_n\} \vdash_\mathcal{I} x$. Let $\mathcal{S}$ be that tree, and replace the subtree at t_j by $\mathcal{S}$. Thus one nested use of one of the $A_i \to x_i$ has been eliminated. By applying this to each of $t_1, \ldots, t_k$, we can reduce m, the maximum number of nested uses of one of the $A_i \to x_i$, by 1.

Repeated applications of the above will eliminate all of the uses of the original $A_i \to x_i$ from the tree for $\mathcal{P} \cup \{A_1 \to x_1\} \cup \cdots \cup \{A_n \to x_n\} \vdash_\mathcal{I} q$, resulting in a tree for $\mathcal{P} \vdash_\mathcal{I} q$. ∎

Lemma 5.6.6 (Tree for restart) *If $\mathcal{P} \vdash_\mathcal{B} \varphi$ succeeds using the restart rule then there is a successful computation tree for $\mathcal{P} \cup \{\varphi \to \bot\} \vdash_\mathcal{I} \varphi$.*

Proof. Each time the restart rule is used in the computation of φ from $\mathcal{P}$, its use can be replaced by the use of $\varphi \to \bot$. Thus $\mathcal{P} \cup \{\varphi \to \bot\} \vdash_\mathcal{I} \varphi$ succeeds without restart, and hence has a computation tree. Compare with the proof of soundness of restart in Theorem 6.5.1. ∎

Lemma 5.6.7 (Cut elimination) *If $\mathcal{P} \cup \{\psi\} \vdash_\mathcal{B} A$ and $\mathcal{P} \cup \{\psi \to \bot\} \vdash_\mathcal{B} A$ then $\mathcal{P} \vdash_\mathcal{B} A$.*

Proof. First we can assume that $\psi = (\alpha \to y)$ and that $y \neq \bot$. Otherwise we can prove the theorem for α, since it can easily be shown that if $\mathcal{P} \cup \{(\alpha \to \bot) \to \bot\} \vdash_\mathcal{B} A$ then $\mathcal{P} \cup \{\alpha\} \vdash_\mathcal{B} A$.

5.6. PROOF OF COMPLETENESS

Our proof is by induction. To explain how we do the induction we need some notation. Let $\mathcal{T} = \langle T, \prec_T, 0, g \rangle$ be a tree for $\mathcal{P}' \cup \{\psi \to \bot\} \vdash_{\mathcal{I}} A$, where $\mathcal{P}' = \mathcal{P} \cup \{A \to \bot\}$. Let $\mathcal{S} = \langle S, \prec_S, 1, h \rangle$ be a tree for $\mathcal{P}' \cup \{\psi\} \vdash_{\mathcal{I}} A$. We assume of course that $S \cap T = \varnothing$.

We use induction on the number n, being the maximal number of nested uses of $\psi \to \bot$ (as a transition formula) in the tree $\mathcal{T}$.

Case $n = 0$

$\psi \to \bot$ is not used in the tree $\mathcal{T}$. Hence $\mathcal{P}' \vdash_{\mathcal{I}} A$ and so $\mathcal{P} \vdash_{\mathcal{B}} A$, by the converse of the previous lemma.

Case n

We assume that if $\mathcal{T}$ has at most $n-1$ nested occurrences of use of $\psi \to \bot$ then $\mathcal{P}' \vdash_{\mathcal{I}} A$. We want to show the same for case n. We accomplish this task by using the trees $\mathcal{T}$ and $\mathcal{S}$ and the tree cut theorem to construct a tree $\mathcal{T}^* = \langle T^*, \prec^*, 0, g^* \rangle$ for $\mathcal{P}' \cup \{\psi \to \bot\} \vdash_{\mathcal{I}} A$ which falls under case $n - 1$, namely the induction hypothesis. $\mathcal{T}^*$ will have at most $n - 1$ nested uses of $\psi \to \bot$, and we will therefore conclude that $\mathcal{P} \vdash_{\mathcal{B}} A$.

The rest of this proof is a detailed analysis of the trees $\mathcal{T}$ and $\mathcal{S}$ and a grafting of various modified parts of these trees into the desired tree $\mathcal{T}^*$.

Let us begin:

Consider the tree $\mathcal{T}$. Let $t_1, \ldots, t_k \in T$ be the last points from the top in which the transition formula in the tree is $\psi \to \bot$, i.e. $(\alpha \to y) \to \bot$. Assume that at the node t_j the goal is a_j; that is, we have $g(t_j) = \langle \mathcal{P}'(t_j) \cup \{(\alpha \to y) \to \bot\}, a_j \rangle$. The transition formula is $(\alpha \to y) \to \bot$ for all j. We thus must have immediate successors s_j of t_j in the tree with $g(s_j) = \langle \mathcal{P}'(t_j) \cup \{(\alpha \to y) \to \bot\}, \alpha \to y \rangle$ and in the trees $\mathcal{T}_{s_j}$, the wff $(\alpha \to y) \to \bot$ is never used!

Figure 5.2 illustrates the situation.

The points $t_{k+1}, \ldots, t_{k_1}$ are endpoints of $\mathcal{T}$ where in the computation path to them, $\psi \to \bot$ is used no more $(n-1)$ nested times. Let $\mathcal{T}'$ be the part of the tree $\mathcal{T}$ down to these points.

Thus we conclude:

(*) $\quad \mathcal{P}'(t_j) \vdash_{\mathcal{I}} \alpha \to y$.

We now use the fact that $\mathcal{P}' \cup \{\alpha \to y\} \vdash_{\mathcal{I}} A$ has the tree $\mathcal{S}$ to conclude (**) below. First, if $\alpha \to y$ is not used in the tree $\mathcal{S}$ then the tree $\mathcal{S}$ with $\alpha \to y$ deleted shows that $\mathcal{P}' \vdash_{\mathcal{I}} A$ succeeds. Since $\mathcal{P}' \vdash_{\mathcal{I}} A$

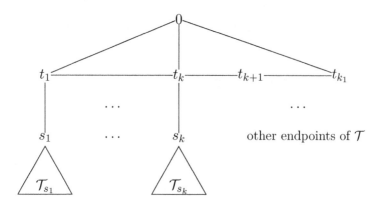

Figure 5.2:

implies $\mathcal{P}' \vdash_\mathcal{B} A$ and $\mathcal{P}' \vdash_\mathcal{B} A$ implies $\mathcal{P} \vdash_\mathcal{B} A$, we get that $\mathcal{P} \vdash_\mathcal{B} A$ succeeds and there is nothing to prove.

If $\alpha \to y$ is used in the tree $\mathcal{S}$, let $r_1, \ldots, r_{k^*}$ be the top (i.e. first) nodes in $\mathcal{S}$ in which $\alpha \to y$ is used. We thus have that (since $y \neq \bot$)

$$f(r_j) = \langle \mathcal{P}'(r_j) \cup \{\alpha \to y\}, y \rangle$$

We can assume that $\alpha \to y$ is not a member of $\mathcal{P}'(r_j)$. Thus

(**) $\mathcal{P}'(r_j) \cup \{\alpha \to y\} \vdash_\mathcal{I} y$.

We also recall that we assumed that the nodes r_j, $j = 1, \ldots, k^*$, are the first nodes (from the root $1 \in \mathcal{S}$) in which $\alpha \to y$ is used. Thus the computation path down to r_j uses transition formulae from $\mathcal{P}'(r_j)$ only.

Let $\mathcal{S}' = \{s \in \mathcal{S} \mid \text{there is no point } s' \in \mathcal{S}, s' \prec_\mathcal{S} s \text{ and } \alpha \to y \text{ is used at } s'\}$. Thus $r_1, \ldots, r_{k^*} \in \mathcal{S}^*$ as well as possibly many other points. Let $\prec'$ be the inherited ordering on $\mathcal{S}'$ (inherited from $\langle \mathcal{S}, \prec_\mathcal{S} \rangle$). Copies of $\langle \mathcal{S}', \prec' \rangle$ as well as a suitable g^* will be defined later and used as building blocks to construct the desired tree $\mathcal{T}^*$. The nodes $r_1, \ldots, r_{k^*}$ are not endpoints since ψ is used at these nodes so the computation must proceed. Let $\bar{r}_1, \ldots, \bar{r}_{k^*}$ be their immediate successors and let $\mathcal{S}_{\bar{r}_1}, \ldots, \mathcal{S}_{\bar{r}_{k^*}}$ be the trees below these successors. See Figure 6.3.

The points $r_1, \ldots, r_{k^*}$ in Figure 5.3 are the first points where ψ is used. The points $r_{k^*+1}, \ldots, r_{k_1^*}$ are endpoints. In the path leading to $r_1, \ldots, r_{k_1^*}$, ψ is not used. Thus $\mathcal{S}'$ is the set of all points in $\mathcal{S}$ above $\{r_1, \ldots, r_{k_1^*}\}$.

5.6. PROOF OF COMPLETENESS 151

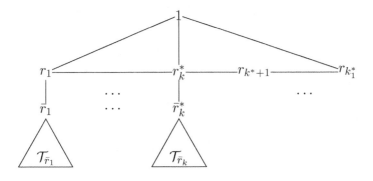

Figure 5.3:

We shall make use later of the trees $\mathcal{S}_{\bar{r}_j}$.

Consider now the tree $\mathcal{T}$ and the node t_j. We have $g(t_j) = \langle \mathcal{P}'(t_j) \cup \{(\alpha \to y) \to \bot\}, a_j \rangle$ at node t_j; $(\alpha \to y) \to \bot$ was used to move to node s_j. We want to succeed with a_j at node t_j by using other clauses from the data and not the clause $(\alpha \to y) \to \bot$. The clause we want to use is $A \to \bot \in \mathcal{P}'(t_j)$. If we do that we will have to ask next for a tree for $\mathcal{P}'(t_j) \vdash_{\mathcal{I}} A$.

Since $\mathcal{P}' \subseteq \mathcal{P}'(t_j)$, we can follow the computation steps of the tree $\mathcal{S}$ (with initial goal A), down to the points $r_1, \ldots, r_{k^*}$ (where $\alpha \to y$ is first used) but no further because $\alpha \to y$ is not available. At the points $r_1, \ldots, r_{k^*}$ the data is $\mathcal{P}'(t_j) \cup \mathcal{P}'(r_1), \ldots, \mathcal{P}'(t_j) \cup \mathcal{P}'(r_{k^*})$, respectively, and the goal is y. From (*) and (**) above we get for each j and i that

- $\mathcal{P}'(t_j) \cup \mathcal{P}'(r_i) \cup \{\alpha \to y\} \vdash_{\mathcal{I}} y$
- $\mathcal{P}'(t_j) \cup \mathcal{P}'(r_i) \vdash_{\mathcal{I}} \alpha \to y$.

Hence by Lemma 6.6.5 we get

(***) $\mathcal{P}'(t_j) \cup \mathcal{P}'(r_i) \vdash_{\mathcal{I}} y$.

Let $\overline{\mathcal{T}}_{t_j, r_i}$ be a tree for (***).

We now have all the components to construct a tree for $\mathcal{P}' \cup \{\psi \to \bot\} \vdash_{\mathcal{I}} A$, which uses $\psi \to \bot$ not more than $n-1$ nested times. We now define the various pieces:

1. Let $T' = \{x \in T \mid x \text{ is above or equal to any of } t_1, \ldots, t_{k_1}\}$. T' is the tree T less all points strictly below $t_1, \ldots, t_{k_1}$. Recall that $t_1, \ldots, t_{k_1}$ are the last points in which $(\alpha \to y) \to \bot$ is used at most

$(n-1)$ nested times. If we can continue the computation at $t_1,\ldots,t_k$, without using $(\alpha \to y) \to \bot$, we will have reduced the nested number of uses of this formula.

2. For each t_j, $j = 1,\ldots,k$, form a copy of $(S', \prec')$ which we denote by S'_{t_j}. We let $S'_{t_j} = \{(t_j, s) \mid s \in S'\}$ and let $(t_j, s_1) \prec'_{t_j} (t_j, s_2)$ iff $s_1 \prec' s_2$.

The above means that we made a special copy of S' for the point t_j in the tree, for $j = 1,\ldots,k$.

3. Among the points of S'_{t_j} are the endpoints $(t_j, r_1),\ldots,(t_j, r_{k^*})$. We graft at these points the trees $\bar{T}_{t_j,r_1},\ldots,\bar{T}_{t_j,r_{k^*}}$ respectively, where $\bar{T}_{t_j,r_i} = \langle \bar{T}_{t_j,r_i}, \prec_{t_j,r_i}, (t_j, r_i), g_{t_j,r_i} \rangle$. We now have T^* as

$$T^* = T' \cup \bigcup_{j \leq k} \left(S'_{t_j} \cup \bigcup_{i \leq k^*} \bar{T}_{t_j,r_i} \right)$$

The ordering $\prec^*$ on T^* is defined as follows: $x \prec^* y$ iff

1. $x, y \in T'$ and $x \prec_T y$

2. $x, y \in S'_{t_j}$ and $x \prec'_{t_j} y$

3. $x, y \in \bar{T}_{t_j,r_i}$ and $x \prec_{t_j,r_i} y$

4. $t_j \prec (t_j, 1)$

5. $x \in T$, $x \prec_T t_j$, and $y \in S'_{t_j} \cup \bigcup_{i \leq k^*} \bar{T}_{t_j,r_i}$

6. $x \in S'_{t_j}$, $x \prec'_{t_j} (t_j, r_i)$, and $y \in \bigcup_{i \leq k^*} \bar{T}_{t_j,r_i}$.

Note that (t_j, r_i) is in both $\bar{T}_{t_j,r_i}$ and S'_{t_j}.

Figure 5.4 describes the tree T^*.

The root of T^* is 0 and the function g^* of T^* is as follows:
$g^*(y) = $ (1) $g(y)$, for $y \in T'$
    (2) $\langle \mathcal{P}'(t_j) \cup \mathcal{P}(s), a_s \rangle$, for $y \in S'_{t_j}$, $y = (t_j, s)$
    and $h(s) = \langle \mathcal{P}(s), a_s \rangle$
    (3) $g_{t_j,r_i}(y)$, for $y \in \bar{T}_{t_j,r_i}$

The function g^* shows that we start at 0 and compute according to the tree $\mathcal{T}$ down to the points $t_1,\ldots,t_k,t_{k+1},\ldots,t_{k_1}$. At the points $t_1,\ldots,t_k$, $\psi \to \bot$ is used in the tree $\mathcal{T}$ for the nth nested time. The goal

5.6. PROOF OF COMPLETENESS

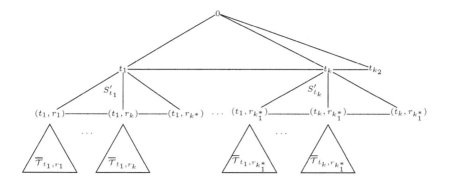

Figure 5.4:

at t_j is a_j. We cannot use $\psi \to \bot$ any more for the goal a_j. We therefore use $A \to \bot$ instead and the next goal is A. To compute A we proceed as in the tree $\mathcal{S}$ which at the point t_j we rename as the copy $\mathcal{S}'_{t_j}$, down to the points $r_1, \ldots, r_{k^*}, \ldots, r_{k_1^*}$, renamed $(t_j, r_1), \ldots, (t_j, r_{k_1^*})$. The tree $\mathcal{S}'_{t_j}$ is used for the data $\mathcal{P}' \cup \{\psi\}$. $\mathcal{P}'$ is available at points t_j of the tree $\mathcal{T}$ but not ψ, but we can continue up to $r_1, \ldots, r_{k^*}, r_{k^*+1}, \ldots, r_{k_1^*}$ (named $(t_j, r_i), i = 1, \ldots, k_1^*$) because ψ is not used. At the points $(t_j, r_1), \ldots, (t_j, r_{k_1^*})$, ψ is used for the first time and the goal is y. We cannot continue as in $\mathcal{S}$ but we continue as in $\mathcal{T}_{(t_j, r_i)}$ and succeed.

The above construction of $\mathcal{T}^*$ completes the inductive step and the lemma is proved. ∎

Final proof of Theorem 5.6.1. Let $\psi_1, \psi_2, \psi_3, \ldots$ be a list of the wffs of the logic. Define a set of assumptions $\mathcal{P}_\infty$ by induction as follows:

$$\mathcal{P}_0 = \mathcal{P}$$
$$\mathcal{P}_{i+1} = \mathcal{P}_i \cup \{\psi_i\} \text{ if } \mathcal{P}_i \cup \{\psi_i\} \not\vdash_B \varphi$$
$$\mathcal{P}_i \cup \{\psi_i \to \bot\} \text{ otherwise}$$

Note that if both $\mathcal{P}_i \cup \{\psi_i \to \bot\} \vdash_B \varphi$ and $\mathcal{P}_i \cup \{\psi\} \vdash_B \varphi$ then $\mathcal{P}_i \vdash_B \varphi$. This was shown in Theorem 6.6.7.

Let

$$\mathcal{P}_\infty = \bigcup_{i=0}^{\infty} \mathcal{P}_i$$

Now we know that if $\mathcal{P}_i \not\vdash_B \varphi$ (which is certainly true for $i = 0$), then $\mathcal{P}_{i+1} \not\vdash \varphi$ also. By induction we have that $\mathcal{P}_\infty \not\vdash_B \varphi$. Success in the computation of φ from $\mathcal{P}_\infty$ will only involve a *finite* part of $\mathcal{P}_\infty$, and hence will be successful from some $\mathcal{P}_i$. If we define an assignment h on the atoms by

$$h(a) = \top \quad \text{iff} \quad \mathcal{P}_\infty \vdash_B a$$

we can show by induction on the structure of the wffs φ that under the assignment h, we have for any φ

$$h(\varphi) = \top \quad \text{iff} \quad \mathcal{P}_\infty \vdash_B \varphi$$

There are three cases to be considered:

1. φ is atomic
   This is given by definition.

2. $\varphi = \alpha \wedge \beta$
   $h(\varphi) = \top$ iff $h(\alpha) = \top$ and $h(\beta) = \top$ (from Definition 1.1.2). Now by induction, $h(\alpha) = \top$ iff $\mathcal{P}_\infty \vdash_B \alpha$, and $h(\beta) = \top$ iff $\mathcal{P}_\infty \vdash_B \beta$. Hence $h(\varphi) = \top$ iff $\mathcal{P}_\infty \vdash_B \alpha$ and $\mathcal{P}_\infty \vdash_B \beta$. Now $\mathcal{P}_\infty \vdash_B \alpha$ and $\mathcal{P}_\infty \vdash_B \beta$ iff $\mathcal{P}_\infty \vdash_B \alpha \wedge \beta$. Thus $h(\varphi) = \top$ iff $\mathcal{P}_\infty \vdash_B \alpha \wedge \beta$.

3. $\varphi = \alpha \to q$
   We prove the two directions of the 'iff' separately.

   (a) Assume that $\mathcal{P}_\infty \vdash_B \alpha \to q$; then certainly if $\mathcal{P}_\infty \vdash_B \alpha$ then $\mathcal{P}_\infty \vdash_B q$ by Lemma 5.6.5. Hence by the induction hypothesis, $h(\alpha) = \top$ implies $h(q) = \top$, i.e. $h(\alpha \to q) = \top$.

   (b) Now assume that $\mathcal{P}_\infty \not\vdash \alpha \to q$, i.e. that $\mathcal{P}_\infty \cup \{\alpha\} \not\vdash q$. We know that $\mathcal{P}_\infty$ either contains α or $\alpha \to \bot$. If $\mathcal{P}_\infty$ contains α then we get $\mathcal{P}_\infty \vdash_B \alpha$ and $\mathcal{P}_\infty \not\vdash q$, i.e. $h(\alpha \to q) = \bot$. The other possibility, that $\mathcal{P}_\infty$ contains $\alpha \to \bot$, cannot arise since $\mathcal{P}_\infty \cup \{\alpha\}$ would be inconsistent and that would imply that $\mathcal{P}_\infty \cup \{\alpha\} \vdash_B q$.

Now—provided that $\mathcal{P}_\infty$ is consistent—every assignment h that is a model for $\bigwedge \mathcal{P}$ (which of course assigns $\top$ to all the atoms provable from $\mathcal{P}$ by the backward rules) assigns $\bot$ to all the formulae which are not provable from $\mathcal{P}$. Thus whenever $\mathcal{P}_\infty \not\vdash \varphi$, there is an assignment h for which $h(\bigwedge \mathcal{P} \to \varphi) = \bot$, i.e. $\bigwedge \mathcal{P} \to \varphi$ is not a tautology. ∎

5.7 Properties of intuitionistic logic fragment of $\{\wedge, \rightarrow, \bot\}$

We saw in the previous section that the backward computation without restart satisfies the cut rule, namely

$$\mathcal{P} \vdash_\mathcal{I} A_i, i = 1, \ldots, k, \text{ and } \mathcal{P} \cup \{A_1, \ldots, A_n\} \vdash_\mathcal{I} q \text{ imply } \mathcal{P} \vdash_\mathcal{I} q$$

We can also prove by induction on the structure of the goal B that

$$\mathcal{P} \vdash_\mathcal{I} B \text{ if } B \in \mathcal{P}$$

It is also clear that $\vdash_\mathcal{I}$ is monotonic and that it satisfies the deduction theorem (by definition) since the rule

$$\mathcal{P} \vdash_\mathcal{I} B \rightarrow q \text{ iff } \mathcal{P} + B \vdash_\mathcal{I} q$$

is one of the computation rules.

$\vdash_\mathcal{I}$ is also clearly transitive (transitivity follows, in fact, from the cut elimination theorem). It is therefore clear that we have that $\vdash \subseteq \vdash_\mathcal{I}$, where $\vdash$ is the consequence relation of Example 5.1.2.

Theorem 5.7.1 (Axiomatization of $\vdash_\mathcal{I}$) *Let $\vdash$ be the system of Example 6.1.2. Then $\vdash_\mathcal{I} = \vdash$.*

Proof. We show that if $\mathcal{P} \not\vdash_\mathcal{I} B$ then there exists an intuitionistic Kripke model (see Definition 2.2.3) in which $\mathcal{P}$ holds but not B. Define $\mathcal{P}_1 \equiv \mathcal{P}_2$ if for all $A \in \mathcal{P}_2$, $\mathcal{P}_1 \vdash_\mathcal{I} A$ and for all $A \in \mathcal{P}_1, \mathcal{P}_2 \vdash_\mathcal{I} A$.

Let $\mathcal{P}/\equiv$ be the equivalence class of $\mathcal{P}$. Let $\mathcal{P}_1/\equiv \, \leq \mathcal{P}_2/\equiv$ be defined as for all $A \in \mathcal{P}_1, \mathcal{P}_2 \vdash_\mathcal{I} A$ (we also write it as $\mathcal{P}_2 \vdash_\mathcal{I} \mathcal{P}_1$). It is easy to show that $\leq$ is well defined. Let $T = \{\mathcal{P}/\equiv \mid \mathcal{P} \not\vdash_\mathcal{I} \bot \text{ and } \mathcal{P} \text{ a set of wffs}\}$.

Define $h(\mathcal{P}/\equiv, q)$ to be 1 iff $\mathcal{P} \vdash_\mathcal{I} q$. Again one can show h is well defined.

Consider now the model $(T, \leq, h)$.

We show that for any $B, \mathcal{P}$ we have that $h(\mathcal{P}/\equiv, B) = 1$ iff $\mathcal{P} \vdash_\mathcal{I} B$. The proof is by structural induction on B.

The crucial case is that of $B \rightarrow C$.

1. Assume $\mathcal{P} \vdash_\mathcal{I} B \rightarrow C$. Then of course $\mathcal{P} \cup B \vdash_\mathcal{I} C$.

   We show that for any $\mathcal{P}'/\equiv \, \geq \mathcal{P}/\equiv$ such that $\mathcal{P}' \vdash_\mathcal{I} B$, we have $\mathcal{P}' \vdash_\mathcal{I} C$. But if $\mathcal{P}' \vdash_\mathcal{I} B$ then $\mathcal{P}' \vdash \mathcal{P} \cup \{B\}$ and hence $\mathcal{P}' \vdash_\mathcal{I} C$. Hence $h(\mathcal{P}/\equiv, B \rightarrow C) = 1$.

2. Assume $h(\mathcal{P}/\equiv, B \to C) = 1$. Then, for any $\mathcal{P}'/\equiv\ \geq \mathcal{P}'/\equiv$ and, in particular, for $\mathcal{P}' = \mathcal{P} \cup \{B\}$ we have (by the induction hypothesis) that if $\mathcal{P}' \vdash_{\mathcal{I}} B$ then $\mathcal{P}' \vdash_{\mathcal{I}} C$. Therefore $\mathcal{P} \cup \{B\} \vdash_{\mathcal{I}} C$ and so $\mathcal{P} \vdash_{\mathcal{I}} B \to C$.

Thus we have seen that if $\mathcal{P} \not\vdash_{\mathcal{I}} B$ then there is a Kripke countermodel and so by Exercise 6.1.3, $\mathcal{P} \not\vdash B$. ∎

The above theorem is very strong. If the database $\mathcal{P}$ is comprised of clauses of the form $A_i \to q_i$ and we know that if $\mathcal{P} \mathrel{\mid\!\sim} q$ then by the theorem we also have $\mathcal{P} \vdash_{\mathcal{I}} q$, but for the computation to succeed, it *must* be the case that $q = q_i$ for some i and $\mathcal{P} \vdash_{\mathcal{I}} A_i$.

This is not true in classical logic. For example, we have $(a \to b) \to a \vdash a$ in classical logic but we do *not* have $(a \to b) \to a \vdash a \to b$.

The above strong cut property can be used to prove interpolation for intuitionistic logic.

Theorem 5.7.2 (Interpolation) *The following holds for the fragment of intuitionistic logic with $\{\wedge, \to, \bot\}$. If $A \mathrel{\mid\!\sim} B$ then there is a C built up from the common atoms of A and B and $\bot$ such that $A \mathrel{\mid\!\sim} C$ and $C \mathrel{\mid\!\sim} B$.*

Proof. We prove that, for a finite database $\mathcal{P}$ of the fragment and for a formula B, if $\mathcal{P} \vdash_{\mathcal{I}} B$ then for some $C, \mathcal{P} \vdash_{\mathcal{I}} C$ and $C \vdash_{\mathcal{I}} B$.

We already know that $\vdash_{\mathcal{I}}$ is the same as $\mathrel{\mid\!\sim}$ of Example 6.1.2 (Theorem 6.7.1).

To find C, we use induction on the complexity of the computation.

Length 1
If $B = q$ is atomic, then $q \in \mathcal{P}$ or $\bot$ is in $\mathcal{P}$ and the interpolant is $C = q$, or $C = \bot$ respectively.

Length $n+1$

1. Assume $\mathcal{P}$ is the set of clauses $\{A_j \to p_j\}$ and the goal is

$$B = \bigwedge_k (B_k \to y_k) \to q$$

Then $\mathcal{P} \vdash_{\mathcal{I}} B$ iff $\{A_j \to p_j\} \cup \{B_k \to y_k\} \vdash_{\mathcal{I}} q$.

Since the goal-directed computation succeeds, the computation must go through one of the clauses. This clause is of the form either $B_m \to y_m$ or $A_m \to p_m$. We check each case.

5.7. FRAGMENT OF $\{\wedge, \rightarrow, \bot\}$

Case 1: q is equal to y_m or $y_m = \bot$, for some m. Clearly $y_m \vdash q$ in this case. Since the computation succeeds and continues by having B_m as a goal, we get

$$\mathcal{P} \vdash_{\mathcal{I}} \bigwedge_k (B_k \rightarrow y_k) \rightarrow B_m$$

B_m is a conjunction, say $B_m = \bigwedge_j B_{m,j}$.
Hence, for each j

$$\mathcal{P} \vdash_{\mathcal{I}} \bigwedge_k (B_k \rightarrow y_k) \rightarrow B_{m,j}$$

By the induction hypothesis there exists a C_j in the common language such that

$$\mathcal{P} \vdash_{\mathcal{I}} C_j \text{ and } C_j \vdash_{\mathcal{I}} \bigwedge_k (B_k \rightarrow y_k) \rightarrow B_{m,j}$$

Let $C = \bigwedge_j C_j$.
We have $\mathcal{P} \vdash_{\mathcal{I}} C$ and $C \vdash_{\mathcal{I}} \bigwedge_k (B_k \rightarrow y_k) \rightarrow B_m$. But $\bigwedge_k (B_k \rightarrow y_k) \vdash_{\mathcal{I}} B_m \rightarrow y_m$ since $B_m \rightarrow y_m$ is one of the conjuncts; therefore we get $C \vdash_{\mathcal{I}} y_m$ and since $y_m \vdash q$ we get $C \vdash_{\mathcal{I}} B$.

Case 2: $q = p_m$ or $p_m = \bot$ for some m. Again the computation succeeds with A_m. We note that in this case p_m is in the common language. We can therefore rewrite what we have into

$$\bigwedge_k (B_k \rightarrow y_k) \vdash \bigwedge \mathcal{P} \rightarrow A_m$$

Therefore there exists an interpolant C_1 such that

$$\bigwedge_k (B_k \rightarrow y_k) \vdash_{\mathcal{I}} C_1 \text{ and } \mathcal{P}, C_1 \vdash_{\mathcal{I}} A_m$$

Therefore, since $p_m \vdash q$ we get

$$(*) \quad (C_1 \rightarrow p_m) \vdash_{\mathcal{I}} \bigwedge_k (B_k \rightarrow y_k) \rightarrow q$$

On the other hand, since $\mathcal{P} \vdash A_m \rightarrow p_m$ and we get

$$\mathcal{P}, C_1 \vdash_{\mathcal{I}} p_m$$

then

$$(**) \quad \mathcal{P} \vdash_{\mathcal{I}} C_1 \rightarrow p_m$$

Thus from (*), (**) it is clear that $C = C_1 \rightarrow p_m$ is the desired interpolant.

2. If B is a conjunction $B = B_1 \wedge \cdots \wedge B_n$ then we find interpolant C_i for B_i and the interpolant for B is $\bigwedge_i C_i$.

This completes the induction and the theorem is proved. ∎

Remark 5.7.3 (Interpolation for classical logic) *The previous theorem can be used to give us a syntactical proof of interpolation for classical logic. We can use interpolation for the intuitionistic fragment, since we have: $A \vdash B$ in classical logic iff $A \hspace{2pt}\vert\hspace{-3pt}\sim \neg B \to B$ in the intuitionistic fragment.*

5.8 Summary of systems and completeness theorems

We now list in Table 6.1 the logical systems we have so far considered, together with their various presentations.

5.9 Worked examples

Example 5.9.1 (Independence of Hilbert axioms) *Show independence of each of the axioms of classical propositional logic (Example 6.2.6). This means that for each of axioms 1–11 show that it is not provable in the logic defined by modus ponens and the remaining axioms.*

Solution (cf. [Novikov, 1964])
Independence of axioms 3–11 is proved by making non-standard (but still two-valued) interpretations of some logical connectives.
   The argument is as follows:

(I) We show that all axioms except one (say, A) are true under any assignment.

(II) We show that A is not true under some assignment.

(III) We show that if formulae $X, X \to Y$ are always true then Y is always true.

Then we obtain that what is provable without A is always true, whereas A is sometimes false. Therefore A is independent.

5.9. WORKED EXAMPLES

Table 5.1:

Logic	Language	Semantic interpretation	Reasoning rules	Comments
Classical propositional logic	Connectives $\neg, \wedge, \vee, \rightarrow, \top, \bot$	Two-valued truth tables Consequence relation defined in Definition 1.2	Forward reasoning rules in Chapter 3. Box consequence in Definition 3.4.5	Exercise 3.4.6 gives soundness
As above	As above	As above	Backward reasoning rules in Section 4.2 with restart	Rules described intuitively with examples
As above	As above	As above	Diminishing resource backward reasoning rules of Section 4.3, with restart	Soundness described in Exercise 4.3.8
Fragment of classical logic without disjunction and negation	$\wedge, \rightarrow, \top, \bot$	As above	The relevant rules for the connectives present in the fragment	Section 6.6 proves completeness
The $\{\wedge, \rightarrow, \bot\}$ fragment of Intuitionistic logic	Connectives are $\wedge, \rightarrow, \bot$	Kripke models given in Definition 2.2.3 semantic consequence $\vdash_\mathcal{I}$	Backward rules for the language without restart	Section 6.6 gives completeness
As above	As above	As above	Diminishing resource for the fragment with bounded restart	Completeness proved in [Gabbay and Olivetti, 2000]
As above	As above	As above	As in Example 5.1.2	Completeness proved in Exercise 5.1.3 and Theorem 5.7.1
Many-valued logic L_n or L_∞	Connectives $\wedge, \vee, \rightarrow, \neg$	As in Definition 2.1.2	Not given in this book	
Temporal many-valued logic	$\wedge, \vee, \twoheadrightarrow, \neg$	As in Definition 2.2.1	Not known	
Classical logic	$\wedge, \vee, \neg, \rightarrow$	As above	Hilbert formulation Section 6.2	See [Niddich, 1962]
Intuitionistic logic	As above	As above	Hilbert formulation	As above
Many-valued logic	As above	As above	As above	As above
Classical logic	$\wedge, \vee, \rightarrow, \neg$	As above	As above	As above

When showing independence of 3–11, we will interpret only one connective in a non-standard way. Implication will be standard, and thus (III) holds trivially.

In Example 1.3.3.4 we showed that all substitution instances of tautologies are also tautologies. The same proof applies to any many-valued tautologies used in this independence proof.

Now let us describe particular interpretations and check (I), (II).

- Axiom 6. Only $\wedge$ will be non-standard.
  Axioms 1–5, 9–11 remain true, because $\to$ is not involved there.
  Axiom 7: $(p \wedge q) \to q$ is true because $p \wedge q \equiv q$. But axiom 6: $(p \wedge q) \to p$ fails when $p = \bot, q = \top$.
  Axiom 8: $(r \to p) \to ((r \to q) \to (r \to p \wedge q))$ becomes equivalent to $(r \to p) \to ((r \to q) \to (r \to q))$, which is also $\top$.

p	q	$p \wedge q$
$\top$	$\top$	$\top$
$\bot$	$\top$	$\top$
$\top$	$\bot$	$\bot$
$\bot$	$\bot$	$\bot$

- Axiom 7. Similarly, take the interpretation

p	q	$p \wedge q$
$\top$	$\top$	$\top$
$\bot$	$\top$	$\bot$
$\top$	$\bot$	$\top$
$\bot$	$\bot$	$\bot$

  Now $p \wedge q \equiv p$, and axiom 6 becomes always true, and axiom 7 is sometimes false.

  Axiom 8 is equivalent to

  $$(r \to p) \to ((r \to q) \to (r \to p))$$

  which is a particular case of axiom 1, and thus it is true.

- Axiom 8. Take the interpretation in which $p \wedge q \equiv \bot$. Then axioms 6, 7 are obviously true.

5.9. WORKED EXAMPLES

Axiom 8 becomes
$$(r \to p) \to ((r \to q) \to (r \to \bot))$$
and this can be made $\bot$, if $p = q = r = \top$.

- Axiom 3 is dual to axiom 6. Take the interpretation $p \vee q \equiv q$ leaving other connectives unchanged. Then axiom 4 is equivalent to $q \to q$, which is true. Axiom 5 is equivalent to $(p \to r) \to ((q \to r) \to (q \to r))$ which is also $\top$. But $p \to p \vee q$ is $\bot$, when $p = \top, q = \bot$.

- Axiom 4 is dual to axiom 7. Then we interpret $p \vee q$ as p. Axiom 3 becomes $p \to p$, axiom 5 becomes $(p \to r) \to ((q \to r) \to (p \to r))$ which is $\top$. But $q \to p \vee q$ is $\bot$, if $p = \bot, q = \top$.

- Axiom 5 is dual to axiom 8. Now we interpret $(p \vee q)$ as $\top$.

  Then Axioms 3, 4, are obviously $\top$ and axiom 5 becomes $(p \to r) \to ((q \to r) \to (\top \to r))$.

  This can be made $\bot$, if $p = q = r = \bot$.

- Axiom 9. Now we interpret $\neg p$ as p. Then axioms 1–8 remain true. Axioms 10, 11 become $p \to p$, which is $\top$. Axiom 9 is equivalent to $(p \to q) \to (q \to p)$, and it is $\bot$, if $p = \bot, q = \top$.

- Axiom 10. Now we interpret $\neg p$ as $\bot$. Axiom 9 is equivalent to $(p \to q) \to (\bot \to \bot)$, which is $\top$. Axiom 11 is obviously true. Axiom 10 becomes $p \to \bot$, which is $\bot$ if $p = \top$.

- Axiom 11. In this case we interpret $\neg p$ as $\top$. Axiom 9 is equivalent to
  $$(p \to q) \to (\top \to \top) \equiv \top$$
  Axiom 10 is obviously $\top$.

  Axiom 11 is equivalent to $\top \to p$, and it is $\bot$ if $p = \bot$.

- Axiom 2. This case is more difficult. Now we interpret our connectives in Łukasiewicz three-valued logic. Axioms 3, 4, 6, 7, 10, 11 are obviously true. Axiom 1 is also true:

  if $q \leq p$ then $(p \to p) = 1$ and so $p \to (q \to p) = 1$;
  if $q > p$ then $(q \to p) = 1 - q + p \geq p$, and thus $p \to (q \to p) = 1$.

It is also easy to check axiom 9:

if $p \leq q$ then $\neg q \leq \neg p$, and so $(\neg q \to \neg p) = 1$, $(p \to q) \to (\neg q \to \neg p) = 1$; if $p > q$ then $1 - q > 1 - p$, $p \to q = 1 - p + q$, $\neg q \to \neg p = 1 - (1 - q) + (1 - p) = 1 + q - p$, and thus $(p \to q) = (\neg q \to \neg p)$, and the axiom is true.

Axiom 2 can be made non-true if we take $p = q = 1/2, r = 0$. Indeed, then $(p \to q) = 1, (p \to r) = (q \to r) = 1/2$, $p \to (q \to r) = (1/2) \to (1/2) = 1$, $(p \to q) \to (p \to r) = 1 \to (1/2) = (1/2)$, and the whole formula is $1 \to (1/2) = 1/2$.

Axioms 5, 8 can be examined via truth tables but there are $3^3 = 27$ rows to check, and we leave this to the reader. It turns out that these axioms are always true. Finally, *modus ponens* derives tautologies from tautologies, because in Łukasiewicz logic $A = A \to B = 1$ only if $B = 1$.

- Axiom 1. Now to show independence we can take four truth values $\{0, 1/3, 2/3, 1\}$ where $\wedge, \vee, \neg$ are interpreted in the same way as in Łukasiewicz logic:

$$x \wedge y = \min(x, y), x \vee y = \max(x, y),$$
$$\neg x = 1 - x$$

and

$$x \to y = \begin{cases} 1, & \text{if } x \leq y \\ 0, & \text{if } x > y \end{cases}$$

Obviously, axioms 3, 4, 6, 7, 10, 11 are true in this interpretation.

To check axiom 5, note that implication takes values 1 or 0, and thus either $(q \to r) \to (p \vee q \to r) = 1$, in which case the axiom is true, or $(q \to r) \to (p \vee q \to r) = 0$. The latter is the case only if $q \to r = 1, p \vee q \to r = 0$, i.e. if $q \leq r$, $\max(p, q) > r$. Hence in this case $p = \max(p, q) > r$, and $p \to r = 0$. The whole implication is again $0 \to 0 = 1$.

To check axiom 8, we use a similar argument. If $(r \to q) \to (r \to p \wedge q) = 0$, then $r \to q = 1, r \to p \wedge q = 0$, and thus $r \leq q, r > \min(p, q)$. Hence $p = \min(p, q) < r$, and $r \to p = 0$. Therefore the whole implication is $0 \to 0 = 1$.

For axiom 9 we notice that $p \to q = \neg q \to \neg p$, because $p \leq q$ iff $1 - q \leq 1 - p$.

5.9. WORKED EXAMPLES

Now consider axiom 2:

$$(p \to (q \to r)) \to ((p \to q) \to (p \to r))$$

First, note that if $p \leq q \leq r$, then the consequent is 1, and the whole implication is 1.

If $p > q$, then $p \to q = 0$, and again the consequent is 1.

If $q > r$, then $q \to r = 0$, and the premise becomes 0, except the only case when $p = 0$. If the premise is 0, the whole implication is 1. If $p = 0$ then the consequent is $(0 \to q) \to (0 \to r) = 1 \to 1 = 1$, and again the whole is 1.

Axiom 1 can be made non-true if we put $p = 1/3$, $q = 1$: then $q \to p = 0, p \to (q \to p) = 0$.

Finally we observe that *modus ponens* when applied continues to derive tautologies, because $A = 1$ and $A \to B = 1$ only if $B = 1$.

Example 5.9.2 (Consequence relation for a Hilbert system)
Let H be a Hilbert system and let $\mathrel{\vdash}_H$ be the consequence relation defined in Definition 6.2.8. Show it has the following properties:

1. $A_1, \ldots, A_n, B \mathrel{\vdash}_H B$.

2. *If* $A_1, \ldots, A_n \mathrel{\vdash}_H B$ *then* $A_1, \ldots, A_n, X \mathrel{\vdash}_H B$.

3. *If* $A_1, \ldots, A_n, X \mathrel{\vdash}_H B$ *and* $C_1, \ldots, C_m \mathrel{\vdash}_H X$ *then* $A_1, \ldots, A_n, C_1, \ldots, C_m \mathrel{\vdash}_H B$.

Solution

1. To prove this we use the single proof sequence (B).

2. Use the same proof sequence in this case.

3. Let $(D_1, \ldots, D_k)$ with $D_k = B$ be the proof sequence for proving B and let $(D'_1, \ldots, D'_k)$ be the sequence for proving X; then $(D'_1, \ldots, D'_k, D_1, \ldots, D_k)$ is a proof sequence for B from $A_1, \ldots, A_n, C_1, \ldots, C_m$.

Example 5.9.3 (The deduction theorem for a Hilbert system)
Let H be a Hilbert system and let $\mathrel{\vdash}_H$ be the consequence relation defined in Definition 6.2.8.

Show that $\vdash_H$ *satisfies the deduction theorem:*

$$A_1, \ldots, A_n, B \vdash_H C \text{ iff } A_1, \ldots, A_n \vdash_H B \twoheadrightarrow C$$

if and only if the following formulae are theorems of H:

$$A \twoheadrightarrow (B \twoheadrightarrow A)$$
$$(A \twoheadrightarrow (B \twoheadrightarrow C)) \twoheadrightarrow ((A \twoheadrightarrow B) \twoheadrightarrow (A \twoheadrightarrow C))$$

Solution

1. Assume $\vdash_H$ satisfies the deduction theorem. Then since $A, B \vdash_H A$ we get that

$$\varnothing \vdash_H A \twoheadrightarrow (B \twoheadrightarrow A)$$

   and similarly since

$$A, A \twoheadrightarrow B, A \twoheadrightarrow (B \twoheadrightarrow C) \vdash_H C$$

   we get

$$\varnothing \vdash_H (A \twoheadrightarrow (B \twoheadrightarrow C)) \twoheadrightarrow ((A \twoheadrightarrow B) \twoheadrightarrow (A \twoheadrightarrow C))$$

   Thus both of these formulae are generated by H.

2. Assume the above formulae are in H. We show the deduction theorem for $\vdash_H$.

   (a) Assume $A_1, \ldots, A_n \vdash_H B \twoheadrightarrow C$, and show that $A_1, \ldots, A_n, B \vdash_H C$. Let $(D_1, \ldots, D_k)$ be a proof sequence for $B \twoheadrightarrow C$; then the following is a proof sequence for C:

$$(D_1, \ldots, D_k, B, C)$$

   (b) Assume $A_1, \ldots, A_n, B \vdash_H C$; we show that $A_1, \ldots, A_n \vdash_H B \twoheadrightarrow C$. Let $(D_1, \ldots, D_k)$ be a proof sequence for C. We use induction on k.

   Case $k = 1$
   In this case $D_k = C$ and hence one of the following subcases holds:

   (a) C is a theorem of H.
   (b) $C = B$.

5.9. WORKED EXAMPLES

(c) $C = A_i$, for some $0 \leq i \leq n$.

For (a) and (c) the proof sequence for $B \twoheadrightarrow C$ is $(C, C \to (B \twoheadrightarrow C), B \twoheadrightarrow C)$.

For (b) the proof sequence is $(B \twoheadrightarrow C)$. We use the fact that $B \twoheadrightarrow B$ is a theorem of H as shown in Exercise 6.2.5.

Case $k > 1$

In this case we have four possibilities for $D_k = C$, (a), (b) and (c): as before and the following case:

(d) For some $k_1, k_2 < k$ we have $D_{k_2} = D_{k_1} \to C$.

By the induction hypothesis we have proof sequences $(E_1^i, \ldots, E_{m_i}^i)$ for proving $B \to D_{k_i}$ from $A_1, \ldots, A_n$, for $i = 1, 2$.

The following is a proof sequence for $B \twoheadrightarrow C$.

$$E_1^1, \ldots, E_{m_1}^1, E_1^2, \ldots, E_{m_2}^2,$$
$$(B \twoheadrightarrow (D_{k_1} \twoheadrightarrow C)) \twoheadrightarrow ((B \twoheadrightarrow D_{k_1}) \twoheadrightarrow (B \twoheadrightarrow C)),$$
$$(B \twoheadrightarrow D_{k_1}) \twoheadrightarrow (B \twoheadrightarrow C), B \twoheadrightarrow C$$

Example 5.9.4 Let $\mathrel{\mid\kern-0.3em\sim}$ be the consequence relation defined at the end of Example 6.2.6, namely $A_1, \ldots, A_n \mathrel{\mid\kern-0.3em\sim} B$ iff $A_1 \to \cdots \to (A_n \to B)\ldots)$ is generated. Show that $\mathrel{\mid\kern-0.3em\sim}$ satisfies the conditions listed in Example 5.9.2.

Solution

$\mathrel{\mid\kern-0.3em\sim}$ satisfies the deduction theorem by Example 5.9.3, since the needed axioms are available. Hence $A_1 \to \cdots \to (A_n \to B)\ldots)$ is generated iff $A_1, \ldots, A_n \mathrel{\mid\kern-0.3em\sim}_H B$ and therefore by the proof in Example 5.9.3 the properties are satisfied.

6
INTRODUCING PREDICATE LOGIC

So far in this book we have concentrated on the classical *propositional* logic, with some deviations along the way to show the possibilities offered by changing the interpretation placed on the truth functors, especially implication ($\rightarrow$). This has meant that we have only considered perhaps half of the functionality of logics, since the power of a logic may be considered to lie in two distinct aspects. The first is its expressiveness as a language for stating assumptions and conclusions. The second is the set of rules it has for checking whether the conclusions follow from the assumptions. The first part of this book concentrated on the latter.

The propositional logics we have dealt with so far are weak in *expressive power*—the smallest semantic units which can receive truth values in propositional logic are *sentences*. Yet we are often concerned with the various pieces of information which make up a sentence. For example, 'the boy with the bicycle was Bethan's brother' contains several data which cannot be represented distinctly in propositional logic. We would be reduced to using a single proposition, b say, to represent the entire sentence. To increase the expressive power of the logic, we need to add the means to talk about the semantic units contained within sentences—this entails using a *predicate* logic. Fortunately, we do not have to abandon the work we have already done, since all the rules of propositional logic concerning connectives and reasoning are valid for predicate logic.

6.1 Simple sentences

The language of predicate logic has a much richer syntax than that of propositional logic. Consider the following deduction:

$$\frac{\text{all girls are good}}{\text{Ann is good}}$$
$$\text{Ann is a girl}$$

In propositional logic we would represent each of the three sentences of the deduction by a proposition such as p, q and r, as the smallest units to which we can assign a truth value. Thus our deduction must be written as

$$\frac{p,\ q}{r}$$

which of course is not valid. We need to use predicate logic to represent the structure of the sentences, and thus indicate the connection (the property of someone being a girl) between the sentence about Ann, and the sentence about being good.

As its name might suggest, predicate logic enables us to describe properties of individual objects, and the relationship between several objects. Thus we can encode the English sentence 'Ann is a girl' by identifying the property ('is a girl') and the object which possesses that property ('Ann'). We might write this as `is-a-girl(Ann)` in predicate logic, or possibly just `girl(Ann)`, to distinguish the predicate—which is outside the brackets—from its arguments which are within the brackets. The English sentence 'Ian loves Ann' with a predicate of 'loves' denoting the relationship between the two objects 'Ian' and 'Ann' might be encoded as `loves(Ian,Ann)`.[1]

Just as we can increase our expressive power by moving from propositional logic to the predicate logic we will consider in the rest of this book, we can further increase the expressive power to reach what are known as higher order logics. To distinguish it from these even more powerful logics, the predicate logic we shall be dealing with is also known as *first-order logic*.

---

[1] Notice that 'Ann is a girl' might also be represented by a binary relationship, such as `is-a(Ann, girl)`.

6.1. SIMPLE SENTENCES

6.1.1 Building simple sentences

The basic building blocks of a predicate language are the set of names of the objects we shall talk about (the *constants*), and the set of names of relationships which we shall use to connect the constants (the *predicates*). Let us take as an example a predicate language describing a classroom of children. The set of constants is {Ann, Brenda, Carol, David, Edward}, and the set of predicates is {boy, girl, sits-next-to}. We need to know how many arguments each relationship has, i.e. how many objects are involved in the relationship. This is sometimes known as the *arity* of the relationship. boy and girl name unary relationships; thus they describe a property of a single object. sits-next-to names a binary relationship, which means that it describes a relationship between two objects.

With this knowledge we can write atomic sentences about the classroom, e.g. boy(David) and sits-next-to(Ann, Carol). An atomic sentence consists of a predicate applied to as many arguments, taken from the set of constants, as the predicate's arity. Thus the following are all well formed atomic sentences in our predicate logic of the classroom:

boy(Edward)

sits-next-to(Brenda, David)

girl(Carol)

boy(Ann)

We said above that all the connective rules of propositional logic are valid for predicate logic. Thus we can combine atomic sentences into more complex compound sentences using the same connectives ($\wedge$, $\vee$, $\neg$, $\rightarrow$) as we did in propositional logic with the atomic propositions. So the following sentences are all well formed in predicate logic:

$\neg$boy(David)

girl(Carol) $\vee$ boy(Carol)

sits-next-to(Ann, Brenda) $\rightarrow$ girl(Brenda)

$\neg$($\neg$boy(Edward) $\wedge$ $\neg$girl(Edward))

Predicates and constants cannot be used in place of each other. Furthermore, as mentioned above, the number of arguments must be the same as the arity of the predicate. The following sentences are therefore *not* well formed, for the reasons given:

Ann(Carol)	(constant **Ann** used as predicate)
boy(girl)	(predicate **girl** used as constant)
boy(David, Edward)	(incorrect number of arguments)
Brenda(girl)	(constant used as predicate and vice versa)
sits-next-to(Edward)	(incorrect number of arguments)

We can express functional relationships between objects by applying a *function name* to the objects' names. Like predicates, function names have a specific arity. For example, if best-friend is a function of arity 1, then best-friend(Ann) is another object, functionally dependent on Ann. Functions can be applied repeatedly as in

best-friend(best-friend(Carol)).

Because the result of applying a function to an object (or group of objects) is another object, they can be used in the arguments to predicates, such as sits- next-to(Ann,best-friend(Ann)). The same division between predicates and constants also applies to function names, namely that they may not be used in place of each other. Notice that best-friend(Ann) is not an atomic sentence but an object, and must have a predicate applied to it, such as girl(best-friend(Ann)).

Exercise 6.1.1

1. *Translate the following English sentences about the classroom into predicate logic:*

    (a) Brenda is a girl and she sits next to Ann.

    (b) Ann doesn't sit next to Edward.

    (c) David is either a boy or a girl.

    (d) Edward is a boy and sits next to Carol's best friend.

2. *Translate the following predicate logic sentences into English:*

    (a) girl(Carol) → ¬boy(Carol)

    (b) sits-next-to(Ann, Brenda) ∧ sits-next-to(Brenda, Ann)

    (c) ¬(boy(David) ∧ girl(David))

    (d) sits-next-to(Brenda, David) → ¬girl(David)

6.1.2 Truth of simple sentences

So far we have only talked about sentences being well formed. Without some way of determining the truth of sentences, predicate logic would only be of limited interest. For propositional logic we used the notion of an interpretation which stated which of the atomic sentences (i.e. the propositions) were true or false, and then extended the interpretation to deal with the various connectives. That is basically what we shall do for predicate logic, but since our atomic sentences are no longer propositions but have structure of their own, we will have to change how we assign truth values to them. The extension of the assignment to cover compound sentences, constructed using the connectives, remains the same as for propositional logic.

The basic elements of an interpretation are a *domain*—a non-empty set of objects—and a set of *relationships* over the domain. (Contrast this with the definition of the predicate language, which is concerned with the *names* of objects and relationships.) We may also have a set of functions over the domain. The domain can be any non-empty set. Often well-understood sets such as the natural numbers ($\mathbb{N}$) are used. For our example we shall use 'real' children, i.e. the objects are the children in our hypothetical classroom. We define an interpretation which maps every element of our set of names to a child in the classroom. We are at liberty to do this in any way we please—the children may not be called Ann, Brenda, Carol, David and Edward. There may only be one child in the class, in which case our interpretation would map all the names to that child. Alternatively we could define an interpretation mapping the names to a completely different domain—such as the dates of birth of British prime ministers, or the cities of the United States.

(As a brief aside, consider the problems facing authors who need to distinguish between the names of objects and the objects themselves, when they have only symbols (i.e. names) at their disposal. In an attempt to stave off confusion, we will write the names of objects thus: **pen**, **book**, **easel**; and refer to the objects themselves thus: *pen*, *book*, *easel*. We also assume that each name refers to a unique object.)

For entirely subjective reasons, let us map the names to a set of five children, whom we shall label A, B, C, D and E:

$$
\begin{array}{llll}
\textbf{Ann} & \mapsto & A \qquad & \textbf{David} \mapsto D \\
\textbf{Brenda} & \mapsto & B \qquad & \textbf{Edward} \mapsto E \\
\textbf{Carol} & \mapsto & C &
\end{array}
$$

Just as we map names of objects to objects in the domain, we map names of relationships to relationships over the domain. For example, we map

$$\begin{array}{rcl} \text{boy} & \mapsto & \{D, E\} \\ \text{girl} & \mapsto & \{A, B, C\} \\ \text{sits-next-to} & \mapsto & \{(A, C), (D, E)\} \end{array}$$

To interpret a sentence such as boy(David), we look at the predicate boy, and see whether the mapping of the argument (David) of the predicate is a member of the map of the predicate. In other words, is the map of David a member of the map of boy, i.e. is $D \in \{D, E\}$? Of course it is; therefore boy(David) is true in this interpretation.

sits-next-to(Ann, Carol) is interpreted by seeing if the mapping of the pair of objects (Ann, Carol) is a member of the mapping of sits-next-to. This reduces to seeing if $(A, C) \in \{(A, C), (D, E)\}$. Hence sits-next-to(Ann, Carol) is also true in this interpretation. However, girl(Edward) is false in the interpretation, because the mapping of Edward (i.e. E) is not a member of the mapping of girl (i.e. $\{A, B, C\}$).

Finding the truth of boy(Carol) ∨ girl(Carol) requires us to use the extended interpretation which handles connectives. Recall from Chapter 1 that $A \lor B$ is interpreted as true if either A or B is interpreted as true. This is valid for predicate logic as well, so

$$\text{boy(Carol)} \lor \text{girl(Carol)}$$

is interpreted as true if either boy(Carol) or girl(Carol) is interpreted as true. This reduces to C being a member of either $\{A, B, C\}$ or $\{D, E\}$, which it is.

Suppose that we have the following function over the domain:

$$\{A \mapsto B,\ B \mapsto C,\ C \mapsto A,\ D \mapsto E,\ E \mapsto D\}$$

we can use this as the mapping of the function best-friend thus:

$$\text{best-friend} \mapsto \{A \mapsto B,\ B \mapsto C,\ C \mapsto A,\ D \mapsto E,\ E \mapsto D\}$$

The interpretation of the sentence girl(best-friend(Ann)) depends therefore on whether the mapping of best-friend(Ann) is a member of the mapping of girl. The mapping of best-friend(Ann) is found by taking the mapping of Ann and applying the mapping of the function best-friend to it. The mapping of Ann is A, and under the mapping of best-friend given above, A becomes B; hence the mapping of best-friend(Ann) is B. Therefore the interpretation of the sentence girl(best-friend(Ann)) reduces to whether B is a member of the mapping of girl, which it is. Thus girl(best-friend(Ann)) is true in the example interpretation.

Note that formally we can interpret syntactical expressions like 'sit next to' as any set of pairs in the domain. However, since 'sit next to' has a meaning in English and it implies symmetry, it is unwise to ignore that and not interpret it as a symmetrical relation.

Exercise 6.1.2

1. *Evaluate the truth value of the following sentences using the interpretation above, together with the extensions to the interpretation to deal with connectives:*

    (a) girl(Brenda) $\rightarrow$ ¬boy(Brenda)

    (b) boy(David) $\wedge$ girl(David) $\rightarrow$ sits-next-to(David, Edward)

    (c) sits-next-to(Edward, David)

    (d) sits-next-to(best-friend(Edward), Edward)

2. *Does the truth value of sentence (a) above change if the interpretation is changed so* boy *maps to* $\{B, D, E\}$ *leaving* girl *still mapped to* $\{A, B, C\}$?

3. *How many atomic sentences can one construct in our classroom language with five constants and three predicates?*

4. *How many interpretations can be distinguished by our language? (Two interpretations are distinguishable if there is a sentence which is true in one case and false in the other.)*

6.2 Variables and quantifiers

Recall the argument with which we began this chapter:

$$\frac{\text{all girls are good}}{\frac{\text{Ann is a girl}}{\text{Ann is good}}}$$

While we can now happily write predicate sentences for 'Ann is a girl' and 'Ann is good', we do not yet have enough notation in the language to translate 'all girls are good'. Each atomic sentence we have written so far has been about a specific object name, or a coupling of specific object names. We lack the ability to describe general cases.

Suppose we wish to state that every child in our classroom was good. We introduce a new predicate **good**, and write

$$\text{good(Ann)} \wedge \text{good(Brenda)} \wedge \text{good(Carol)} \wedge \\ \text{good(David)} \wedge \text{good(Edward)}$$

This can get quite cumbersome, so let us introduce a shorthand:

$$\bigwedge_{x \in \{\text{Ann, Brenda, Carol, David, Edward}\}} \text{good(x)}$$

where the large conjunction symbol behaves in a similar way to the $\sum$ summation symbol in mathematics. The variable x stands for each of the elements of the indicated set. Let us suppose that a new child joins the class, with the name of **Fred**. We must therefore write

$$\bigwedge_{x \in \{\text{Ann, Brenda, Carol, David, Edward, Fred}\}} \text{good(x)}$$

This would become increasingly tiresome as **George**, **Harry**, **Ian** and friends all join the class, so we make a final notation change and write

$$\forall x. \ \text{good(x)}$$

This is read as good(x) for every object name in the language. The $\forall$ (pronounced 'for all') is called the *universal quantifier*. Now we can change the set of object names as often as we wish without changing the sentence describing the property. Let us return to our sentence 'all girls are good'. For our original class we could write

$$\begin{array}{ll} & (\ \text{girl(Ann)} & \rightarrow \quad \text{good(Ann)}\ ) \\ \wedge & (\ \text{girl(Brenda)} & \rightarrow \quad \text{good(Brenda)}\ ) \\ \wedge & (\ \text{girl(Carol)} & \rightarrow \quad \text{good(Carol)}\ ) \\ \wedge & (\ \text{girl(David)} & \rightarrow \quad \text{good(David)}\ ) \\ \wedge & (\ \text{girl(Edward)} & \rightarrow \quad \text{good(Edward)}\ ) \end{array}$$

We write the implication for **David** and **Edward** as well, because our interpretation may map these 'boys' names' to girls in the real classroom. This large conjunction can be written more compactly using the quantifier $\forall$:

6.2. VARIABLES AND QUANTIFIERS

$\forall$x. girl(x) $\rightarrow$ good(x)

Recall that when we originally introduced the connectives, we agreed a precedence hierarchy which allowed us to omit many brackets from our logical sentences. We add $\forall$ at the bottom of the hierarchy so that it is below $\rightarrow$. Therefore the sentence above is syntactically equivalent to

$\forall$x. (girl(x) $\rightarrow$ good(x))

The universal quantifier $\forall$ has a counterpart $\exists$, known as the *existential quantifier*. Just as $\forall$x. good(x) takes each object name, puts it as an argument to good and makes one *conjunction*, $\exists$x. good(x) puts each object name as an argument to good and makes one *disjunction*. In the classroom example, $\exists$x. good(x) would produce

good(Ann)$\vee$ good(Brenda)$\vee$ good(Carol)$\vee$
good(David)$\vee$ good(Edward)[2]

The interpretation of a predicate formula is extended to evaluate the truth or falsity of quantifier sentences very simply. $\forall$x. φ(x), where φ(x) stands for some (possibly complex) sentence involving the variable x, is true in an interpretation if and only if φ(x) is true in the interpretation *for all* replacements of x in φ(x) by an object name. Similarly, $\exists$x. φ(x) is true in an interpretation if and only if φ(x) is true in the interpretation *for some* (i.e. at least one) replacement of x in φ(x) by an object name. We are assuming that every object has a syntactical name.

Let $\varphi(z)$ be a formula with the free variable z and assume x and y do not appear in φ. It is clear from our interpretation that $\forall x.\varphi(x)$ and $\forall y.\varphi(y)$ mean the same. This means that we can change bound variables to completely new ones.

Example 6.2.1 *In our classroom interpretation,* $\forall$x.boy(x) *is not true, since there are names (*Ann, Brenda *and* Carol*) which can replace x in* boy(x) *which would make* boy(x) *false in the given interpretation. On the other hand,* $\exists$x.boy(x) *is true, since replacing x with either* David *or* Edward *would make* boy(x) *true.*

---

[2]If the domain is infinite, we cannot replace $\forall$ and $\exists$ by ordinary $\wedge$ and $\vee$ as we are doing here, unless we allow infinitely long sentences.

6.2.1 Scope of quantifiers

Sentences can have more than one quantifier in them. Since the x in ∀x is simply a variable, we can use symbols other than x. Thus we can write ∀y, ∃z and so forth. Hence

```
(∀x. boy(x)) ∨ (∃y. girl(y))
```

is a well-formed sentence. As in mathematics, variables can be renamed at will, provided we obey certain *scoping* rules.

In formulae such as $\forall x.\varphi$, the subformula φ is the *scope* of the quantifier ∀x. Suppose that φ contains no other occurrence of ∀x. Then if x occurs in φ, then x is said to be *free* in φ. However, it becomes *bound* by the ∀x, so that it is not free in $\forall x.\varphi$. For example, in ∀x. boy(x), the variable x is free in the subformula boy(x), but bound in the formula as a whole. If the scope of a quantifier for variable x contains another quantifier for x, e.g. ∀x. (boy(x)∧∀x.girl(x)) then the free/bound status of x within the subformula boy(x)∧∀x.girl(x) depends on where it occurs in the subformula. Any occurrence within the scope of the subformula's quantifier is bound; otherwise it is free. Thus the x which is the argument to boy is free within the subformula, and the x which is the argument to girl is bound. The x which is the argument to boy is bound by the outer ∀x so that it is not free in the formula as a whole.

The scoping rules do not depend on which quantifier is used, so that for example,

```
∃x. (boy(x) ∧ ∀x. girl(x))
```

still has the x in boy(x) bound by the outer quantifier (now ∃x) and the x in girl(x) bound by the inner quantifier ∀x. In the following formulae, x is bound, but y is free.

```
boy(y) ∧ ∀x. good(x)
∃x. (good(x) ∧ girl(y))
```

We need to clarify a point about the meaning of the quantifiers. Consider the formula

```
∀x. girl(x) → ∃x. good(x)
```

From what we have already said, this is true if and only if for every replacement of x in the subformula girl(x) → ∃x. good(x) the subformula is true. In fact, if we were to replace every occurrence of x, we would override the ∃x quantifier, and distort the intended meaning of the sentence. We instead replace only the *free* occurrences of x in the subformula. Thus ∀x. girl(x) → ∃x. good(x) is true (for the original class) iff

$$
\begin{aligned}
&(\text{girl(Ann)} &&\to \exists x. \text{good(x)}) \\
\wedge\ &(\text{girl(Brenda)} &&\to \exists x. \text{good(x)}) \\
\wedge\ &(\text{girl(Carol)} &&\to \exists x. \text{good(x)}) \\
\wedge\ &(\text{girl(David)} &&\to \exists x. \text{good(x)}) \\
\wedge\ &(\text{girl(Edward)} &&\to \exists x. \text{good(x)})
\end{aligned}
$$

is true.

To decide the truth of the subformula ∃x. good(x) we need to expand the subformula to a disjunction by replacing each *free* occurrence of x by an object name. Thus the entire formula is equivalent to

$$
\begin{aligned}
&(\text{girl(Ann)} &&\to (\text{good(Ann)} \vee \text{good(Brenda)} \vee \cdots \vee \text{good(Edward)})) \\
\wedge\ &(\text{girl(Brenda)} &&\to (\text{good(Ann)} \vee \text{good(Brenda)} \vee \cdots \vee \text{good(Edward)})) \\
\wedge\ &(\text{girl(Carol)} &&\to (\text{good(Ann)} \vee \text{good(Brenda)} \vee \cdots \vee \text{good(Edward)})) \\
\wedge\ &(\text{girl(David)} &&\to (\text{good(Ann)} \vee \text{good(Brenda)} \vee \cdots \vee \text{good(Edward)})) \\
\wedge\ &(\text{girl(Edward)} &&\to (\text{good(Ann)} \vee \text{good(Brenda)} \vee \cdots \vee \text{good(Edward)}))
\end{aligned}
$$

If we have a formula ∀x. φ(x) and we wish to replace the x with y, we can do so provided we only change the *free* occurrences of x in φ(x) with y. For example, ∀x. (good(x) ∨ boy(x)) is syntactically equivalent to ∀y. (good(y) ∨ boy(y)). The formula (∀x. girl(x)) ∨ ∃x. boy(x) cannot be changed to use y throughout in one step, as the x in boy(x) is not free in girl(x) ∨ ∃x. boy(x). We must first change the subformula ∃x. boy(x) to ∃y. boy(y), and then change (∀x. girl(x)) ∨ ∃y. boy(y) to (∀y. girl(y)) ∨ ∃y. boy(y).

Exercise 6.2.2 *Consider a language with the following predicate symbols:* friend(x,y), *(meaning x is a friend of y)*, beautiful(x), man(x), *and* woman(x) *and just one constant which is* John.

1. *Write the following English sentences in predicate logic:*

    (a) *No one is both a man and a woman*

    (b) *There are beautiful women*

    (c) *No friend of John is beautiful*

    (d) *John is a beautiful man*

(e) *No one is a friend of anyone who is not a friend of himself*

(f) *Every man is a friend of some beautiful woman but not vice versa*

(g) *A friend of a friend is a friend*

(h) *If all one's friends are beautiful then one is a woman*

(i) *Everyone is either a man or a woman.*

2. *Translate the following predicate logic sentences into English:*

   (a) ∃x. ∃x. [woman(x)] → woman(x)]

   (b) ∀x. ∃y. [friend(x, y) → friend(y, x)]

   (c) ∀x. [∃y. [friend(y, y)] → man(x)].

3. *Find the free and bound variables of the following expression:*

   ∃x.[[[∃x. friend(x, y)] → friend(y, y)] ∧ friend(x,y) ∧ woman(x) ∧ [woman(y) → woman(x)] ∧ ∀y. friend(y, y)]

4. *Consider the language with the numbers as constants* $\{0,1,2,3,\ldots\}$ *and the predicates* x<y *(x is smaller than y) and* prime(x) *(x is a prime number). Express the following (if possible):*

   (a) *there is no greatest number*

   (b) *there exists no greatest prime number*

   (c) *there exists a first element*

   (d) y *is a prime number*

   (e) z *does not divide 9*

   (f) *John loves a prime number.*

5. *Paradox of the barber*
   In a certain village, there is a barber who shaves all those in the village who do not shave themselves. Is this possible, i.e. is it consistent?

6.2.2 Quantifier rules and equivalences

We are finally in a position to represent the argument at the start of the chapter. The English sentence 'all girls are good' can be written in predicate logic as ∀x. (girl(x)→ good(x)). Thus the whole argument can be written

6.2. VARIABLES AND QUANTIFIERS

$$\frac{\forall x.\ (\text{girl}(x) \to \text{good}(x))\qquad \text{girl}(\text{Ann})}{\text{good}(\text{Ann})}$$

Remember that $\forall x.\varphi(x)$ means (the possibly infinite expression) $\varphi(a_1) \wedge \varphi(a_2) \wedge \varphi(a_3) \wedge \ldots$ where $a_1, a_2, a_3, \ldots$ are the object names we are dealing with. So since **Ann** is one of the names we are dealing with, the $\forall x.\ \varphi(x)$ in the argument above means

$$\varphi(\text{Ann}) \wedge \varphi(a_1) \wedge \varphi(a_2) \wedge \varphi(a_3) \wedge \ldots$$

So we have

$$\begin{array}{ll} & (\ \text{girl}(\text{Ann}) \to \text{good}(\text{Ann})\ ) \\ \wedge & (\ \text{girl}(a_1) \to \text{good}(a_1)\ ) \\ \wedge & (\ \text{girl}(a_2) \to \text{good}(a_2)\ ) \\ \wedge & (\ \text{girl}(a_3) \to \text{good}(a_3)\ ) \\ & \quad\vdots \end{array}$$

The ($\wedge$E) rule which we used when we introduced the forward propositional rules is still valid here, and allows us to split up the big conjunction, so that we have

$$\begin{array}{l} \text{girl}(\text{Ann}) \to \text{good}(\text{Ann}) \\ \text{girl}(a_1) \to \text{good}(a_1) \\ \text{girl}(a_2) \to \text{good}(a_2) \\ \text{girl}(a_3) \to \text{good}(a_3) \\ \quad\vdots \\ \text{girl}(\text{Ann}) \end{array}$$

as the assumptions. The ($\to$E) rule now allows us to deduce

$$\frac{\text{girl}(\text{Ann}) \to \text{good}(\text{Ann}) \qquad \text{girl}(\text{Ann})}{\text{good}(\text{Ann})}$$

So we can see that our argument is valid.

In the above we went from $\forall x.\ \varphi(x)$ to $\varphi(\text{Ann})$, by means of the ($\wedge$E) rule. We can write this as a forward deduction rule in its own right, namely ($\forall$E):

$$\frac{\forall x.\ \varphi(x)}{\varphi(a)}$$

where **a** is some object name in the language. There is a deduction rule for the existential quantifier as well, based on the (∨I) rule:

$$\frac{A}{A \vee B}$$

Recall that ∃x. $\varphi(\mathbf{x})$ is syntactically equivalent to (the possibly infinite expression) $\varphi(\mathbf{Ann}) \vee \varphi(a_1) \vee \varphi(a_2) \vee \varphi(a_3) \vee \ldots$. Therefore if we know $\varphi(a_i)$ for some name a_i, by (∨I) we know $\varphi(a_1) \vee \varphi(a_2) \vee \ldots \vee \varphi(a_i) \vee \ldots$ and thus we know ∃x. $\varphi(\mathbf{x})$. This is the (∃I) rule:

$$\frac{\varphi(\mathbf{a})}{\exists \mathbf{x}.\ \varphi(\mathbf{x})}$$

The existential quantifier is known as the 'dual' of the universal quantifier because ∀x. ¬$\varphi(\mathbf{x})$ is equivalent to ¬∃x. $\varphi(\mathbf{x})$. To see this, remember that ¬$(A \vee B)$ is equivalent to ¬$A \wedge \neg B$. Hence ¬∃x. $\varphi(\mathbf{x})$, which can be expanded to ¬$(\varphi(a_1) \vee \varphi(a_2) \vee \varphi(a_3) \vee \ldots)$, is equivalent to ¬$\varphi(a_1) \wedge \neg \varphi(a_2) \wedge \neg \varphi(a_3) \wedge \ldots$. This, of course, is ∀x. ¬$\varphi(\mathbf{x})$. Hence

$$\forall \mathbf{x}.\ \neg \varphi(\mathbf{x}) \equiv \neg \exists \mathbf{x}.\ \varphi(\mathbf{x})$$

By a similar argument, we have

$$\neg \forall \mathbf{x}.\ \varphi(\mathbf{x}) \equiv \exists \mathbf{x}.\ \neg \varphi(\mathbf{x})$$

Further manipulation of the expansions of ∀ and ∃ gives us a set of equivalences which can be used to move quantifiers around in formulae without changing the meaning; some of these equivalences are presented in Figure 6.1. Note that we assume that the variable being quantified over (for illustration purposes in Figure 6.1 we use **x**) does not appear as a free variable in the formula γ.

We shall give an illustration of how to show that the equivalences of Figure 6.1 are valid. Take for example the formula ∀x. $[\varphi(\mathbf{x}) \vee \gamma]$, where γ contains no free occurrences of **x**. This formula can be written (by expanding the ∀x) as

$$[\varphi(a_1) \vee \gamma] \wedge [\varphi(a_2) \vee \gamma] \wedge [\varphi(a_3) \vee \gamma] \wedge \ldots$$

for each object $a_1, a_2, a_3, \ldots$ in the domain. Now recall the propositional equivalence

$$[p \vee r] \wedge [q \vee r] \equiv [p \wedge q] \vee r$$

6.2. VARIABLES AND QUANTIFIERS

$$\forall x.\ \neg\varphi(x) \equiv \neg\exists x.\ \varphi(x)$$
$$\neg\forall x.\ \varphi(x) \equiv \exists x.\ \neg\varphi(x)$$
$$\forall x.\ [\varphi(x) \vee \gamma] \equiv [\forall x.\ \varphi(x)] \vee \gamma$$
$$\exists x.\ [\varphi(x) \vee \gamma] \equiv [\exists x.\ \varphi(x) \vee \gamma]$$
$$\forall x.\ [\varphi(x) \wedge \gamma] \equiv [\forall x.\ \varphi(x)] \wedge \gamma$$
$$\exists x.\ [\varphi(x) \wedge \gamma] \equiv [\exists x.\ \varphi(x)] \wedge \gamma$$
$$\forall x.\ [\varphi(x) \to \gamma] \equiv [\exists x.\ \varphi(x)] \to \gamma$$
$$\exists x.\ [\varphi(x) \to \gamma] \equiv [\forall x.\ \varphi(x)] \to \gamma$$
$$\forall x.\ [\gamma \to \varphi(x)] \equiv \gamma \to \forall x.\ \varphi(x)$$
$$\exists x.\ [\gamma \to \varphi(x)] \equiv \gamma \to \exists x.\ \varphi(x)$$
$$\forall x.\ [\varphi(x) \wedge \psi(x)] \equiv [\forall x.\ \varphi(x)] \wedge [\forall x.\ \psi(x)]$$
$$\exists x.\ [\varphi(x) \vee \psi(x)] \equiv [\exists x.\ \varphi(x)] \vee [\exists x.\ \psi(x)]$$

Figure 6.1: Quantifier equivalences

This enables us to write the expansion of the $\forall x$ as

$$[\varphi(a_1) \wedge \varphi(a_2) \wedge \varphi(a_3) \wedge \ldots] \vee \gamma$$

which can be compressed by contracting back to the $\forall x$

$$[\forall x.\ \varphi(x)] \vee \gamma$$

as required. Similar proofs hold for the remaining equivalences. Note that the following are *inequivalences*:

$$\forall x.\ [\varphi(x) \vee \psi(x)] \not\equiv [\forall x.\ \varphi(x)] \vee [\forall x.\ \psi(x)]$$
$$\exists x.\ [\varphi(x) \wedge \psi(x)] \not\equiv [\exists x.\ \varphi(x)] \wedge [\exists x.\ \psi(x)]$$

Consider the first inequivalence, and consider the following concrete example:

[∀x. man(x) ∨ ∀x. woman(x)]   →   ∀x. [man(x) ∨ woman(x)]

It is clearly true. If every person is a man or every person is a woman, then every person is either a man or a woman. The converse is not, however, true—if every person is a man or a woman, there is no basis for concluding that every person is a man or every person is a woman.

Therefore:

$$\forall x. [\texttt{man(x)} \lor \texttt{woman(x)}] \not\equiv [\forall x. \texttt{man(x)} \lor \forall x. \texttt{woman(x)}]$$

A similar argument can be made for the second inequivalence—if there is a man, and there is a woman, one cannot conclude that there is someone who is both a man and a woman.

Exercise 6.2.3 *Prove that these two quantifier equivalences from Figure 6.1 are true:*

1. $\forall x. [\varphi(x) \land \psi(x)] \equiv [\forall x. \varphi(x)] \land [\forall x. \psi(x)]$
2. $\exists x. [\varphi(x) \lor \psi(x)] \equiv [\exists x. \varphi(x)] \lor [\exists x. \psi(x)]$.
3. *Use quantifier equivalences from Figure 7.1 to show that the formula* $\exists x. \forall y. (\varphi(x) \to \varphi(y))$ *is always true.*

6.2.3 Prenex normal form

The importance of the quantifier equivalences is that they enable us to write every wff in an equivalent form with all the quantifiers on the outside. Thus a formula such as

$$\forall x. [\texttt{p(x)} \to \exists y. [\texttt{q(y)} \land \forall z. \texttt{r(y,z)}]]$$

can be rewritten as the equivalent formula

$$\forall x. \exists y. \forall z. [\texttt{p(x)} \to \texttt{q(y)} \land \texttt{r(y,z)}]$$

Formulae with all their quantifiers grouped together outside the propositional connectives are said to be in *prenex* normal form. The prenex formula can be abstractly presented as

$$\mathcal{Q}_1 x_1. \mathcal{Q}_2 x_2. \cdots . \mathcal{Q}_n x_n . \varphi(x_1, x_2, \ldots, x_n)$$

where each $\mathcal{Q}_i$ is either $\forall$ or $\exists$, and $\varphi(x_1, x_2, \ldots, x_n)$ is a quantifier-free formula, known as the *matrix*, involving the variables $x_1, x_2, \ldots, x_n$. It is usual for the matrix to be in disjunctive normal form (Definition 1.2.3). The quantifiers $\mathcal{Q}_1 x_1. \mathcal{Q}_2 x_2. \cdots . \mathcal{Q}_n x_n$ are known as the *prefix*. It is permissible for n to be zero, so that quantifier-free formulae are in prenex normal form. We now show that there is an equivalent prenex formula for each predicate formula.

Theorem 6.2.4 *Every formula α of predicate logic can be transformed to another formula β in prenex normal form such that $\alpha \Leftrightarrow \beta$ is valid.*

6.2. VARIABLES AND QUANTIFIERS

Proof: The proof is an induction over the structure of the formula α. There are two base cases, when α is atomic or α is $\bot$; in either case α is quantifier free and hence by definition in prenex normal form. α is thus its own prenex equivalent.

There are six induction cases, although four of these are redundant, being definable from the two we deal with here, and the base cases:

$\alpha = \varphi \rightarrow \psi$. By the induction hypothesis we can assume that φ and ψ are already in prenex normal form, or have been replaced by their prenex equivalents. If necessary we can further rewrite the formulae to ensure that the sets of variables $\{x_1, \ldots, x_n\}$ and $\{y_1, \ldots, y_m\}$ are disjoint. Hence

$$\varphi = Q_1 x_1.Q_2 x_2.\cdots.Q_n x_n.\varphi'$$
$$\psi = \mathcal{R}_1 y_1.\mathcal{R}_2 y_2.\cdots.\mathcal{R}_m y_m.\psi'$$

where the Q_i and $\mathcal{R}_j$ are quantifiers. By repeated application of the following equivalences from Figure 6.1

$$\forall x. [\varphi(x) \rightarrow \gamma] \equiv [\exists x. \varphi(x)] \rightarrow \gamma$$
$$\exists x. [\varphi(x) \rightarrow \gamma] \equiv [\forall x. \varphi(x)] \rightarrow \gamma$$

we can move all the quantifiers $Q_1, \ldots, Q_n$ from under the $\rightarrow$ to outside it (as ψ contains none of the variables $x_1, \ldots, x_n$). We now have

$$Q'_1 x_1.Q'_2 x_2.\cdots.Q'_n x_n.[\varphi' \rightarrow \mathcal{R}_1 y_1.\mathcal{R}_2 y_2.\cdots.\mathcal{R}_m y_m.\psi']$$

where Q'_i is $\forall$ if Q_i is $\exists$ and vice versa. By use of two more equivalences from Figure 6.1

$$\forall x. [\gamma \rightarrow \varphi(x)] \equiv \gamma \rightarrow \forall x. \varphi(x)$$
$$\exists x. [\gamma \rightarrow \varphi(x)] \equiv \gamma \rightarrow \exists x. \varphi(x)$$

we can move $\mathcal{R}_1, \ldots, \mathcal{R}_m$ from under the $\rightarrow$ as well, leaving only the quantifier-free formulae φ' and ψ' under the $\rightarrow$:

$$Q'_1 x_1.Q'_2 x_2.\cdots.Q'_n x_n.\mathcal{R}_1 y_1.\mathcal{R}_2 y_2.\cdots.\mathcal{R}_m y_m.[\varphi' \rightarrow \psi']$$

which is the prenex equivalent for α. (The $\mathcal{R}$ quantifiers are not changed by the equivalences.)

$\alpha = \exists x.\varphi$. We assume that φ is of the form $\mathcal{Q}_1 x_1.\mathcal{Q}_2 x_2.\cdots.\mathcal{Q}_n x_n.\varphi'$, by the induction hypothesis. Hence $\alpha = \exists x.\mathcal{Q}_1 x_1.\mathcal{Q}_2 x_2.\cdots.\mathcal{Q}_n x_n.\varphi'$ which is in prenex normal form. ∎

Many automatic theorem-proving techniques rely on formulae being in prenex normal form, and we shall also exploit it. Once we have translated a set of formulae into this form, we can then convert the matrix for each formula into the $\mathcal{B}$-clausal form of Definition 5.4.2, because each matrix is simply atoms joined together by propositional connectives, and thus all the familiar manipulations that we used to transform arbitrary propositional formulae into $\mathcal{B}$-clauses can be used on the matrix.

Exercise 6.2.5 *Write the following formulae in prenex normal form (p, q are predicate symbols):*

1. $\exists x.\forall y.p(x,y) \land \exists x.\forall y.q(x,y)$

2. $\exists x.\forall y.p(x,y) \to \forall x.\exists y.q(x,y)$

6.2.4 Equality

There is a special predicate, $x \approx y$, which is interpreted as identity. Thus $a \approx b$ if and only if they name the same element in the domain. The symbol '$\approx$' is a formal symbol for equality. From now on we use the more common '$=$' symbol.

6.3 Formal definitions

We have now completed the informal presentation of predicate logic, and we now provide formal definitions of the material covered above. Because of the interdefinability of the various connectives, the following definition provides a minimal set of connectives. Exercise 6.3.3 involves the derivation of some of the other connectives.

Definition 6.3.1 (Syntax of predicate logic) *The formulae of the predicate logic language $\mathcal{L}$ are built up from the following symbols:*

- *a set $\mathcal{L}_{\mathrm{pred}}$ of predicate symbols, each with an associated arity, which is a positive integer,*

- *a set $\mathcal{L}_{\mathrm{cons}}$ of constant symbols,*

6.3. FORMAL DEFINITIONS

- a set $\mathcal{L}_{\text{var}}$ of variable symbols,
- a set $\mathcal{L}_{\text{func}}$ of function symbols, each with an associated arity, which is a positive integer,
- a quantifier $\exists$,
- classical connectives $\bot$ and $\to$. The other classical connectives are definable from $\bot$ and $\to$.

The set of terms $\mathcal{L}_{\text{term}}$ is given by the following rules:

- any member of $\mathcal{L}_{\text{cons}}$ is a term in $\mathcal{L}_{\text{term}}$, with no variables;
- any member x of $\mathcal{L}_{\text{var}}$ is a term in $\mathcal{L}_{\text{term}}$, with variable x (itself);
- if f is a member of $\mathcal{L}_{\text{func}}$ with arity n, and $t_1, \ldots, t_n$ are terms in $\mathcal{L}_{\text{term}}$, then $t = f(t_1, \ldots, t_n)$ is a term in $\mathcal{L}_{\text{term}}$. The set of variables of t is the union of all the sets of variables of $t_1, \ldots, t_n$.

We can now define the well-formed formulae (wffs) of $\mathcal{L}$ by the rules:

- $\bot$ is a wff of $\mathcal{L}$, with no free variables;
- if p is a member of $\mathcal{L}_{\text{pred}}$ with arity n, and $t_1, \ldots, t_n$ are terms in $\mathcal{L}_{\text{term}}$, then $p(t_1, \ldots, t_n)$ is a wff in $\mathcal{L}$, with the free variables being all the variables of $t_1, \ldots, t_n$;
- if φ and ψ are wffs in $\mathcal{L}$ then so is $\varphi \to \psi$, with the free variables being the union of those free in φ and ψ;
- if v is a variable in $\mathcal{L}_{\text{var}}$ and φ is a wff in $\mathcal{L}$, then $\exists v.\varphi$ is a wff in $\mathcal{L}$. The free variables of $\exists v.\varphi$ are those of φ less the variable v.

Each symbol in $\mathcal{L}$ must be interpreted in order for wffs of $\mathcal{L}$ to be given a truth value. For propositional logic, we simply assigned truth values to the propositions via a function h, and then extended h to propagate the truth values from the propositions through the formulae. As outlined in Section 6.1.2, the change from atomic propositions to predicates with structured terms being the simplest wffs means that the basic model becomes more complicated.

Definition 6.3.2 (Semantics of predicate logic) *The formulae are given truth values with respect to an* interpretation *or a* model $\mathcal{M} = \langle \mathcal{D}, \pi_{\text{cons}}, \pi_{\text{func}}, \pi_{\text{pred}} \rangle$, *with the four components:*

- $\mathcal{D}$, a non-empty domain of objects,

- π_{cons}, a mapping from members of $\mathcal{L}_{\text{cons}}$ to $\mathcal{D}$,

- π_{func}, mapping each member of $\mathcal{L}_{\text{func}}$ to a function mapping $\mathcal{D}^n$ to $\mathcal{D}$, for each $p \in \mathcal{L}_{\text{func}}$ $\pi_{\text{func}}(p) : \mathcal{D}^n \mapsto \mathcal{D}$, where n is the arity of the member of $\mathcal{L}_{\text{func}}$, and

- π_{pred}, mapping each member of $\mathcal{L}_{\text{pred}}$ to $\mathcal{D}^n \mapsto \{\top, \bot\}$, where n is the arity of the member of $\mathcal{L}_{\text{pred}}$.

We also need to interpret the free variables in the formulae. This is done by defining a variable assignment V, which is a mapping from $\mathcal{L}_{\text{var}}$ to $\mathcal{D}$. We need the notation $V_{[v \mapsto d]}$ to mean the assignment V' satisfying $V'(x) = V(x)$, for $x \neq v$ and $V'(v) = d$. Given this, we can interpret all the terms of $\mathcal{L}$ by means of a term mapping π_{term}, based on V, π_{cons} and π_{func}, which maps all members of $\mathcal{L}_{\text{term}}$ to $\mathcal{D}$. For t in $\mathcal{L}_{\text{term}}$:

- for all members c of $\mathcal{L}_{\text{cons}}$, $\pi_{\text{term}}(c) = \pi_{\text{cons}}(c)$;

- for all members v of $\mathcal{L}_{\text{var}}$, $\pi_{\text{term}}(v) = V(v)$;

- for all members f of $\mathcal{L}_{\text{func}}$ with arity n, $\pi_{\text{term}}(f(t_1, \ldots, t_n)) = \pi_{\text{func}}(f)(\pi_{\text{term}}(t_1), \ldots, \pi_{\text{term}}(t_n))$;

- $\pi_{\text{term}}(t)$ is called the value of t in the model, under the assignment V.

Finally we can define the truth of a wff φ of $\mathcal{L}$, with respect to an interpretation $\mathcal{M}$ and a variable assignment V. This is written as $\langle \mathcal{M}, V \rangle \vDash \varphi$, read as '$\varphi$ holds in $\langle \mathcal{M}, V \rangle$', or '$\langle \mathcal{M}, V \rangle$ is a model of φ' and given by

$\langle \mathcal{M}, V \rangle \nvDash \bot$
$\langle \mathcal{M}, V \rangle \vDash p(t_1, \ldots, t_n)$ iff $\pi_{\text{pred}}(p)(\pi_{\text{term}}(t_1), \ldots, \pi_{\text{term}}(t_n)) = \top$
$\langle \mathcal{M}, V \rangle \vDash \varphi \rightarrow \psi$ iff $\langle \mathcal{M}, V \rangle \vDash \varphi$ implies $\langle \mathcal{M}, V \rangle \vDash \psi$
$\langle \mathcal{M}, V \rangle \vDash \exists v. \varphi$ iff   there exists $d \in \mathcal{D}$ and $\langle \mathcal{M}, V_{[v \mapsto d]} \rangle \vDash \varphi$

Let $\varphi(x_1, \ldots, x_n)$ be a formula with the free variables $x_1, \ldots, x_n$. It is common to use the notation $\mathcal{M} \vDash \varphi(a_1, \ldots, a_n)$ to represent

$$\langle \mathcal{M}, V_{[x_i \mapsto a_i | i=1, \ldots, n]} \rangle \vDash \varphi.$$

6.3. FORMAL DEFINITIONS

If the formula contains no free variables, then an arbitrary mapping (sometimes called empty *mapping) can be used as the initial variable assignment. We use the notation* [ ] *for such a mapping and we write* $\langle \mathcal{M}, [\,] \rangle \models \varphi$.

We write $\models \varphi$ *to indicate that 'φ holds in all models* $\langle \mathcal{M}, V \rangle$'.

Exercise 6.3.3

1. *Given that* $\neg \alpha \equiv \alpha \to \bot$, $\alpha \vee \beta \equiv (\alpha \to \bot) \to \beta$, $\alpha \wedge \beta \equiv (\alpha \to (\beta \to \bot)) \to \bot$ *and* $\forall x.\, \varphi \equiv (\exists x.\, (\varphi \to \bot\,)) \to \bot$, *derive definitions for*

    (a) $\langle \mathcal{M}, V \rangle \models \neg \varphi$

    (b) $\langle \mathcal{M}, V \rangle \models \varphi \vee \psi$

    (c) $\langle \mathcal{M}, V \rangle \models \varphi \wedge \psi$

    (d) $\langle \mathcal{M}, V \rangle \models \forall x.\, \varphi$

2. *Let the language be based on the following sets:*

    - $\mathcal{L}_{\text{cons}} = \{\texttt{a, b, c, d}\}$
    - $\mathcal{L}_{\text{var}} = \{\texttt{x, y, z}\}$
    - $\mathcal{L}_{\text{func}} = \{\texttt{f,g}\}$, $\texttt{f}$ *with arity 1 and* $\texttt{g}$ *with arity 2*
    - $\mathcal{L}_{\text{pred}} = \{\texttt{p,q,r}\}$, $\texttt{p}$ *with arity 2,* $\texttt{q}$ *with arity 1 and* $\texttt{r}$ *with arity 3.*

    Let the model be given by $\mathcal{M} = \langle \mathcal{D}, \pi_{\text{cons}}, \pi_{\text{func}}, \pi_{\text{pred}} \rangle$:

    - $\mathcal{D} = \{1, 2, 3, 4, 5, 6, 7\}$,
    - π_{cons}: $\{\texttt{a} \mapsto 1, \texttt{b} \mapsto 3, \texttt{c} \mapsto 5, \texttt{d} \mapsto 7\}$
    - π_{func}: $\{\texttt{f} \mapsto \{1 \mapsto 2, 2 \mapsto 3, 3 \mapsto 5, 4 \mapsto 7, 5 \mapsto 7, 6 \mapsto 7, 7 \mapsto 7\}, \texttt{g} \mapsto \{(u,v) \mapsto u+v \text{ if } u+v \leq 7, \text{ otherwise } (u,v) \mapsto 1\}\}$
    - π_{pred}:

    $$\{\ \texttt{p} \mapsto \{(2,3) \mapsto \top, (3,2) \mapsto \top\},$$
    $$\texttt{q} \mapsto \{2 \mapsto \top, 4 \mapsto \top, 6 \mapsto \top\},$$
    $$\texttt{r} \mapsto \{(1,3,5) \mapsto \top, (2,4,6) \mapsto \top\}\ \}$$

    All other combinations are assigned $\bot$ *by* π_{pred}.

Work out which of the following formulae are true in the above model, with the initial variable assignment []:

(a) p(f(a),b)
(b) r(a,f(b),c)
(c) ∃x.(¬q(x) → p(f(a),g(x,2)))
(d) ∀x.(¬q(x) → p(f(a),g(x,2)))

Exercise 6.3.4 *Consider the language based on the following sets:*

- $\mathcal{L}_{\text{cons}} = \varnothing$
- $\mathcal{L}_{\text{var}} = \{x, y, z\}$
- $\mathcal{L}_{\text{func}} = \varnothing$
- $\mathcal{L}_{\text{pred}} = \{s, p\}$, *both with arity 3.*

Let the model be given by $\mathcal{N} = \langle \mathbb{N}, \pi_{\text{cons}}, \pi_{\text{func}}, \pi_{\text{pred}} \rangle$ *where*

- $\mathbb{N} = \{0, 1, 2, \ldots, \}$ *is the set of natural numbers*
- $\pi_{\text{cons}}, \pi_{\text{func}}$ *are empty*
- $\pi_{\text{pred}}(s)(a, b, c) = \top$ *iff* $a + b = c$,
  $\pi_{\text{pred}}(p)(a, b, c) = \top$ *iff* $a \cdot b = c$.

(a)–(f): Write down a formula with a single free variable x stating that

(a) $x = 0$ (d) x *is even*

(b) $x = 1$ (e) x *is prime*

(c) $x = 2$ (f) $x > 27$.

(g), (h): Write down a formula with two free variables x, y stating that

(g) $x = y$

(h) $x < y$

(i) *Write down a sentence stating that there are infinitely many primes.*

Exercise 6.3.5 *Check whether the following formulae have models:*

(a) $\exists x. \forall y. (Q(x, x) \land \neg Q(x, y))$

(b) $\exists x.\exists y.(P(x) \wedge \neg P(y))$

(c) $\exists x.\forall y.(Q(x,y) \to \forall z.R(x,y,z))$

(d) $P(x) \to \forall y.P(y)$.

Exercise 6.3.6 *Check whether the following formulae are valid:*

(a) $\exists x.P(y) \to \forall x.P(x)$

(b) $\neg(\exists x.P(x) \to \forall x.P(x))$

(c) $\exists x.\forall y.Q(x,y) \to \forall y.\exists x.Q(x,y)$

(d) $\forall x.\exists y.Q(x,y) \to \exists y.\forall x.Q(x,y)$.

6.4 Worked examples

The purpose of the worked examples of this section is to demonstrate the expressive power of predicate logic as a language for specification. This capability plays a crucial role in all computer science applications.

Example 6.4.1 (Euler's problem 1782) *There are 36 officers of six regiments (nos 1, 2, ..., 6) and of six ranks (nos 1, 2, ..., 6), so that every pair of numbers is presented by some officer. The problem is to arrange them in a 6×6 square, such that at every row and at every column ranks and regiments of officers are different. This means that we have to find a six-element model of some set of sentences. This example presents a simple set of sentences for which it is difficult to check whether it has a model.*

1. Write down these sentences, using the following language:
    - $\mathcal{L}_{\text{cons}} = \emptyset$
    - $\mathcal{L}_{\text{var}} = \{x, y, z, t, u, v, w\}$
    - $\mathcal{L}_{\text{func}} = \{rk, rg\}$
    - $\mathcal{L}_{\text{pred}} = \{\ =\ \}$.

    ($rk(a,b)$ means the rank of the officer standing at the a-th row and at the b-th column, $rg(a,b)$ means the regiment of this officer.)

    [Euler's conjecture was that such models do not exist. This was proved over a century later, in 1900.]

2. *Construct a three-element model of the same set of sentences.*

Solution

1. $\varphi_1 : \forall x.\forall y.\exists z.\exists t.(rk(z,t) = x \wedge rg(z,t) = y)$

    This means: 'every pair (rank, regiment) is presented by some officer'.

    $\varphi_2 : \forall x.\forall y.\forall z.(rk(x,y) = rk(x,z) \rightarrow y = z)$

    this means: 'ranks at every row are distinct'.

    $\varphi_3 : \forall x.\forall y.\forall z.(rk(y,x) = rk(z,x) \rightarrow y = z)$

    ('Ranks at every column are distinct').

    $\varphi_4 : \forall x.\forall y.\forall z.(rg(x,y) = rg(x,z) \rightarrow y = z)$

    $\varphi_5 : \forall x.\forall y.\forall z.(rg(y,x) = rg(z,x) \rightarrow y = z)$

    (φ_4 and φ_5 are similar to φ_2, φ_3).

2. One of the possible solutions is

1,1	2,3	3,2
3,3	1,2	2,1
2,2	3,1	1,3

    (The pair of elements in the above square in the third row and first column (i.e. position (3, 1)) is '2,2'. This means that $rk(3,1) = 2, rg(3,1) = 2$. Similarly, $rk(1,2) = 2, rg(1,2) = 3$, etc.)

Example 6.4.2 (Affine planes) *Consider the language with*

- $\mathcal{L}_{\text{cons}} = \emptyset$
- $\mathcal{L}_{\text{var}} = \{x, y, \ldots\}$
- $\mathcal{L}_{\text{func}} = \emptyset$
- $\mathcal{L}_{\text{pred}} = \{=, P, L, \in\}$ *with unary* P, L, *binary* $=, \in$.

$P(x)$ is read as 'x is a point', $L(x)$ as 'x is a line'; $\in (x, y)$ is read as 'x belongs to y', or as 'y contains x'.

Write down sentences, stating that

(α) *Everything is either a point or a line, but not both.*

(β) *Every line contains at least two points.*

6.4. WORKED EXAMPLES

(γ) Everything containing a point is a line.

(δ) Everything belonging to a line is a point.

(ε) Every two distinct points belong to some line, and this line is unique.

(ξ) Given a point x and a line y, not containing x, one can find a unique line containing x which has no common points with y.

(ζ) There exist at least two lines.

Solutions

(α) $\forall x.[(P(x) \vee L(x)) \wedge \neg(P(x) \wedge L(x))]$

(β) $\forall x.(L(x) \rightarrow \exists y.\exists z.(\in (y,x) \wedge \in (z,x) \wedge \neg y = z))$

(γ) $\forall x.\forall y.(P(x) \wedge \in (x,y) \rightarrow L(y))$

(δ) $\forall x.\forall y.(L(x) \wedge \in (y,x) \rightarrow P(y))$

(ε) $\forall x.\forall y.(P(x) \wedge P(y) \wedge \neg x = y \rightarrow \exists z.[L(z) \wedge \in (x,z) \wedge \in (y,z) \wedge \forall t.(L(t) \wedge \in (x,t) \wedge \in (y,t) \rightarrow t = z)])$

(ξ) $\forall x.\forall y.[P(x) \wedge L(y) \wedge \neg \in (x,y) \rightarrow \exists z.[L(z) \wedge \in (x,z) \wedge \neg \exists t.(\in (t,z) \wedge \in (t,y)) \wedge \forall u[L(u) \wedge \in (x,u) \wedge \neg \exists t.(\in (t,u) \wedge \in (t,y)) \rightarrow u = z]]]$

(ζ) $\exists x.\exists y.(L(x) \wedge L(y) \wedge \neg x = y)$.

Example 6.4.3 (Affine planes, continued[3]) *A model of the set of formulae* $\Delta = \{\alpha, \beta, \gamma, \delta, \varepsilon, \xi, \zeta\}$ *of the previous example is called an affine plane.*

1. Construct an affine plane with four points.

2. Show that in every finite affine plane all lines contain an equal number of points.

3. Show that in any affine plane containing a line with n points, every point belongs to exactly $n+1$ lines.

---
[3] For more information on the logical approach to affine geometry, see [Blumenthal, 1980].

6. INTRODUCING PREDICATE LOGIC

4. *Show that the number of points in any finite affine plane is a full square.*[4]

Solutions

Here is the picture:

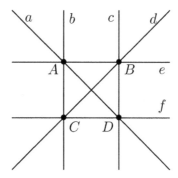

There are four points (A, B, C, D) and six lines (a, b, c, d, e, f).
Here is the description of the model according to Definition 7.3.1

$D = \{A, B, C, D, a, b, c, d, e, f\}$;
$\pi_{\text{pred}}(P)(X) = \top$ iff $X \in \{A, B, C, D\}$;
$\pi_{\text{pred}}(L)(X) = \top$ iff $X \in \{a, b, c, d, e, f\}$;
$\pi_{\text{pred}}(\in)(X, y) = \top$ iff
$\quad (X, y) \in \{(A, a), (D, a), (A, b), (C, b), (B, c), (D, c), (B, d), (C, d), (A, e),$
$\quad (B, e), (C, f), (D, f)\}$;
$\pi_{\text{pred}}(=)(X, Y) = \top$ iff $X = Y$.

The truth of formulae $(\alpha), (\beta), (\gamma), (\delta), (\zeta)$ in this model is quite clear. (ε) holds because for any two distinct points there is a unique line containing them, namely:

$\quad a$ for A, D; $\quad d$ for B, C;
$\quad b$ for A, C; $\quad e$ for A, B;
$\quad c$ for B, D; $\quad f$ for C, D.

---

[4]However, it is not yet known completely which full squares can occur as numbers of points of affine planes. Nobody knows, for example, if there exist affine planes with 100 points.

6.4. WORKED EXAMPLES

(ξ) can be checked for all possible values x, y for which the premise is true:

$$\begin{aligned}
&\text{if } x = A, y = d \quad \text{then } z = a, \\
&\text{if } x = A, y = c \quad \text{then } z = b, \\
&\text{if } x = A, y = f \quad \text{then } z = e, \\
&\text{if } x = B, y = a \quad \text{then } z = d, \\
&\text{if } x = B, y = b \quad \text{then } z = c, \\
&\text{if } x = B, y = f \quad \text{then } z = e, \\
&\text{if } x = C, y = a \quad \text{then } z = d, \\
&\text{if } x = C, y = c \quad \text{then } z = b, \\
&\text{if } x = C, y = e \quad \text{then } z = f, \\
&\text{if } x = D, y = e \quad \text{then } z = f, \\
&\text{if } x = D, y = d \quad \text{then } z = a, \\
&\text{if } x = D, y = b \quad \text{then } z = c.
\end{aligned}$$

2. To deal with affine planes, we somewhat simplify our notations. According to tradition, we will denote lines by small letters, and points by capitals. We write $B \in a$ instead of $\in (B, a)$ and $B \notin a$ instead of $\neg \in (B, a)$, as is usually done.

Also we write $a \| b$ (read as 'a is parallel to b') instead of

$$\neg \exists X. (X \in z \land X \in b)$$

Now consider an affine plane $\mathcal{M}$; let a, a' be two distinct lines in $\mathcal{M}$. First we notice that there exists A'_1 such that

$$\mathcal{M} \vDash A'_1 \in a' \land A'_1 \notin a \qquad (2.1)$$

To show this, assume the contrary. Since $\mathcal{M} \vDash (\beta)$, one can find two points (say, B_1, B_2) such that $\mathcal{M} \vDash B_1 \neq B_2 \land B_1 \in a' \land B_2 \in a'$.
By our assumption, we also have that

$$\mathcal{M} \vDash B_1 \in a \land B_2 \in a$$

Now the formula (ε) provides that $a = a'$, which is a contradiction. Thus (2.1) holds for some A'_1. Likewise, there exists A_1 such that

$$\mathcal{M} \vDash A_1 \in a \land A_1 \notin a' \qquad (2.2)$$

Now let $A_1, \ldots, A_n$ be all points of a, and let us show that a' contains exactly n points.

Owing to (ε), there exists a unique line b_1, such that

$$\mathcal{M} \vDash A_1 \in b_1 \wedge A_1' \in b_1 \tag{2.3}$$

($A_1 \neq A_1'$ is clear from (2.1).)

Then if we take any A_i, $1 < i \leq n$, it follows that

$$\mathcal{M} \vDash A_i \notin b_1 \tag{2.4}$$

because otherwise

$$\mathcal{M} \vDash A_1 \in b_1 \wedge A_i \in b_1 \wedge A_1 \in a \wedge A_i \in a \wedge A_1 \neq A_i$$

which implies $a = b_1$, owing to (ε). but $a \neq b_1$, because $A_1' \in b_1$, $A_2' \notin a$. So (2.4) holds.

Now since $\mathcal{M} \vDash (\xi)$, there exists a unique line b_i such that

$$\mathcal{M} \vDash A_i \in b_i \wedge b_i \| b_1 \tag{2.5}$$

Further on, we observe that

$$\mathcal{M} \nvDash b_i \| a' \tag{2.6}$$

Indeed, suppose the contrary; then (by (2.1), (2.5), (2.3))

$$\mathcal{M} \vDash a' \| b_i \wedge b_1 \| b_i \wedge A_1' \in a' \wedge A_1' \in b_1 \wedge A_1' \notin a$$

Hence by (ξ) we get that $a' = b_1$. But this is impossible because $A_1 \in b_1$, $A_1 \notin a'$ (by (2.2), (2.3)).

Thus (2.6) holds.

From (2.6) it follows that a', b_i have a common point. This point is unique, because otherwise $a' = b_i$ (owing to (ε)), which is impossible, because $b_i \| b_1$, $a' \nparallel b_1$ (since they have a common point A_1').

So let A_i' be this point:

$$\mathcal{M} \vDash A_i' \in b_i \wedge A_i' \in a'$$

In this way each point at a corresponds to a unique point at a', and it remains to prove that $A_i \mapsto A_i'$ is a one-to-one correspondence.

This is clear if we find the inverse function. But we can construct it in the same way. Namely, given a point D, such that $D \in a'$, $D \neq A_1'$, we can construct a unique line l, such that $\mathcal{M} \vDash D \in l \wedge l \| b_1$.

6.4. WORKED EXAMPLES

This gives us the unique C, such that

$$\mathcal{M} \models C \in l \land C \in a$$

Since $C \in a$, it follows that $C = A_j$, for some j. But then $l = b_j$, because

$$\mathcal{M} \models l \| b_1 \land b_j \| b_1 \land C \in l \land C \in b_j$$

And therefore $D = A'_j$, because

$$\mathcal{M} \models D \in l \land D \in a'$$

This mapping, $A_i \mapsto A'_i$, is one-to-one, implying that $A'_1, \ldots, A'_n$ are exactly all points of a'.

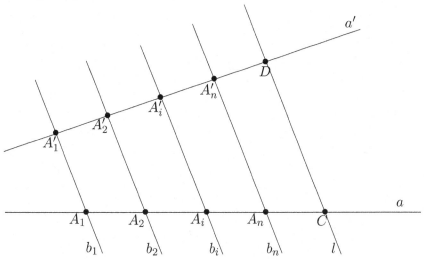

3. First we prove that in any affine plane $\mathcal{M}$

$$\mathcal{M} \models \forall A. \exists a. A \notin a \tag{3.1}$$

In fact, by (ζ) we can find two different lines (say, b and c) in $\mathcal{M}$. If $A \notin b$ or $A \notin c$, there is nothing to prove. If $A \in b$ and $A \in c$, we take B, C such that

$$\mathcal{M} \models B \in b \land C \in c \land B \neq A \land C \neq A \tag{3.2}$$

Such B, C exist because $\mathcal{M} \models (\varepsilon)$. Then $B \neq C$ because otherwise we had

$$\mathcal{M} \models A \in b \land A \in c \land B \in b \land B \in c \land A \neq B \land b \neq c$$

which contradicts (ε). Applying (ε) again, we find a line a, such that

$$\mathcal{M} \vDash B \in a \land C \in a$$

Then $A \notin a$ because otherwise $A \in a \land B \in a \land A \in b \land B \in b$ which implies $a = b$, owing to (ε); likewise, we obtain that $a = C$, and then $b = c$, which is a contradiction.

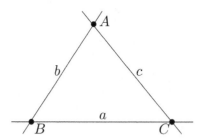

Now we can prove the main statement. By item (2), the line a contains exactly n points, say $A_1, \ldots, A_n$. Obviously for every i, $A \neq A_i$ (since $A \notin a$). Then we can use (ε) again and get that for any i $(1 \leq i \leq n)$

$$\mathcal{M} \vDash \exists a_i.(A \in a_i \land A_i \in a_i) \tag{3.3}$$

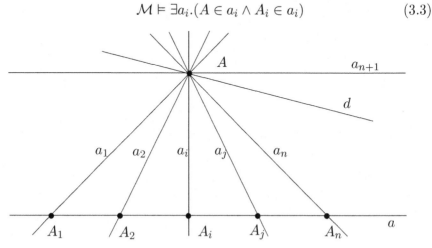

Next, we notice that

$$\text{if } i \neq j \text{ then } a_i \neq a_j \tag{3.4}$$

In fact, $a_i = a_j$ implies (by (3.3))

$$\mathcal{M} \vDash A_i \in a_i \land A_j \in a_i \land A_i \in a \land A_j \in a$$

6.4. WORKED EXAMPLES

and thus $a_i = a$ by (ε), since $A_i \neq A_j$. But then $A \in a$, which brings us to a contradiction. Thus (3.4) holds.

So we have found n different lines containing A. There is also a line a_{n+1}, such that $A \in a_{n+1}, a_{n+1} \| a$ (by (ξ)). Obviously, $a_{n+1} \neq a_i$ for any $i \leq n$ because a_i and a have a common point A_i.

It remains to show that

$$\mathcal{M} \models \forall d.(A \in d \to d = a_1 \lor d = a_2 \lor \cdots \lor d = a_{n+1}) \qquad (3.4)$$

Really, if $A \in d$ and $d \not\| a$ then $d = a_{n+1}$, owing to (ξ). And if $A \in d$ and $d \| a$ then d and a must have a common point, so for some $i \leq n$

$$\mathcal{M} \models A_i \in d \land A \in d \land A_i \in a_i \land A \in a_i$$

This implies $a_i = d$ (by (ε)), because $A_i \neq A$.

Therefore $a_1, \ldots, a_{n+1}$ are exactly all lines containing A.

4. Let $\mathcal{M}$ again be an affine plane, with a line a, containing exactly n points: $A_1, \ldots, A_n$. As above, we can construct a line b_1, such that

$$\mathcal{M} \models A \in b_1 \land a \neq b_1 \qquad (4.1)$$

Then by (ξ) we get for any $i > 1$

$$\mathcal{M} \models \exists b_i.(A_i \in b_i \land b_i \| b_1) \qquad (4.2)$$

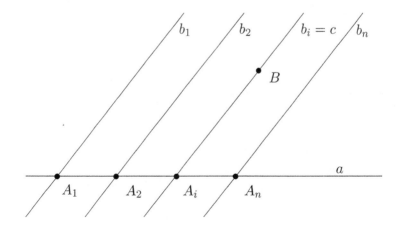

We notice that $b_i \| b_j$ whenever $i \neq j$, because otherwise for some C we had
$$\mathcal{M} \models C \in b_i \wedge C \in b_j \wedge b_i \neq b_j \wedge b_i \| b_1 \wedge b_j \| b_1$$
in contradiction with (ξ).

By item (2) each b_i consists of n points, and in total we get $n \cdot n = n^2$ distinct points in $\mathcal{M}$.

Finally, we observe that we have calculated all points in $\mathcal{M}$. In fact, if we take any point $B \notin a$, and if $B \notin b_1$, then by (ξ) there exists a line c, such that
$$\mathcal{M} \models B \in c \wedge c \| b_1 \tag{4.3}$$
Hence $c \not\| a$, because otherwise we had
$$\mathcal{M} \models b_1 \| c \wedge a \| c \wedge A_1 \in b_1 \wedge A_1 \in a \wedge b_1 \neq a$$
in contradiction with (ξ).

Thus c and a have a common point, i.e. for some i,
$$\mathcal{M} \models A_i \in c$$
Now (ξ) yields that $b_i = c$, because
$$\mathcal{M} \models A_i \in b_i \wedge A_i \in c \wedge b_i \| b_1 \wedge c \| b_1$$
Therefore, every point belongs to some of the lines $b_1, \ldots, b_n$.

Example 6.4.4 *Show that the following formula φ is false in any finite model, but is true in some infinite model:*
$$\varphi = \forall x. \neg Q(x,x) \wedge \forall x. \forall y. \forall z. (Q(x,y) \wedge Q(y,z) \to Q(x,z)) \wedge \forall x. \exists y. Q(x,y)$$

Solution

1. Assume that φ is true in some model $\mathcal{M}$ with the domain D. We have to show that D is infinite.

   First take any $d_0 \in D$. Since $\mathcal{M} \models \varphi$, it follows that $\mathcal{M} \models \forall x. \exists y. Q(x,y)$ and hence (by Exercise 7.3.3) $\mathcal{M} \models \exists y. Q(d_0, y)$. Then $\mathcal{M} \models Q(d_0, d_1)$ for some $d_1 \in D$.

   Now we can find a sequence $d_0, d_1, \ldots \in D$ by induction as follows. If $d_0, \ldots, d_n$ are already chosen, we note that $\mathcal{M} \models \exists y. Q(d_n, y)$. And

6.4. WORKED EXAMPLES

then $\mathcal{M} \models Q(d_n, d_{n+1})$ for some $d_{n+1} \in D$. Finally we notice that all elements d_n in the sequence are different. In fact $d_n \neq d_{n+1}$ because otherwise we had $\mathcal{M} \models Q(d_n, d_n)$ which contradicts $\mathcal{M} \models \forall x. \neg Q(x, x)$. Also $d_n \neq d_{n+m}$ for any $m > 1$.

To see this, it is sufficient to show that for any $m \geq 1$

(a) $\mathcal{M} \models Q(d_n, d_{n+m})$, and recall that $\mathcal{M} \models \neg Q(d_n, d_n)$. The claim (a) is proved by induction over m:

- $Q(d_n, d_{n+1})$ by our construction.
- If $\mathcal{M} \models Q(d_n, d_{n+m})$ is proved, we notice that

$$\mathcal{M} \models Q(d_{n+m}, d_{n+m+1})$$

by the construction. Since $\mathcal{M} \models \varphi$, we have also that

$$\mathcal{M} \models \forall x. \forall y. \forall z. (Q(x, y) \wedge Q(y, z) \rightarrow Q(x, z))$$

and thus

(b) $\mathcal{M} \models Q(d_n, d_{n+m}) \wedge Q(d_{n+m}, d_{n+m+1}) \rightarrow Q(d_n, d_{n+m+1})$, by Exercise 7.3.3.

The premise in (b) is true by our construction and the inductive hypothesis, and therefore $\mathcal{M} \models Q(d_n, d_{n+m+1})$.

This completes the proof of (a), and therefore we have found an infinite subset in D.

2. Consider a model $\mathcal{M}$ with the domain $\mathbb{N}$ and π_{pred} such that

$$\pi_{\text{pred}}(Q)(a, b) = \top$$

iff $a < b$. Then

- $\mathcal{M} \models \forall x. \neg Q(x, x)$, because $\neg a < a$ holds for any a.
- $\mathcal{M} \models \forall x \forall y. \forall z. (Q(x, y) \wedge Q(y, z) \rightarrow Q(x, z))$, because '<' is a transitive relation in $\mathbb{N}$.
- $\mathcal{M} \models \forall x. \exists y. Q(x, y)$, because for any a we have $\mathcal{M} \models \exists y. Q(a, y)$, and the latter holds because we can take, say, $y = a + 1$.

Example 6.4.5 *Find a formula with equality which holds in a model iff*

1. *the domain is one-element;*

2. the domain is more than n-element;

3. the domain is n-element for some finite n.

Solution

1. Consider the formula $\varphi_1 : \forall x.\forall y.(x = y)$.

   We have
   $$\mathcal{M} \vDash \varphi_1 \text{ iff for any } a \in D,\ \mathcal{M} \vDash \forall y.(a = y)$$
   iff for any $a, b \in D$, $\mathcal{M} \vDash a = b$. The latter means that all elements of the domain coincide, i.e. that the domain is one-element.

2. We have to say that there exist at least $(n+1)$ distinct elements in the domain. Consider the formula
   $$\psi_{n+1} : \exists x_1.\exists x_2.\cdots.\exists x_{n+1}.\left(\bigwedge_{i \neq j} \neg(x_i = x_j)\right)$$
   In fact, $\mathcal{M} \vDash \psi_{n+1}$ iff there exist $d_1, \ldots, d_{n+1} \in D$ such that $d_i \neq d_j$ whenever $i \neq j$, i.e. all $d_1, \ldots, d_{n+1}$ are different.

3. Let φ_n be the formula
   $$\exists x_1.\cdots.\exists x_n.\left[\left(\bigwedge_{i \neq j} \neg(x_i = x_j)\right) \wedge \forall y.\left(\bigvee_{i=1}^{n}(y = x_i)\right)\right]$$
   Then $\mathcal{M} \vDash \varphi_n$ iff there exist $d_1, \ldots, d_n \in D$ such that
   $$\mathcal{M} \vDash \bigwedge_{i \neq j} \neg(d_i = d_j) \wedge \forall y.\left(\bigvee_{i=1}^{n}(y = d_i)\right)$$
   The first conjunct says that all the d_i are distinct, and the second holds iff for any $d \in D$
   $$\mathcal{M} \vDash (d = d_1 \vee \cdots \vee d = d_n)$$
   i.e. iff every $d \in D$ is among $d_1, \ldots, d_n$.

   Thus $\mathcal{M} \vDash \varphi_n$ iff $D = \{d_1, \ldots, d_n\}$ for some $d_1, \ldots, d_n$.

6.4. WORKED EXAMPLES

Example 6.4.6

1. Show that the following formula is false in any two-element model, but true in some three-element model:

$$\exists x.\exists y.\exists z.(P(x) \land Q(x) \land P(y) \land \neg Q(y) \land \neg P(z))$$

2. Find a formula without equality which is false in some five-element model, but true in any model of less number of elements.

Solution

1.
$$\mathcal{M} \models \exists x.\exists y.\exists z.(P(x) \land Q(x) \land P(y) \land \neg Q(y) \land \neg P(z))$$

iff for some $d_1, d_2, d_3 \in D$

$$\mathcal{M} \models P(d_1) \land Q(d_1) \land P(d_2) \land \neg Q(d_2) \land \neg P(d_3)$$

Then

$d_1 \neq d_3$	since	$\mathcal{M} \models P(d_1) \land \neg P(d_3)$
$d_1 \neq d_2$	since	$\mathcal{M} \models Q(d_1) \land \neg Q(d_2)$
$d_2 \neq d_3$	since	$\mathcal{M} \models P(d_2) \land \neg P(d_3)$

Thus the formula cannot be true in a two-element model.

On the other hand, it has a three-element model. That is, we can take $\mathcal{M}_0$ with the domains $\{d_1, d_2, d_3\}$ such that:

$\pi_{\text{pred}}(P)(d) = \top$ iff $d = d_1$ or $d = d_2$;

$\pi_{\text{pred}}(Q)(d) = \top$ iff $d = d_1$.

Then obviously

$$\mathcal{M}_0 \models P(d_1) \land Q(d_1) \land P(d_2) \land \neg Q(d_2) \land \neg P(d_3)$$

and thus the formula is true.

2. The idea is the same. We can write a formula saying that a model contains at least five different elements, and then take its negation. So let

$$\varphi = \exists x_1.\exists x_2.\exists x_3.\exists x_4.\exists x_5.(P(x_1) \land Q(x_1) \land R(x_1) \land \\ \neg P(x_2) \land Q(x_2) \land R(x_2) \land P(x_3) \land \neg Q(x_3) \land R(x_3) \land \\ \neg P(x_4) \land \neg Q(x_4) \land R(x_4) \land \neg R(x_5))$$

Similarly to (a) we have:

$\mathcal{M} \models \varphi$ only if D has at least five elements. Now take the formula $\neg\varphi$. Then $\mathcal{M} \models \neg\varphi$ if $card\ D < 5$ ('card D' means 'the number of elements in D'). On the other hand, $\neg\varphi$ is false (i.e. φ is true) in some five-element model. That is, let $D = \{d_1, d_2, d_3, d_4, d_5\}$

$$\pi_{\text{pred}}(P)(d) = \top \text{ iff } d = d_1 \text{ or } d = d_3; \pi_{\text{pred}}(Q)(d) = \top$$
$$\text{iff } d = d_1 \text{ or } d = d_2; \pi_{\text{pred}}(R)(d) = \top \text{ iff } d \neq d_5$$

and consider a model $\mathcal{M}_0$ with these D, π_{pred}.

Then

$$\mathcal{M}_0 \models P(d_1) \wedge Q(d_1) \wedge R(d_1) \wedge \neg P(d_2) \wedge Q(d_2) \wedge R(d_2) \wedge$$
$$P(d_3) \wedge \neg Q(d_3) \wedge R(d_3) \wedge \neg P(d_4) \wedge \neg Q(d_4) \wedge R(d_4) \wedge \neg R(d_5)$$

Hence $\mathcal{M}_0 \models \varphi$.

Example 6.4.7 *Let $\varphi(x_1, \ldots, x_n)$ be a quantifier-free formula in a language without functional symbols and constants. Prove that if $\forall x_1. \cdots. \forall x_n.\varphi(x_1, \ldots, x_n)$ holds in any n-element model, then it holds in any model.*

Solution
Assume that $\forall x_1. \cdots. \forall x_n.\varphi(x_1, \ldots, x_n)$ holds in any n-element model. Consider an arbitrary model $\mathcal{M}$, such that $card\ \mathcal{M} \neq n$, and show that $\mathcal{M} \models \forall x_1. \cdots. \forall x_n.\varphi(x_1, \ldots, x_n)$ as well. Then two cases are possible, (1) and (2) below.

1. $card\ \mathcal{M} < n$. Let $D = \{a_1, \ldots, a_m\}$ ($m < n$) be the domain of $\mathcal{M}$. Then we can put $(n-m)$ new elements into D, in such a way that our formula would keep its truth value. So we introduce new elements $b_1, \ldots, b_{n-m}$ and make them behave in the same way as a_1, say. That is, we do the following:

    - Take $D' = \{a_1, \ldots, a_m, b_1, \ldots, b_{m-n}\}$.
    - Define the mapping $c \mapsto c^0$ from D' to D sending every b_j to a_1 and every a_i to itself.
    - If $\pi_{\text{pred}}(P)$ is an interpretation of a k-place predicate symbol P in $\mathcal{M}$, we define its new interpretation $\pi'_{\text{pred}}(P)$:

6.4. WORKED EXAMPLES

(a1) $\pi'_{\text{pred}}(P)(c_1, \ldots, c_k) = \pi_{\text{pred}}(P)(c_1^0, \ldots c_k^0)$.

Now let $\mathcal{M}'$ be the model based on D', π'_{pred}. Then for any $c_1, \ldots, c_s \in D'$, and for any quantifier-free formula $\psi(x_1, \ldots, x_s)$, we have

(a2) $\mathcal{M}' \vDash \psi(c_1, \ldots, c_s)$ iff $\mathcal{M} \vDash \psi(c_1^0, \ldots, c_s^0)$.

This is proved by induction on ψ. The base is given by (a1) because

$$\mathcal{M}' \vDash P(c_1, \ldots, c_k) \text{ iff } \pi'_{\text{pred}}(P)(c_1, \ldots, c_k) = \top \text{ iff}$$
$$\pi_{\text{pred}}(P)(c_1^0, \ldots, c_k^0) = \top \text{ iff } \mathcal{M} \vDash P(c_1^0, \ldots, c_k^0)$$

Other cases are all alike, so we consider only $\psi = \psi_1 \vee \psi_2$:

$$\mathcal{M}' \vDash \psi_1(c_1, \ldots, c_k) \vee \psi_2(c_1, \ldots, c_k)$$
$$\text{iff } \mathcal{M}' \vDash \psi_1(c_1, \ldots, c_k) \text{ or } \mathcal{M}' \vDash \psi_2(c_1, \ldots, c_k)$$
$$\text{iff } \mathcal{M} \vDash \psi_1(c_1^0, \ldots, c_k^0) \text{ or } \mathcal{M} \vDash \psi_2(c_1^0, \ldots, c_k^0)$$
$$\text{iff } \mathcal{M} \vDash \psi_1(c_1^0, \ldots, c_k^0) \vee \psi_2(c_1^0, \ldots, c_k^0)$$

From (a2) we obtain

$$\mathcal{M} \vDash \forall x_1. \cdots . \forall x_n. \varphi(x_1, \ldots, x_n)$$

iff for any $c_1, \ldots, c_n \in D$, $\mathcal{M}' \vDash \varphi(c_1, \ldots, c_n)$ (since $c_i^0 = c_i$).
But the latter holds because $\mathcal{M}' \vDash \forall x_1. \cdots . \forall x_n. \varphi(x_1, \ldots, x_n)$ by our assumption.
Therefore $\mathcal{M} \vDash \forall x_1. \cdots . \forall x_n. \varphi(x_1, \ldots, x_n)$.

2. card $\mathcal{M} > n$. Let D again be the domain of $\mathcal{M}$, π_{pred} be the interpretation function. To show that $\mathcal{M} \vDash \forall x_1. \cdots . \forall x_n. \varphi(x_1, \ldots, x_n)$ we have to consider arbitrary $d_1, \ldots, d_n \in D$ and show

(b1) $\mathcal{M} \vDash \varphi(d_1, \ldots, d_n)$.

We take these $d_1, \ldots, d_n$ and restrict our $\mathcal{M}$ to the new domain $D' = \{d_1, \ldots, d_n\}$. More precisely, we define

(b2) $\pi'_{\text{pred}}(P)(c_1, \ldots, c_k) = \pi_{\text{pred}}(P)(c_1, \ldots, c_k)$ (for any $c_1, \ldots, c_k \in D'$).
Let $\mathcal{M}'$ be the model formed by D', π'_{pred}. Take any quantifier-free $\psi(x_1, \ldots, x_s)$. Then the following holds:

(b3) for any $c_1, \ldots, c_s \in D', \mathcal{M}' \vDash \psi(c_1, \ldots, c_s)$ iff $\mathcal{M} \vDash \psi(c_1, \ldots, c_s)$.

This is proved by induction on ψ similarly to (a2), with (b2) as a starting point.

Now $\mathcal{M}' \vDash \forall x_1.\cdots.\forall x_n.\varphi(x_1,\ldots,x_n)$ by (1) because *card* $\mathcal{M}' \leq n$ (note that some of the $d_1,\ldots,d_n$ may be identical).

Hence $\mathcal{M}' \vDash \varphi(d_1,\ldots,d_n)$ and thus $\mathcal{M} \vDash \varphi(d_1,\ldots,d_n)$ by (b3). That is, we have proved (b1).

7

FORWARD AND BACKWARD PREDICATE RULES

Let us summarize our knowledge so far. In propositional logic we can use backward rules to answer the question

$$\text{data} \quad \vdash \quad \text{query}$$

i.e. does the query follow logically from the data? We have seen three techniques for solving this problem. In Chapter 1, we used an automatable method based on constructing a truth table for an implication from the data to the query, and checking that the implication was a tautology. In Chapter 3, we presented the 'natural deduction' forward reasoning rules, which require that one make guesses when reasoning, and are hence not easily automatable. These forward rules were then developed into the backward rules for the propositional logic, in Chapter 4.

The truth table method always gives us an answer. We also know how many steps we need to get the answer. The backward method may loop but is in principle as mechanical as the truth table method. The forward rules method requires ingenuity and one is not guaranteed to find an answer. The following basic notions are shared by all three methods:

- Consistency of a set of formulae (same as satisfiability), meaning all formulae in the set can be made true in a single interpretation.

- Tautology (or logical theorem), meaning a formula which is always true.

These notions carry over to the predicate logic of the previous chapter. As we stated there, all the propositional rules of reasoning are valid in predicate logic, since predicate logic extends propositional logic with richer means of expression. The basis of this greater expressiveness is the use of quantifiers, for which we require additional reasoning rules. In the previous chapter we encountered some quantifier rules, mainly having to do with pulling quantifiers out of a formula. This chapter will introduce further quantifier reasoning rules, so that we are able to get an answer to a question from data, this time expressing both the data and the query in predicate logic:

$$\text{predicate data} \quad \vdash \quad \text{predicate query}$$

This will require a combination of the quantifier rules, together with the rules at the end of Chapter 4.

7.1 Reasoning with variables

Our first step is to rewrite the data and query in a ready to compute form. We have to do the translation of $A \vee B$ as $(A \to \bot) \to B$, and $\neg A$ as $A \to \bot$ as we did for the propositional case. We can also translate $\exists x.\, A(x)$ as $\neg \forall x.\, \neg A(x)$ but still the presence of quantifiers everywhere in the formula of the database is not good enough for ease of computation. Our first task is to stylize all data and queries into the prenex normal form, by putting all quantifiers at the front:

$$(\forall \mathrm{x}.\exists \mathrm{y}.\forall \mathrm{z}.\ldots)[\text{matrix}]$$

with no quantifiers in the matrix. Recall that the string of quantifiers is called the prefix. In the previous chapter we looked at rules which allow us to rewrite any formula into the prenex normal form, with all quantifiers at the front. We can use the propositional rewrite rules to write the quantifier-free matrix in a form ready for computation, i.e. in the form of $\mathcal{B}$-clauses (with atomic heads etc.) Thus for example we can have a database such as

1. ∀x.∃y. [A(x, y) → Q(x, x)]

2. A(a, a)

with a query of [∃z. A(z, a)]→Q(a, a). The above is the best formulation we can give using our rewrite rules. All quantifiers are at the

7.1. REASONING WITH VARIABLES

front, and the quantifier-free matrix is rewritten in $\mathcal{B}$-clausal form with atomic heads. For computing purposes this is not enough. The existential quantifiers present in the prefix are a big nuisance. We would like to be able to write the data with *universal* quantifiers only. Now we know that $\exists y. \varphi(y) \Leftrightarrow \neg \forall y. \neg \varphi(y)$, so we might replace each existential quantifier in the prefix by a negated universal quantifier. In the case of sentence 1 above we could write

$\forall$x. $\neg\forall$y. $\neg$[A(x, y)$\rightarrow$Q(x, x)]

This does not solve the problem because we now have a negation stuck in the middle of the quantifier prefix. We want to have a *universal prenex* form of

$$(\forall x.\forall y.\forall z.\ldots)[\text{matrix}]$$

We shall achieve this by using *Skolem functions*, which we shall cover later in this chapter.

Let us assume for a moment that we can always rewrite data in the above universal prenex form. The big advantage, as far as computation is concerned, is that theoretically it allows us to reduce the problem of whether a query succeeds from some data from predicate logic into propositional logic. Here is an example of how it is done, which should give an intuitive notion of what happens.

Example 7.1.1 *From the items of data:*

1. *A person is happy if he loves someone*

2. *Carol loves David*

we must show that the query 'Carol is happy' follows. We translate the English sentences into predicate logic, using the predicates loves(x,y) *for* x *loves* y *and* happy(x) *for* x *is happy. The data becomes:*

1. $\forall$x. [[$\exists$y. loves(x, y)] $\rightarrow$ happy(x)]

2. loves(Carol, David)

and we must show the sentence happy(Carol). *To begin computation we have to rewrite, and first we push the quantifiers outside. The first sentence in the data becomes*

$\forall$x.$\forall$y. [loves(x, y) $\rightarrow$ happy(x)]

by applying the quantifier rule

$$\forall x.[\varphi(x) \to \gamma] \equiv [\exists x.\varphi(x)] \to \gamma$$

from Figure 6.1 to move the $\exists y$ from the antecedent to become $\forall y$ outside the implication, since y does not appear in happy(x). *We are fortunate that we now have this sentence in the universal prenex form, and we do not need to concern ourselves with Skolem functions for now. The second sentence in the data contains no quantifiers, and is thus already in the universal prenex form, as is the conclusion we wish to prove. The data is thus*

1. $\forall x.\forall y.$ [loves(x, y) $\to$ happy(x)]

2. loves(Carol, David)

and we must show the sentence happy(Carol).

The reduction of this predicate formula to the propositional case is done by substituting all the possible constant names that we have, for the universally quantified variables in the first formula in the data.

Thus since the data contains only universal quantifiers

$$\forall x.\forall y.[matrix]$$

we write down all the possible combinations $\theta_1, \ldots, \theta_n$ of names in the data and replace the universally quantified sentence with

$$[matrix]\theta_1 \wedge \cdots \wedge [matrix]\theta_n$$

where θ_i is a variable assignment such as [x $\mapsto$ David, y $\mapsto$ Carol]. *Our first data sentence therefore is replaced by n variable-free sentences—in this case, since there are just two names in the example, there are four different ways to assign the names to the two variables; hence $n = 4$:*

1.

	x $\mapsto$	y $\mapsto$		
1	Carol	Carol	loves(Carol, Carol)	$\to$ happy(Carol)
2	Carol	David	loves(Carol, David)	$\to$ happy(Carol)
3	David	Carol	loves(David, Carol)	$\to$ happy(David)
4	David	David	loves(David, David)	$\to$ happy(David)

2. loves(Carol, David)

7.1. REASONING WITH VARIABLES

The question then is whether we can use propositional rules, especially backward ones, to get `happy(Carol)`. *Clearly in this case we can, by matching the query with the head of implication 1.2, and asking for* `loves(Carol, David)` *which succeeds.*

The process of replacing variables by constants, as seen in the preceding example, and by other variables, is known as *substitution*. In Definition 6.3.2, variable assignments were defined as mappings from $\mathcal{L}_v$ to $\mathcal{D}$. Substitutions are a little more general than that, being *syntactic* rather than semantic operations.

Definition 7.1.2 (Substitution) *A* substitution *is a finite set of pairs (*bindings*) of the form v/t where v is a variable and t is a term. A typical substitution is thus*

$$\sigma = \{v_1/t_1, \ldots, v_n/t_n\}$$

We also write $\sigma(v_i) = t_i$, $i = 1, \ldots, n$.

The application *of a substitution σ to a formula A is the simultaneous replacement of every free occurrence of each v_i in σ by t_i, denoted $A\sigma$ and known as a* substitution instance.

In order for the substitution to be done correctly, we must assume that the free variables in the terms t_i are different from the bound variables of the formula A, otherwise the bound variables of A need to be renamed.

There are a number of properties of substitutions which we shall either require or prefer. A required property is that the substitution be *functional*. This means that the variables v_i in the substitution are distinct. Thus $\sigma_1 = \{x/a, y/b, x/c\}$ is not functional, whereas $\sigma_2 = \{x/a, y/b, z/c\}$ is. A preferred property is that of *idempotency*, where for any formula A, we have $A\sigma = (A\sigma)\sigma$. For this to be the case, none of the variables v_i may appear in any of the terms t_j, unless $v_i \mapsto v_i$. Thus $\sigma_1 = \{x/a, y/b\}$ is idempotent, whereas $\sigma_2 = \{x/a, y/x\}$ is not.

Substitutions may be joined together, by means of a composition operator ∘, defined as follows. Given two substitutions, both functional and idempotent,

$$\sigma_1 = \{u_1/s_1, \ldots, u_m/s_m\}$$
$$\sigma_2 = \{v_1/t_1, \ldots, v_n/t_n\}$$

the composition of σ_1 and σ_2, written $\sigma_1 \circ \sigma_2$, is given by

$$\sigma_1 \circ \sigma_2 = \{u_1/s_1\sigma_2, \ldots, u_m/s_m\sigma_2\} \cup \{v_j/t_j \mid v_j/t_j \in \sigma_2 \\ \text{and } v_j \notin \{u_1, \ldots, u_m\}\}$$

Composition possesses the following properties:

1. $\sigma_1 \circ \sigma_2$ is functional

2. $A(\sigma_1 \circ \sigma_2) = (A\sigma_1)\sigma_2$ for any formula A

3. $(\sigma_1 \circ \sigma_2) \circ \sigma_3 = \sigma_1 \circ (\sigma_2 \circ \sigma_3)$.

Because of the asymmetric definition of $\circ$, note, however, that in general $\sigma_1 \circ \sigma_2 \neq \sigma_2 \circ \sigma_1$, i.e. $\circ$ is not commutative.

7.1.1 Introduction to unification

Returning to the problem of computing from predicate sentences, the next step is to ask: if we have several clauses in the universal prenex form, each with a number of universally quantified variables, and many constants in the language, do we have to write a possibly (or practically) infinite list of clauses just to perform the computation?

The answer is no. Any proof that a query follows from some data must contain a finite number of inferences, or applications of rules, so that in any proof, we can only use a finite number of the possibly infinite number of substitutions of constants for variables. In fact, we need only substitute exactly those constants which we will need to carry out the proof. To do this we do not begin by making substitutions, but retain the universal prenex form of the clauses until the moment during the proof when we discover that we need to use the clause. It is only at this stage that we generate an instance of the clause with constants substituted for the variables.

This process is best illustrated by an example, so let us check again whether Carol is happy. Our data is

1. `loves(x, y)` $\rightarrow$ `happy(x)`

2. `loves(Carol, David)`

We have chosen to omit the universal quantifiers from the first clause, since all variables appearing in the data clauses must be universally quantified. There is an implicit agreement that any possible combination

7.1. REASONING WITH VARIABLES

of constants can be substituted for the variables x and y in the clause. To demonstrate that happy(Carol) follows from the data, we again try to match the query against the head of some rule in the data. There is only one head which could match the query, namely happy(x), provided that the variable x was substituted for by Carol. We make that substitution, but leave the choice of the substitution for y until later. The new clause is added to the data

   loves(Carol, y) → happy(Carol)

and is used as before, generating a subquery loves(Carol, y) which, we must demonstrate, follows from the extended data. This new query is different from all the others we have previously seen, in that it contains a variable, y. Without worrying about the meaning of this variable for the moment, let us continue with the proof. We need to find a head in the data which will match loves(Carol, y). Fortunately such a head exists, loves(Carol, David), provided we can make a substitution of David for y. Because we left the choice of a substitution for y open when we generated the new clause loves(Carol, y) → happy(Carol), we may now generate a further clause

   loves(Carol, David) → happy(Carol)

as a retrospective justification of our choice to substitute David for y now. Our proof is complete, so we have demonstrated happy(Carol) by only making the substitution [x ↦ Carol, y ↦ David].

The process of generating the substitutions [x ↦ Carol] and [y ↦ David] is known as *unification*, because when we have two sentences such as happy(Carol) and happy(x), we wish to blend (or unify) the two sentences together, so that they may be regarded as identical. Once the two sentences are unified, we may use the propositional reasoning rules on them, so that being able to unify sentences is important. Unification is at the heart of almost all the reasoning methods for first-order logic.

Although the two examples of unification in the illustration above involved just making one variable substitution in one sentence to make it identical to the other, ground, sentence, there is much more to unification. Suppose that we have the two sentences

   loves(x, Brenda)   and   loves(brother-of(y), y).

By means of unification, we can make substitutions in these sentences so that they turn into a single unified sentence. Informally, unification proceeds as follows. The only constant we have is **Brenda**, which is the second argument to `loves` in the first sentence. If the two sentences are to be unified, the second argument to `loves` in the second sentence must become equal to **Brenda**; hence we can generate the substitution [y ↦ **Brenda**]. Now we substitute throughout the two sentences for y, generating:

   `loves(x, Brenda)` and `loves(brother-of(Brenda), Brenda)`.

Now we have unified the second arguments of `loves`, we move to the first arguments. In the first sentence the first argument is a variable, x, and in the second sentence it is the term `brother-of(Brenda)`. We must therefore generate the substitution [x ↦ brother-of(Brenda)] and replace x throughout the sentences, leaving us with

   `loves(brother-of(Brenda), Brenda)` and
   `loves(brother-of(Brenda), Brenda)`.

By means of the combined substitution [x ↦ brother-of(**Brenda**), y ↦ **Brenda**], we have unified the two sentences into a single unified sentence.

Definition 7.1.3 (Unifiers) *A substitution σ is a* unifier *of two atomic sentences A and B iff $A\sigma = B\sigma$. A substitution σ is a* most general unifier *of two atomic sentences A and B iff it is a unifier of A and B, and for any other unifier σ' of A and B, there is a substitution θ such that $\sigma' = \sigma \circ \theta$.*

7.1.2 A unification algorithm

The algorithm that we shall give is just one of many for computing unifiers. The idea is to work through the syntactic construction of two atomic sentences to see where they differ, and whether the differences can be overcome by making substitutions for the variables. Should a difference be found which cannot be overcome by a set of variable substitutions, then the sentences are not unifiable, otherwise a set of variable substitutions (the most general unifier) is output.

The unification algorithm works on a stack S of pairs of terms, and goes through one cycle for each pair on the stack. A unifier θ is used to store the substitutions as they arise. We present the algorithm in an informal pseudo-code in Figure 7.1. We assume that the algorithm is to work on two atomic sentences A_1 and A_2. The algorithm begins by

begin
  $S :=$ empty stack
  put all the term pairs from A_1 and A_2 on S
  $\theta := \{\}$
  while S is not empty
    pop the term pair $\langle s_1, s_2 \rangle$ from S
    $e_1 = s_1\theta;\ e_2 = s_2\theta$
    if e_1 and e_2 are both constants and $e_1 \neq e_2$
      then fail
    else if e_1 and e_2 are both functions and e_1 and e_2 have
        different functors
      then fail
    else if either e_1 or e_2 is a function and the other is a constant
      then fail
    else if e_1 and e_2 are both functions and e_1 and e_2 have
        the same functor
      then push pairs of their corresponding arguments onto S
    else if e_1 and e_2 are both variables
      then $\theta := \theta \circ \{e_1/e_2\}$
    else if either e_1 or e_2 is a variable which occurs strictly
        within the other
      then fail
    else if e_1 is a variable
      then $\theta := \theta \circ \{e_1/e_2\}$
    else if e_2 is a variable
      then $\theta := \theta \circ \{e_2/e_1\}$
  end while
  return θ
end

Figure 7.1: The unification algorithm

placing the pairs of corresponding terms from A_1 and A_2 onto the stack S. The unifier θ is initialized to the empty substitution.

The algorithm proceeds to cycle through the following steps for each pair on the stack, until a failure point is reached, or the stack becomes empty: (i) the top pair of terms is popped from S, and the latest version of θ is applied to them, yielding terms e_1 and e_2; (ii) the terms are compared and if there is no way of matching them, failure is generated, otherwise an appropriate binding is added to θ. If e_1 and e_2 are functional terms which might unify, their term pairs are added onto the stack for further comparison.

Example 7.1.4 *We will unify the atomic sentences* p(x,f(x),a) *and* p(y,f(g(z),z)). *The initial stack is*

x, y
f(x), f(g(z))
a, z

The stack is not empty, so we pop the top pair off the stack, and apply θ (currently empty) to the pair, getting x *and* y. *The terms are compared, and since both are variables, we add* $\{x/y\}$ *to θ. The bottom of the cycle is reached, so we again see that the stack is not empty, and pop the top pair of terms off the stack, and apply θ. This time θ is not empty, so applying it to the top pair,* f(x) *and* f(g(z)), *yields* f(y) *and* f(g(z)). *These terms can now be compared, and as they are functional terms with the same functor (*f*), the term pair* y *and* g(z) *are pushed onto the stack:*

y, g(z)
a, z

Going back to the beginning of the cycle, we pop the top pair of terms off the stack, and apply θ, which has no effect, so we are left with comparing y *and* g(z). *Since* y *is a variable and* g(z) *is not, the binding* $\{y/g(z)\}$ *is added to θ. In the final run through the cycle, the pair of terms* a *and* z *are popped off the stack, θ is applied, and the terms compared. Since* z *is a variable and* a *is not, the binding* $\{z/a\}$ *is added to θ. The stack is now empty, and the algorithm ends, returning the composition* $\theta = \{x/y\} \circ \{y/g(z)\} \circ \{z/a\}$, *which is indeed the following simultaneous substitution* $\{x/g(a), y/g(a), z/a\}$.

Exercise 7.1.5 *For each of the following pairs of atomic sentences, decide whether or not they unify, and if the latter, what the most general unifier is.*

7.1. REASONING WITH VARIABLES

1. p(a,f(z)) *and* p(x,f(b))
2. q(f(g(x))) *and* r(c)
3. r(x,x,f(b)) *and* r(g(f(b)),g(y),y)
4. p(a,f(g(x))) *and* p(y,g(z))
5. q(b,x) *and* q(y,f(x))
6. q(f(a,x)) *and* q(f(y,f(z)))

7.1.3 The meaning of variables in clauses

When we were demonstrating by means of unification that Carol was happy, we generated a subquery which contained a variable, loves(Carol, y). Although we did not concern ourselves with y's nature then, we should do so now. Is y a universal or existential variable in loves(Carol, y)? It was generated from the clause

loves(Carol, y)→happy(Carol)

which is in the data as a universal prenex form clause. Hence there is an implicit universal quantifier on y:

$$\forall y.\ [\ \texttt{loves(Carol, y)} \to \texttt{happy(Carol)}] \quad (7.1)$$

so one might assume that y is a universally quantified variable in the query loves(Carol, y). However, one would be wrong. The clause (7.1) can be rewritten via the quantifier equivalences to

$$[\exists y.\ \texttt{loves(Carol, y)}] \to \texttt{happy(Carol)} \quad (7.2)$$

so that when the query happy(Carol) is asked, we must demonstrate that ∃y. loves(Carol, y) holds. From this we can see that the variables in the query are, in fact, *existentially* quantified.

The general form of the data and the query, and the variables they contain, is therefore:

Universal sentences (query):
A(x) → B(x)

B(x) → C(x)

Existential sentence (query):

C(y)

7. FORWARD AND BACKWARD PREDICATE RULES

The convention is that the data clauses and query are regarded as having the following quantification:

∀x. [A(x) → B(x)]

∃y. C(y)

∀x. [B(x) → C(x)]

Remember that because the x is quantified for each clause, the x in A(x) → B(x) is not the same as the x in B(x) → C(x).

The distinction becomes more central when we add clauses to the database from the goal. Consider the following example:

Universal sentence (data)
$A(x) \to B(x)$
$B(x) \to C(x)$
Existential sentence (query)
$A(y) \to C(y)$.

Since the query is an implication, we need to add $A(y)$ to the database and ask the query $C(y)$. Our database is now

1. $A(x) \to B(x)$

2. $B(x) \to C(x)$

3. $A(y)$

and the query is $C(y)$.

The original logical meaning was whether

$$\models \forall x.(A(x) \to B(x)) \land \forall x.(B(x) \to C(x)) \to \exists y.(A(y) \to C(y))$$

The same meaning must be retained after we put $A(y)$ into the database. To see what we have to do let us pull $\exists y$ up front. We get

$$\models \exists y.[\forall x.(A(x) \to B(x)) \land \forall x.(B(x) \to C(x)) \to (A(y) \to C(y))]$$

which is equivalent to

$$\models \exists y.[\forall x.(A(x) \to B(x)) \land \forall x.(B(x) \to C(x)) \land A(y) \to C(y)]$$

The last formula is the conjunction of the data implying (→) the goal.

We thus need to agree on two types of variables: universal variables (like the variable x) and existential variables (like the variable y). A database may have both kinds of variables. The goal has only existential variables. Assuming that Data = $\{A_1, \ldots, A_n\}$ then success in computation of Data ⊢? Goal means: ⊨ (∃ existential variables) $[\bigwedge_i (\forall$ universal variables) $A_i \rightarrow$ Goal].

What do we do then when we want to show a query of the form ∀z. Q(z), in which we must show that Q(z) follows from the data for all possible values of z? The way to do this is to choose a new constant say, k, which has not been mentioned anywhere before, in either the data or the query, and try to show that Q(k) follows from the data. If we succeed, then we have shown that

$$\frac{\mathcal{P}}{\text{Q(k)}}$$

where $\mathcal{P}$ represents the data, is logically valid. This means that all models for $\mathcal{P}$ are models for Q(k). Now since k is not mentioned anywhere in $\mathcal{P}$, any model for $\mathcal{P}$ is a model for Q(k), no matter what element in the domain k is mapped to. Hence any model for $\mathcal{P}$ is a model for Q(k) for all values of k; thus any model for $\mathcal{P}$ is a model for ∀z. Q(z). This gives us

$$\frac{\mathcal{P}}{\forall \text{z. Q(z)}}$$

as required. Therefore we have a new forward rule known as *universal generalization*:

If $\dfrac{\mathcal{P}}{\text{Q(k)}}$ and k is not mentioned in $\mathcal{P}$, then $\dfrac{\mathcal{P}}{\forall \text{z. Q(z)}}$

See the soundness theorem, 8.3.6.

7.1.4 Skolemization

We will now deal with *Skolem functions*. Earlier in this chapter we said that we would need such functions to eliminate existential quantifiers from clauses such as

∀x.∃y. [A(x, y) → Q(x, x)]

so as to present all clauses in the universal prenex form. Consider the formula ϕ = ∀x.∃y.R(x,y). In every model in which this formula is true, for every name a, there must exist a y such that R(a,y) is true. Expanded out, this is

$$[\exists y. R(a_1,y)] \land [\exists y. R(a_2,y)] \land \ldots$$

Of course the y promised by $[\exists y. R(a_i, y)]$ depends on a_i. Let $f(a_i)$ be a function to choose such a y. The function f is known as a Skolem choice function; for each x it chooses y = f(x) such that R(x,y) holds.

We can thus write $\varphi = \forall x. R(x,f(x))$. The relationship between φ and ϕ is that

1. $\varphi \to \phi$ is a theorem of logic,

2. in any model in which ϕ is true, f can be chosen in such a way that φ is true.

The second property yields

For every formula ψ, $\dfrac{\phi}{\psi}$ succeeds if and only if $\dfrac{\varphi}{\psi}$ succeeds

We really should write $\varphi = \forall x. R(x,f_\phi(x))$, since the Skolem function is introduced especially for the formula ϕ. If we had a further formula θ to Skolemize, we would introduce a new Skolem function f_θ.

In practice, the predicate R will have an intended meaning and so the Skolem function f will also be endowed with a meaning. For example, ϕ might mean 'every man is married to some woman', i.e.

$$\phi = \forall x. [\text{man}(x) \to \exists y. (\text{woman}(y) \land \text{married}(x, y))]$$

If we pull the quantifiers to the front, we get

$$\phi = \forall x.\exists y. [\text{man}(x) \to (\text{woman}(y) \land \text{married}(x, y))]$$

Taking the Skolem function f, we have

$$\varphi = \forall x. [\text{man}(x) \to (\text{woman}(f(x)) \land \text{married}(x, f(x)))]$$

We can read f(x) as 'the first wife of x'. In fact there is nothing to stop us from using wife-of(x) instead of f(x):

$$\varphi = \forall x. [\text{man}(x) \to (\text{woman}(\text{wife-of}(x)) \land \text{married}(x, \text{wife-of}(x)))]$$

When we have an existential quantifier within the scope of more than one universal quantifier, the Skolem function takes all of the universally

7.1. REASONING WITH VARIABLES

quantified variables as its arguments. So that to eliminate the existential quantifiers from ϕ where

$$\phi = \forall x.\forall y.\exists z.\, [R(x,\ y) \wedge Q(z)]$$

we do this by taking a function $g(x,y)$ to replace z, giving us

$$\varphi = \forall x.\forall y.\, [R(x,\ y) \wedge Q(g(x,\ y))]$$

This example shows that wholesale automatic Skolemization loses information present in the original formula. We can rewrite ϕ to be

$$\phi' = \forall x.\forall y.\, [R(x,\ y) \wedge \exists z.\, Q(z)]$$

since the z does not appear in $R(x,\ y)$. Similarly, we can remove $Q(z)$ from the scope of the $\forall x$ and $\forall y$ because neither x nor y appears in $Q(z)$. Hence

$$\phi'' = [\forall x.\forall y.\, R(x,\ y)] \wedge [\exists z.\, Q(z)]$$

We can now see that the existential variable does not depend on x or y at all, and if Skolemization were to take place on this rewritten version of ϕ, we could use a 0-place function, i.e. a constant h to replace the quantifier, yielding

$$\varphi'' = [\forall x.\forall y.\, R(x,\ y)] \wedge [Q(h)]$$

This second Skolem formula, φ'', contains the information that the formula is satisfiable in a model with only one element for which Q is true. The original Skolem formula, φ, did not. The information was lost when the original Skolemization took place. This is why we stated above that $\varphi \rightarrow \phi$ is a theorem of logic, and not that $\varphi \leftrightarrow \phi$ is a theorem of logic.

Exercise 7.1.6

1. Skolemize this prenex normal form sentence:

$$\forall x.\exists y.\forall z.\forall u.\exists v.\, [R(x,\ z) \rightarrow \neg Q(u,\ v)]$$

2. Where necessary, pull the quantifiers out of, and Skolemize, the following sentences:

    (a) $\forall x.\, [\exists y.\, A(x,\ y) \rightarrow B(x)]$
    (b) $\neg[\forall x.\, A(x) \wedge \forall y.\, B(y)]$

(c) $\exists x.\exists y.\ [A(x) \wedge \forall z.\ B(z, y)]$

(d) $\forall x.\forall y.\exists z.\forall u.\forall v.\exists t.\ [A(x, y) \to B(x, t) \wedge C(z)]$

The above discussion showed essentially how to eliminate existential quantifiers from a formula: put it in prenex normal form and then Skolemize.

Thus, given the query
$$\gamma \vdash ? \alpha$$
we can Skolemize γ into a formula $\gamma^\forall$ in universal normal form, i.e. $\gamma = \forall x_1.\cdots,\forall x_n.\gamma_0$ where γ_0 is a formula without quantifiers, such that the following holds:

(∗) $\quad \gamma \vdash \alpha$ iff $\gamma^\forall \vdash \alpha$

Our next question is what is the desired ready for computation form of the formula α?

We want to rewrite α into a formula $\alpha^\exists$ of the form $\alpha^\exists = \exists x_1.\cdots.\exists x_n.\alpha_0$, where α_0 is without quantifiers, such that for all wffs β we have

(∗∗) $\quad \beta \vdash \alpha$ iff $\beta \vdash \alpha^\exists$

To do this, observe that
$$\beta \vdash \alpha \text{ iff } \neg\alpha \vdash \neg\beta$$

Skolemize $\neg\alpha$ into a formula $\alpha' = \forall x_1.\cdots.\forall x_n.\alpha'_0$.

Thus we have for all β,
$$\alpha' \vdash \neg\beta \text{ iff } \neg\alpha \vdash \neg\beta$$

and hence for all β
$$\beta \vdash \neg\alpha' \text{ iff } \beta \vdash \alpha$$

but $\neg\alpha' \equiv \exists x_1.\cdots.\exists x_n.\neg\alpha'_0$.

Hence let $\alpha^\exists = \neg\alpha'$ and this is the formula we are looking for.

Example 7.1.7 (Preparing for computation) *In Section 8.3 we shall define the notion of $\mathcal{B}$-computation, for checking whether $A \vdash ?B$, for arbitrary formulae of predicate logic A and B. The $\mathcal{B}$-computation procedure requires that A and B have the form of $\mathcal{B}$-clauses (Definition 8.3.1) and A is universally quantified and B is existentially quantified. We therefore need to know how to rewrite an arbitrary problem $A \vdash ?B$ into an equivalent problem $A^{*\forall} \vdash ?B^{*\exists}$ such that $A^{*\forall}$ is a universally quantified $\mathcal{B}$-clause and $B^{*\exists}$ is an existentially quantified $\mathcal{B}$-clause. We already know how to do this, as follows:*

1. Skolemize A into A' of the form $\forall x_1.\cdots.\forall x_n.A'_0$ where A'_0 contains no quantifiers.

2. A'_0 can be rewritten using propositional translation $*$ into A_0^* as in Definition 4.3.1.
   $A^{*\forall}$ can now be defined as $A^{*\forall} = \forall x_1.\cdots.\forall x_n.A_0^*$.

3. Find $B_1 = (\neg B)^{*\forall} = \forall y_1.\cdots.\forall y_m.B_0^*$ using steps (1) and (2) above. Let $B^{*\exists} = \neg B_1 = \exists y_1.\cdots.\exists y_m.\neg B_0^*$.

We have
$$A \vdash B \text{ iff } A^{*\forall} \vdash B^{*\exists}$$

Here is an example:
$$\forall x.\exists u.R(x,u) \vdash ? \exists u.\forall x.R(x,u)$$

becomes
$$\forall x.R(x,f(x)) \vdash ? \exists u.R(g(u),u)$$

where f and g are Skolem functions.

In comparison
$$\exists u.\forall x.R(x,u) \vdash ? \forall x.\exists u.R(x,u)$$

becomes
$$\forall x.R(x,c) \vdash ? \exists u.R(d,u)$$

where c,d are constants.

7.2 Reasoning forwards

We are now in a position to produce proofs in predicate logic that a conclusion follows from some assumptions (or equivalently, that a query succeeds from some data) using forward reasoning rules. We make use of all the forward reasoning rules for propositional logic, as presented in Chapter 3, as well as the quantifier equivalences of Figure 6.1 of the preceding chapter. The additional rules of reasoning that we have introduced in this chapter are:

- Universal instantiation ($\forall$ elimination)

$$\frac{\forall x.\ A(x)}{A(c)}$$

where c is a constant in the language.

- Universal generalization ($\forall$ introduction)

$$\text{If } \frac{\mathcal{P}}{\mathtt{Q(c)}} \text{ and c is not mentioned in } \mathcal{P}, \text{ then } \frac{\mathcal{P}}{\forall \mathtt{z.\ Q(z)}}$$

- $\exists$ introduction

$$\frac{\mathtt{A(t)}}{\exists \mathtt{x.\ A(x)}}, \text{ t a term of the language}$$

where c is a constant in the language.

- Skolemization ($\exists$ elimination)

$$\text{If } \frac{\mathcal{P}}{\exists \mathtt{x.\ A(x)}} \text{ and k is not mentioned in } \mathcal{P}, \text{ then } \frac{\mathcal{P}}{\mathtt{A(k)}}$$

We annotate the constant k as being introduced as a Skolem constant and we do not allow the use of the ($\forall$-introduction) rule to it.

We shall work with Skolemized sentences, so that there are no existentially quantified variables in the data. Recall that variables in the query are existentially quantified, however. The query is not to be Skolemized in the same way as the data. The universal generalization rule ($\forall$I) gives a hint as to how we remove quantifiers from queries. The ($\forall$I) states that if we can prove a sentence about an individual whose name does not appear in the data, then we can prove the same sentence about all individuals. This leads to the idea of using Skolem constants and functions in the query to represent *universally* quantified variables. This can easily be done by the Skolemization method we have already presented, but by applying it to a negated form of the query, and subsequently negating the Skolemized formula again. For example, suppose that we have the data $\forall \mathtt{z.\ loves(Bethan,z)}$ and we ask the query $\exists \mathtt{x}.\forall \mathtt{y.\ loves(x,y)}$. The data contains no existential quantifiers, and so is represented by the clause

1. loves(Bethan,z)

The query, on the other hand, contains both existential and universal quantifiers, and so must be Skolemized. As it is a query, we first

7.2. REASONING FORWARDS

negate it, to get ∀x.∃y. ¬loves(x,y), so that the universal quantifiers become existential, and vice versa. We can now Skolemize, getting ∀x. ¬loves(x,f(x)). This can then be negated again, to get ∃x. loves(x,f(x)), and thus we have a query in which the universally quantified variable has been replaced by a Skolem function. By convention, we can now omit the existential quantifier from the query, and understand that all variables in the query are existentially quantified. A simple unification between loves(Bethan,z) and loves(x,f(x)) yields the unifier {x/Bethan, z/f(x)} which is our answer.

Example 7.2.1 *Using these rules, we shall attempt to show that the following argument is valid:*

> *Some logicians teach at Imperial College. Anyone who teaches at Imperial College lives in London. All logicians who live in London are good. All good logicians support logic programming. Therefore some good logicians support logic programming.*

We must first translate the sentences above into logic, resulting in
1. ∃x.[logician(x) ∧ teaches-at(x,Imperial-College)]
2. ∀x.[teaches-at(x,Imperial-College) → lives-in-London(x)]
3. ∀x.[logician(x) ∧ lives-in-London (x) → good(x)]
4. ∀x.[logician(x) ∧ good(x) → support-LP(x)]
5. ∃x.[logician(x) ∧ good(x) ∧ support-LP(x)]

These logic sentences are in prenex normal form, but sentence 1 in the data and sentence 5 (the query) are not in universal prenex form. Only sentences in the data must be in universal prenex form, so sentence 1 must be Skolemized, introducing the Skolem constant **k**.

6. logician(k) ∧ teaches-at(k,Imperial-College)

Now we may begin reasoning. We use universal instantiation for **k** *in sentences 2, 3 and 4, which produces the following new sentences:*

7. teaches-at(k,Imperial-College) → lives-in-London(k)
8. logician(k) ∧ lives-in-London (k) → good(k)
9. logician(k) ∧ good(k) → support-LP(k)

Using the now familiar propositional forward rules we can reason from 6, 7, 8 and 9 that

10. `logician(k) ∧ good(k) ∧ support-LP(k)`

Finally, by applying existential introduction to sentence 10 we reach

5. `∃x. [logician(x) ∧ good(x) ∧ support-LP(x)]`

Example 7.2.2 *Here is another example of reasoning with forward rules. Is the following argument valid?*

> *Everything pleasant is either illegal or immoral. Programming in Pascal is unpleasant. Therefore programming in Pascal is neither illegal nor immoral.*

As for the good logicians above, we must translate the argument into logical form, giving us

1. `∀x. [pleasant(x) → illegal(x) ∨ immoral(x)]`
2. `¬pleasant(Pascal)`
3. `¬illegal(Pascal) ∧ ¬immoral(Pascal)`

These sentences are already in universal prenex form, so we can start to compute straight away. Using universal instantiation for the constant `Pascal` *in sentence 1 we get*

4. `pleasant(Pascal) → illegal(Pascal) ∨ immoral(Pascal)`

By now you should have realized that the conclusion does not *follow from the assumptions. One of the drawbacks of the forward reasoning rules method is that it does not indicate when the conclusion does not follow from the assumptions. To prove that the argument is not valid, we must provide a model to serve as a counter-example, i.e. a model which satisfies all the assumptions, but not the conclusion. In the propositional case, using truth tables, we could find a line in the truth table to serve as the counterexample. For the time being, we shall have to guess at a counterexample for invalid predicate arguments.*

Consider the model given by $\mathcal{M} = \langle \mathcal{D}, \pi_{\text{cons}}, \pi_{\text{func}}, \pi_{\text{pred}} \rangle$:

- $\mathcal{D} = \{p\}$

- π_{cons}: $\{\text{Pascal} \mapsto p\}$

- π_{func}: $\varnothing$

- π_{pred}: $\{\text{immoral} \mapsto \{p \mapsto \top\}\}$, *all other predicates are assigned* $\{p \mapsto \bot\}$ *by* π_{pred}.

7.2. REASONING FORWARDS

We should check the assumptions against this model. Is it true in model $\mathcal{M}$ that

  1. ∀x. [pleasant(x) → illegal(x) ∨ immoral(x)]

is satisfied? The answer is yes, since for x = p *(the only element in the domain) we get* $\bot \to \bot \vee \top$, *which reduces to* $\top$. *The second assumption,*

  2. ¬pleasant(Pascal)

is also satisfied, since the constant Pascal *maps to the domain element p, which is mapped to* $\bot$ *by* π_p. *This is then changed to* $\top$ *by the* ¬ *connective. Finally, since the assumptions are satisfied by* $\mathcal{M}$, *we must demonstrate that the conclusion is not satisfied by* $\mathcal{M}$:

  3. ¬illegal(Pascal) ∧ ¬immoral(Pascal)

Only immoral(Pascal) *maps to* $\top$ *in* $\mathcal{M}$, *so that sentence 3 reduces to* $\top \wedge \bot$ *which further reduces to* $\bot$. *Hence* $\mathcal{M}$ *is indeed a counterexample to the above argument.*

Exercise 7.2.3 *Demonstrate that the following argument is valid:*

> *All interrupt commands are undesirable. Some control commands are interrupt commands. Therefore some control commands are undesirable.*

Example 7.2.4 *This example indicates a common misconception about the meaning of the universal quantifier. Suppose that we have to show that the following argument is valid:*

> *All the lecturers like every student taking course 140. No lecturer likes John. Therefore John does not take course 140.*

One translation into universal prenex form is

  1. ∀x.∀y. [student(x) ∧ lecturer(y) ∧ takes(140,x) → likes(y, x)]
  2. ∀x. [lecturer(x) → ¬likes(x, John)]
  ---
  3. ¬[student(John) ∧ takes(140, John)]

To show the conclusion, we will show a rewritten form of it, namely

  3'. [student(John) ∧ takes(140, John)] → $\bot$

which we will attempt to show by adding the antecedent student(John) ∧ takes(140, John) to the assumptions, and try to derive ⊥:

    4.   student(John) ∧ takes(140, John)

Since we are interested in John we should substitute John *in assumptions 1 and 2, and try to derive* ⊥. *Notice that assumption 2 can, like the conclusion, be written to have* ⊥ *as its consequent:*

    2′.   ∀x. [lecturer(x) ∧ likes(x, John) → ⊥]

So to show ⊥, *by assumption 2 it is sufficient to show that for some* x, lecturer(x) ∧ likes(x, John). *You can try and show this for as long as you like, but you will not succeed. The conclusion does not follow from the assumptions. To see this, take a model in which John is a student and John is taking course 140, but there are no lecturers. Assumptions 1 and 2 will be true since the* lecturer(y) *(in the antecedents) will be false for all* y, *but the conclusion 3 will be false.*

A common misconception is that the statement 'every object has some property' implies the existence of at least one object. This is not the case. One might state that 'every common factor of two different prime numbers is even', despite the fact that two different prime numbers have, by definition, no common factor. You may feel that the translation of the lecturer problem from English into logic was incorrect, and that a further assumption, that there are teachers, should be added:

    0.   ∃x. lecturer(x)

The argument can now be demonstrated to be valid. Skolemizing the new assumption 0, with the constant k, *we get*

    5.   lecturer(k)

Applying universal instantiation to assumptions 1 (x=John, y=k) *and 2′* (x=k), *we get*

    6.   student(John) ∧ lecturer(k) ∧ takes(140, John)
         → likes(k, John)
    7.   [lecturer(k) ∧ likes(k, John)] → ⊥

Again trying to show ⊥, *we use the rewritten form of assumption 2, this time in its instantiated form 7, so that we must show* lecturer(k) ∧ likes(k, John). *Now applying* modus ponens *to sentences 4, 5 and 6 gives us* likes(k, John). *With* ∧ *introduction with sentence 5, we get* lecturer(k) ∧ likes(k, John) *and thus the conclusion is proved.*

7.3 Reasoning backward

As in the propositional case, we resolve all our difficulties by using backward rather than forward reasoning; we go from the conclusion to the assumptions. The assumptions are prepared in the universal prenex form, with universal quantifiers outside a quantifier-free matrix. The matrix is in a predicate form of the propositional $\mathcal{B}$-clauses. Recall the definition of $\mathcal{B}$-clauses from Chapter 5:

1. An atom q is a $\mathcal{B}$-clause, as is $\bot$.

2. If $A_1, \ldots, A_n$ are $\mathcal{B}$-clauses and q is atomic or $\bot$ then

$$A_1 \wedge \cdots \wedge A_n \to q$$

   is a $\mathcal{B}$-clause.

3. A query is a conjunction of $\mathcal{B}$-clauses, e.g.

$$A \wedge B \wedge \cdots \wedge C$$

   where A, B, C are $\mathcal{B}$-clauses.

The extension to the predicate case is simple, and involves only a change to the definition of an atomic clause. We divide our variables into two types: universal variables $\{x_1, x_2, x_3, \ldots\}$ and existential variables $\{y_1, u_1, y_2, u_2, \ldots\}$. This division is for the purpose of computation only.

Definition 7.3.1 (Predicate $\mathcal{B}$-clauses)

1. *An atom $p(t_1, \ldots, t_n)$, where p is a predicate symbol and the t_i are terms, is a $\mathcal{B}$-clause, as is $\bot$.*

2. *If $A_1, \ldots, A_n$ are $\mathcal{B}$-clauses and q is atomic or $\bot$ then*

$$A_1 \wedge \cdots \wedge A_n \to q$$

   is a $\mathcal{B}$-clause.

3. *A query is a conjunction of $\mathcal{B}$-clauses whose free variables are all existential, e.g.*

$$A \wedge B \wedge \cdots \wedge C$$

   where A, B, C are $\mathcal{B}$-clauses.

7. FORWARD AND BACKWARD PREDICATE RULES

4. *A database is a set of $\mathcal{B}$-clauses with possibly mixed free variables.*

The propositional backward reasoning rules were formally defined in Definition 5.4.4 as follows:

1. When A is an atom q we have $\langle \mathcal{P}, G \rangle \vdash_\mathcal{B} q$ iff either

    (a) $q \in \mathcal{P}$ or $\bot \in \mathcal{P}$, or
    (b) for some $B \to q \in \mathcal{P}$ or $B \to \bot \in \mathcal{P}$ we have $\langle \mathcal{P}, G \rangle \vdash_\mathcal{B} B$.

2. When A is a conjunction $B \wedge C$, we have $\langle \mathcal{P}, G \rangle \vdash_\mathcal{B} B \wedge C$ iff $\langle \mathcal{P}, G \rangle \vdash_\mathcal{B} B$ and $\langle \mathcal{P}, G \rangle \vdash_\mathcal{B} C$

3. When A is an implication $B \to q$ we have $\langle \mathcal{P}, G \rangle \vdash_\mathcal{B} B \to q$ iff $\langle \mathcal{P} \cup \{B\}, G \rangle \vdash_\mathcal{B} q$

4. The restart rule states that $\langle \mathcal{P}, G \rangle \vdash_\mathcal{B} q$ for an atom q if $\langle \mathcal{P}, G \rangle \vdash_\mathcal{B} G$, where G is the original goal of the computation.

The predicate form of these rules is essentially the same, with a few modifications to handle variables and quantifiers. First, when we have a conjunctive goal, such as $B \wedge C$, in the predicate case B and C may have common variables, such as b(u,y) and c(y,z) where y is a common variable. We must ensure that the variable substitutions which we generate when solving the first subgoal b(u,y) are applied to the second subgoal c(y,z) before we solve it. This ensures that incorrect answers are not generated. For example, if we have the assumptions

$$b(1,2) \quad b(1,3) \quad c(2,3) \quad c(1,4)$$

the subgoal b(u,y) will succeed with the substitutions $\{u/1, y/2\}$ and $\{u/1, y/3\}$. The subgoal c(y,z) will succeed with the substitutions $\{y/2, z/3\}$ and $\{y/1, z/4\}$. However, there is only one correct substitution for the whole goal b(u,y) $\wedge$ c(y,z), which is $\{u/1, y/2, z/3\}$. By applying the two answers for the subgoal b(u,y) to the second subgoal c(y,z), we generate two new subgoals c(2,z) and c(3,z). Only the former succeeds, with answer $\{z/3\}$ which, when composed with the answer to the first subgoal, gives the desired $\{u/1, y/2, z/3\}$.

When the goal is an atomic formula Q(s,t,...) where s and t etc. are terms then we have a number of possibilities:

- Q(p,q,...) is one of the assumptions, where p and q etc. are terms. Provided that Q(s,t,...) and Q(p,q,...) unify, with most general unifier θ, we succeed with answer θ.

- ⊥ is one of the assumptions, in which case we succeed with the empty unifier as the answer.

- $\varphi(\mathtt{p,q,u,v}\ldots) \to \mathtt{Q(p,q,\ldots)}$ is one of the assumptions, where p, q, u, v, etc. are terms. Provided that $\mathtt{Q(s,t,\ldots)}$ and $\mathtt{Q(p,q,\ldots)}$ unify, with most general unifier θ, and the subgoal $\varphi(\mathtt{p,q,u,v}\ldots)$ succeeds with answer σ, we succeed with answer $\theta \circ \sigma$.

- $\varphi(\mathtt{p,q,u,v}\ldots) \to \bot$ is one of the assumptions, where p, q, u, v, etc. are terms. Provided that the subgoal $\varphi(\mathtt{p,q,u,v}\ldots)$ succeeds with answer σ, we succeed with answer σ.

Finally, the restart rule must be modified to take variables into account. Recall that the variables in the goal are *existentially* quantified. When we attempt to show that a goal $\varphi(\mathtt{u})$ succeeds, we are really showing that the goal $\exists \mathtt{u}.\ \varphi(\mathtt{u})$ succeeds. We can justify the restart rule on the basis that $\psi \equiv \neg\psi \to \psi$, so that instead of asking the goal ψ, we ask the goal $\neg\psi \to \psi$. The implication is checked by adding $\neg\psi$ or rather $\psi \to \bot$ to the database, and asking for ψ. If we use the restart rule, we are attempting to show $\bot$ by asking for ψ. Because in predicate logic ψ is of the form $\exists \mathtt{u}.\ \varphi(\mathtt{u})$, we add the rule $(\exists \mathtt{u}.\ \varphi(\mathtt{u})) \to \bot$ to the database and so when we restart, we are asking the goal $\exists \mathtt{u}.\ \varphi(\mathtt{u})$ again. The presence of the existential quantifier is important, as it indicates that the goal involves another new existential variable y different from the variables used so far. After all, $\exists \mathtt{u}.\ \varphi(\mathtt{u})$ is equivalent to $\exists \mathtt{y}.\ \varphi(\mathtt{y})$. Now in the computation leading up to the use of the restart rule, we may have generated bindings for u, but these bindings have no connection with the restarted goal. If we just go on using u, we might generate an incorrect answer by use of the bindings.

Definition 7.3.2 (Predicate $\mathcal{B}$-computation) *We define the success of the computation $\langle \mathcal{P}, G \rangle \vdash_\mathcal{B} A$ with an associated answer substitution θ. θ is a substitution of terms to the existential variables. The terms used in the substitution θ are built up from constants, function symbols and existential variables only.*

1. *When A is an atom $\mathtt{Q(s,t,\ldots)}$ where s and t etc. are terms we have $\langle \mathcal{P}, G \rangle \vdash_\mathcal{B} \mathtt{Q(s,t,\ldots)}$ with answer θ iff one of the following holds*

    (a) *$\mathtt{Q(p,q,\ldots)} \in \mathcal{P}$, where p and q etc. are terms and $\mathtt{Q(s,t,\ldots)}$ and $\mathtt{Q(p,q,\ldots)}$ unify with most general unifier (θ_1, θ), where θ_1*

is the substitution to the universal variables of the data clauses, and θ is the substitution to the existential variables.

(b) $\bot \in \mathcal{P}$, and $\theta = \{\}$.

(c) $\varphi(\mathtt{p},\mathtt{q},\mathtt{u},\mathtt{v}\ldots) \to Q(\mathtt{p},\mathtt{q},\ldots) \in \mathcal{P}$, where $\mathtt{p}, \mathtt{q}, \mathtt{u}, \mathtt{v}$, etc. are terms and $Q(\mathtt{s},\mathtt{t},\ldots)$ and $Q(\mathtt{p},\mathtt{q},\ldots)$ unify with most general unifier (θ'_1, θ'), and $\langle \mathcal{P}, G \rangle \vdash_B \varphi(\mathtt{p},\mathtt{q},\mathtt{u},\mathtt{v}\ldots)$ with answer σ, and $\theta = \theta' \circ \sigma$.

(d) $\varphi(\mathtt{p},\mathtt{q},\mathtt{u},\mathtt{v}\ldots) \to \bot \in \mathcal{P}$, where $\mathtt{p}, \mathtt{q}, \mathtt{u}, \mathtt{v}$, etc. are terms and $\langle \mathcal{P}, G \rangle \vdash_B \varphi(\mathtt{p},\mathtt{q},\mathtt{u},\mathtt{v}\ldots)$ with answer θ.

2. When A is a conjunction $B \wedge C$, we have $\langle \mathcal{P}, G \rangle \vdash_B B \wedge C$ with answer $\theta \circ \sigma$ iff $\langle \mathcal{P}, G \rangle \vdash_B B$ with answer θ and $\langle \mathcal{P}, G \rangle \vdash_B C\theta$ with answer σ.

3. When A is an implication $B \to Q(\mathtt{s},\mathtt{t},\ldots)$ we have $\langle \mathcal{P}, G \rangle \vdash_B B \to Q(\mathtt{s},\mathtt{t},\ldots)$ with answer θ iff $\langle \mathcal{P} \cup \{B\theta\}, G \rangle \vdash_B Q(\mathtt{s},\mathtt{t},\ldots)$ with answer θ.

4. The restart rule states that $\langle \mathcal{P}, G \rangle \vdash_B Q(\mathtt{s},\mathtt{t},\ldots)$ for an atom $Q(\mathtt{s},\mathtt{t},\ldots)$ with answer θ if $\langle \mathcal{P}, G \rangle \vdash_B G$, where G is the original goal of the computation, written with completely new existential variables $y_1, \ldots, y_k$ and with θ extended to $y_1, \ldots, y_k$ as $\theta(y_j) = y_j$.

5. Another way of looking at the restart rule is that every time we use the rule, we restart with new variables $y_1, \ldots, y_k$ and a new substitution for them, obtained by extending θ. Thus assume that at the start of the computation the goal is assumed to be $A(x,t,\ldots)$ with the free existential variables $u_1, \ldots, u_k$. We start with the substitution $\theta^1(u_i) = y_i^1, i = 1, \ldots, k$ where y_i^1 are also existential variables. θ^1 can be refined to an answer θ^1_* at some stage of the computation in which we choose to restart. We restart with the original goal but with completely new substitution $\theta^2(u_i) = y_i^2$, with completely new existential $y_i^2, i = 1, \ldots, k$.

We continue until we succeed. We shall have on record, at the moment of success, variables $y_1^1, \ldots, y_k^1, \ldots, y_1^n, \ldots, y_k^n$ and an extended substitution θ for these variables. If we regard $\theta(y_i^m)$ as a substitution $\theta^m_*(u_i)$ for the variable u_i, we can say we have got n different substitutions $\theta^1_*, \ldots, \theta^n_*$ for the variables $u_1, \ldots, u_k$. We shall see from the soundness theorem that this means that the formula $\bigvee_{i=1}^n G\theta^n_*$ successfully follows from $\mathcal{P}\theta$.

7.3. REASONING BACKWARD

Example 7.3.3 *We shall illustrate the use of the rules by redoing Example 7.2.1 using the backward reasoning method. The argument, when Skolemized, can be represented by*

1. logician(k)
2. teaches-at(k, Imperial-College)
3. teaches-at(x, Imperial-College) → lives-in-London(x)
4. logician(x) ∧ lives-in-London(x) → good(x)
5. logician(x) ∧ good(x) → support-LP(x)
6. logician(u) ∧ good(u) ∧ support-LP(u)

The goal is conjunctive, so we must ensure that we solve all three conjuncts with the correct answer substitution. We begin by asking the subgoal logician(u) *which succeeds immediately from sentence 1, with the substitution* {u/k}, *i.e.* θ(u) = k. *This substitution is then applied to the remaining goal,* good(u) ∧ support-LP(u), *giving* good(k) ∧ support-LP(k). *Again we have a conjunctive goal, so we ask the first subgoal of the conjunction,* good(k). *This will unify with the head of sentence 4. Notice that sentence 4, like the other sentences in the database, uses the variable* x. *Because each sentence is universally quantified, the* x *in sentence 4 is distinct from each of the* x *in the other sentences. (To avoid confusion, we can rename the variable in sentence 4 at this point, to* y, *say.) Thus the unification with the goal produces the substitution* $\theta_1(x) = k$ *which is composed with the earlier substitution. We generate the new subgoal from the antecedent of sentence 4, and apply the substitution, getting* logician(k) ∧ lives-in-London(k). *The first part of this will succeed from sentence 1, and the second part will unify with the head of sentence 3 and generate the subgoal*

teaches − at(k, Imperial − College),

which will succeed from sentence 2.

This only leaves the third of the original conjuncts, support-LP(k) *which unifies with the head of sentence 5, generating the subgoal* logician(k) ∧ good(k). *These can be solved as we have just seen, so that the entire goal succeeds, with the answer substitution* {u/k}, *i.e.* θ(u) = k. *(We omit the bindings for variables generated during the course of the computation.)*

Example 7.3.4 *Let us illustrate all features involved with an example. Let the data be*

$$\forall x_1, x_2.(R(x_1, x_2) \rightarrow Q(x_2))$$

and the goal be
$$\exists u,y.[\neg\neg R(u,y) \to Q(y)]$$
We first rewrite the data and goal as $\mathcal{B}$-clauses. We get the problem
$$R(x_1,x_2) \to Q(x_2) \vdash_{\mathcal{B}}?((R(u,y) \to \bot) \to \bot) \to Q(y)$$
which becomes using the $\to$ rule

1. $R(x_1,x_2) \to Q(x_2)$

2. $(R(u,y) \to \bot) \to \bot$

with the query $\vdash_{\mathcal{B}}?Q(y)$.
   We use clause 1. We substitute $\theta_1(x_1) = y_1, \theta_1(x_2) = y, \theta(y) = y$.
   We get the goal $\vdash_{\mathcal{B}}?R(y_1,y)$ with answer substitution θ. Note that θ_1 had to substitute existential variable terms for the universal variables x_1, x_2 in clause 1.
   Using clause 2 we ask for the goal $\vdash_{\mathcal{B}}?R(u,y) \to \bot$. We add the antecedent to the data as clause 3 and ask for $\bot$

3. $R(u,y)$

with the query $\vdash_{\mathcal{B}}?\bot$.
   We now choose to restart. We ask the original query with completely new existential variables u', y'
$$\vdash_{\mathcal{B}}?((R(u',y') \to \bot) \to \bot) \to Q(y')$$
We add the antecedent to the data as clause 4 and ask for the head

4. $(R(u',y') \to \bot) \to \bot$

with the query $\vdash_{\mathcal{B}}?Q(y')$.
   We now substitute in clause 1 $\theta_2(x_1) = z_1, \theta_2(x_2) = y'$ and $\theta(y') = y'$ and get
$$\vdash_{\mathcal{B}}?R(z_1,y')$$
Substitute $\theta(z_1) = u, \theta(y') = y$ and we can succeed.
   Success means that
$$\vdash \exists u,y.[\forall x_1,x_2.(R(x_1,x_2) \to Q(x_2)) \to (\neg\neg R(u,y) \to Q(y)]$$

7.3. REASONING BACKWARD

Exercise 7.3.5 *Decide, using the predicate backward rules, whether the following arguments are valid. See Example 8.1.7 on how to rewrite the formulae into $\mathcal{B}$-clauses.*

1. $\dfrac{\forall\text{x. } (\text{a(x)} \lor \text{b(x)}) \\ \neg \exists \text{x. a(x)}}{\forall \text{x. b(x)}}$

2. $\dfrac{\exists \text{x. } (\text{a(x)} \to \text{b(x)})}{(\exists \text{x. a(x)}) \to \exists \text{x. b(x)}}$

3. $\dfrac{\exists \text{x.} \forall \text{y. a(x,y)}}{\forall \text{y.} \exists \text{x. a(x,y)}}$

4. $\dfrac{\exists \text{x.} \exists \text{y. r(x,y)} \\ \forall \text{x.} \forall \text{y. r(x,y)} \to \text{r(y,x)} \\ \forall \text{x.} \forall \text{y.} \forall \text{z. r(x,y)} \land \text{r(y,z)} \to \text{r(x,z)}}{\forall \text{x. r(x,x)}}$

5. $\dfrac{\forall \text{x.} \exists \text{y.R(x, y)}}{\exists \text{y.} \forall \text{x.R(x, y)}}$

Theorem 7.3.6 (Soundness of the predicate $\mathcal{B}$-computation)
Consider the $\mathcal{B}$-computation with restart and assume that at some stage we have the query $\langle \mathcal{P}, G \rangle \vdash_\mathcal{B}?A$ and a history of restart substitutions $\theta_*^1, \ldots, \theta_*^n$. Assume that this query now succeeds with substitution θ. Then we have that $\vDash ((\forall \text{ universal variables})(\bigwedge \mathcal{P}) \to \bigvee_{i=1}^n G\theta_*^n \lor A\theta)$, i.e. it is true in every classical interpretation, under any assignment.

Proof. By induction on the computation.

We first show that for a computation without restart we have $\langle \mathcal{P}, G \rangle \vdash_\mathcal{B} A$ iff $\vDash ((\forall \text{ universal variables}) \bigwedge \mathcal{P} \to A\theta)$.

Our proof reduces the predicate case to the propositional case. Let $u_1, \ldots, u_k$ be the existential variables of A and let $t_i = \theta(u_i)$. Let $\mathcal{P}'$ be the result of substituting for the universal variables of the clauses of $\mathcal{P}$ all possible combinations of $t_1, \ldots, t_n$. $\mathcal{P}'$ can be regarded as a finite propositional database and $A\theta$ succeeds propositionally from it.

Hence by propositional soundness we have $\bigwedge \mathcal{P}' \vdash A\theta$.

Therefore

$(\forall \text{ universal variables}) \bigwedge \mathcal{P} \vdash A\theta)$ and so $\vDash ((\forall \text{ universal variables}) \bigwedge \mathcal{P} \to A\theta)$

Assume now that A succeeds with n uses of the restart rule and that the existential variables involved are $y_1^1, \ldots, y_k^1, \ldots, y_1^n, \ldots, y_k^n$. Consider the new computation of $A\theta$ from the new database $\mathcal{P}_1$ where

$$\mathcal{P}_1 = \mathcal{P}\theta \cup \{G(y_1^m, \ldots, y_k^m) \to \bot \mid m \leq n\}\theta$$

$A\theta$ can succeed without restart from this database, because at the mth trigger of restart, we do not use the restart rule but use the clause

$$G(y_i^m)\theta \to \bot$$

We therefore get that $\vDash ((\forall \text{ universal variables}) \bigwedge \mathcal{P} \wedge \bigwedge_m (G(y_i^m)\theta \to \bot) \to A\theta)$.

Since

$$\bigwedge \mathcal{P} \wedge \bigwedge_m (G(y_i^m)\theta \to \bot) \to A\theta$$

is equivalent to

$$\bigwedge \mathcal{P} \to \bigvee_m G(y_i^m)\theta \vee A\theta$$

our inductive case is proved.

Thus the theorem is proved. ∎

Remark 7.3.7 (Completeness, Herbrand theorem) *We are not going to prove completeness in this book. See [Gabbay and Olivetti, 2000, Metcalfe et al., 2009].*

The following well-known theorem follows from the soundness of completeness of our computation.

Herbrand theorem
Let A be a formula of the form $\exists u_1, \ldots, u_n \cdot A_0$, where A_0 is quantifier free and assume that $\vDash A$ in classical logic. Then for some substitutions $\theta^1, \ldots, \theta^m$ of terms to the variables $u_1, \ldots, u_n$ we have $\vDash \bigvee_j A_0 \theta^j$.

See also Example 8.5.2.

Remark 7.3.8 (Hilbert formulation) *It is possible to give a Hilbert formulation for classical and for intuitionistic predicate logics by adding the following schemas for quantifiers to the schemas in Examples 6.2.6 and 6.2.7.*

12. $\forall x. A(x) \to A(t)$, t arbitrary term.

13. $A(t) \to \exists x. A(x)$, t arbitrary term.

7.3. REASONING BACKWARD

14. $\dfrac{\vdash A \to B(x)}{\vdash A \to \forall x.B(x)}$ x not free in A.

15. $\dfrac{\vdash B(x) \to A}{\vdash \exists x.B(x) \to A}$ x not free in A.

7.3.1 Connection with resolution

In this book we have concentrated on reasoning via a set of simple rules, from assumptions to conclusions (reasoning forwards) and from conclusions back to assumptions (reasoning backwards). Another well-known method of checking whether a query follows from a database is *resolution*.

We know that if A follows logically from $\mathcal{P}$, i.e. $\mathcal{P} \vdash A$ succeeds, then any interpretation which makes all the formulae of $\mathcal{P}$ true must also make A true. In other words $\wedge \mathcal{P} \to A$ is a predicate tautology. Thus we cannot make $\mathcal{P}$ true and A false, or in other words we cannot make $\mathcal{P} \cup \{\neg A\}$ all true, i.e. $\mathcal{P} \cup \{\neg A\}$ is not consistent. In our case $\mathcal{P}$ contains universal formulae of the form $\forall$x. B(x). The query A is an existential formula of the form $\exists$y. C(y). If we negate the query, we get $\neg A = \forall$y. $\neg$C(y), which is a universal formula, like those in $\mathcal{P}$. Therefore if we have a computation for checking whether a set of universal sentences is consistent or not, then we can check whether a query A follows from data $\mathcal{P}$. If $\mathcal{P} \cup \{\neg A\}$ can be shown to be inconsistent, then $\mathcal{P} \vdash A$.

Resolution is a method for checking inconsistency. For the purposes of this method it is convenient to write the clauses of $\mathcal{P} \cup \{\neg A\}$ in conjunctive normal form. So for example the acceptable (to our backward computation) clause

(A∧B→C)→D

will have to be rewritten for resolution as

(A∨D) ∧ (B∨D) ∧ (¬C∨D)

which becomes the set of assumptions

A∨D
B∨D
¬C∨D

Suppose we have two clauses in the database of the form

$$\neg A \vee p$$
$$\neg q \vee A'$$

They are logically equivalent to

$$A \rightarrow p$$
$$q \rightarrow A'$$

If we can unify A with A' we can get, by transitivity of $\rightarrow$,

$$q\theta \rightarrow p\theta$$

where θ is the substitution which makes $A'\theta = A\theta$. This can be rewritten as

$$\neg q\theta \vee p\theta$$

Thus the basic computation of resolution is: if the set to be tested contains the formulae

$$\neg A \vee p$$
$$\neg q \vee A'$$

and A and A' can be unified with most general unifier θ, we can add a new clause, the *resolvent*

$$\neg q\theta \vee p\theta$$

to the set.

When we have formulae in the set of the form

$$\neg A \vee p$$
$$A'$$

the resolvent is just $p\theta$. In this way, the formulae

$$\neg C$$
$$C$$

resolve to the *empty clause*, written $\square$. As one might expect from having both C and $\neg C$, the empty clause indicates the presence of an inconsistency. Now $\neg C$ is equivalent to $C \rightarrow \bot$, so that in the backward reasoning method, we would have

7.4. DECIDABILITY

$$\frac{C \to \bot}{C}$$

as assumptions, so that the query $\bot$ would succeed from the formulae, and hence we would deduce that the assumptions were inconsistent.

Resolution is a powerful technique, with many variations having been designed to cope with different logics and applications, and is worthy of a book in its own right. For further details, see [Goubault-Larrecq and Mackie, 1997; Gallier, 1986].

7.4 Decidability

A logic (see Definition 6.1.1) is said to be decidable iff there exists an algorithm (which always terminates) to decide for any two wffs A, B whether $A \mathrel{|\!\!\sim} B$ holds or not.

According to this definition, classical propositional logic is decidable. For any A, B we can check by truth tables whether $A \to B$ is a tautology, as explained in Section 1.1.2. In fact, all the propositional systems mentioned in the table of Section 6.8 are decidable.

The decidability of the classical propositional calculus can also be seen from the diminishing resource backward computation algorithm of Definition 4.3.3. This algorithm is terminating, as the database gets smaller and smaller. The restart rule involves only one scan of the history and needs to be applied only once. Remark 4.3.4 quotes completeness of the algorithm for classical propositional logic (with restart) and for intuitionistic propositional logic (with bounded restart); therefore these logics are decidable.[1]

Classical predicate logic, however, is undecidable. Although there are many algorithms for checking whether $A \vdash B$ holds in classical predicate logic, they are not guaranteed to terminate for arbitrary $A \vdash ? B$. Such systems are called *semi-decidable* or *recursively enumerable*.

Thus classical predicate logic and intuitionistic predicate logic are semi-decidable but not decidable.

---

[1] At this point it is worthwhile to take another look at Example 4.2.5.

7.5 Worked examples

Example 7.5.1
Data
If everybody buys Jane Fonda's Quick Slimming book, then someone is bound to follow the programme successfully, and be liked by all her friends, envied by her neighbours and generally feel more self-confident and loving towards her children.

Anyone liked by all her friends must be good natured.

Goal
Therefore if everybody buys Jane Fonda's book then someone is good natured.

Let us translate, using the obvious predicates:

Data

(1) $\forall x. \text{Buy}(x, J) \rightarrow \exists y. [\text{Follow}(y) \land$
$\forall z. [\text{Friend}(z, y) \rightarrow \text{Like}(z, y)] \land$
$\forall z. [\text{Neighbour}(z, y) \rightarrow \text{Envy}(z, y)] \land$
$\text{Confident}(y) \land$
$\forall z. [\text{Child}(z, y) \rightarrow \text{Love}(y, z)]]$

(2) $\forall x. [\forall y. [\text{Friend}(y, x) \rightarrow \text{Like}(y, x)]. \rightarrow \text{Good}(x)]$.

Goal

(3) $\forall x. \text{Buy}(x, J) \rightarrow \exists x. \text{Good}(x)$

We need a ready for computation form. We pull the quantifiers out and get the (*) formulation as follows:

(1*) $\exists x. \exists y. \forall z_1. \forall z_2. \forall z_3. \text{Buy}(x, J) \rightarrow (\text{Follow}(y) \land$
$[\text{Friend}(z_1, y) \rightarrow \text{Like}(z_1, y)] \land$
$[\text{Neighbour}(z_2, y) \rightarrow \text{Envy}(z_2, y)]$
$\land \text{Confident}(y) \land$
$[\text{Child}(z_3, y) \rightarrow \text{Love}(y, z_3)])$

(2*) $\forall x. \exists y. [(\text{Friend}(y, x) \rightarrow \text{Like}(y, x)) \rightarrow \text{Good}(x)]$

7.5. WORKED EXAMPLES

and the Goal is

(3*) $\exists x.\exists y.[\text{Buy}(x, J) \to \text{Good}(y)]$.

The above equivalences are valid in classical logic. In intuitionistic logic we cannot rewrite as we did.

Skolemizing and decomposing (1*) into several clauses we get the final (#) formulation:

Data

(1#) 1. $Buy(c_1, J) \to Follow(c_2)$
2. $Buy(c_1, J) \to (Friend(x, c_2) \to Likes(x, c_2))$
3. $Buy(c_1, J) \to (Neighbour(x, c_2) \to Envy(x, c_2))$
4. $Buy(c_1, J) \to Confident(c_2)$
5. $Buy(c_1, J) \to (Child(x, c_2) \to Love(c_2, x))$.

(2#) $[\text{Friend}(f(x), x) \to Like(f(x), x)] \to \text{Good}(x)$.

Goal

(3#) $Buy(x, J) \to \text{Good}(y)$.

The above is ready for computation. Notice that after Skolemizing and rewriting we lost a bit of the natural linguistic structure.

The computation

Data#	$\vdash_B$?	Goal#
which reduces to		
(1#), (2#)	$\vdash_B$?	(3#)
which reduces to		
(1#), (2#)	$\vdash_B$?	$Buy(u, J) \to Good(v)$
where u, v are existential variables.		
This reduces to		
(4#), (1#), (2#)	$\vdash_B$?	$Good(v)$
where (4#) = $Buy(u, J)$		
Unify with (2#) and get		
(4#), (1#), (2#)	$\vdash_B$?	$Friend(f(v), v) \to Like(f(v), v))$
which reduces to		
(4#), (1#), (2#), (5#)	$\vdash_B$?	$Like(f(v), v))$

where (5#) $Friend(f(v), v)$ was added to the database.

   Unify with (1#.2). Choose $c_2 = v$. Thus everywhere where v appears it becomes c_2, including the database.
   We now get

Current data	**Current goal**
(1#)	? $Friend(f(c_2), c_2)$
(2#)	
(4#) $Buy(u, J)$	
(5#) $Friend(f(v), v))$, for $v = c_2$	
i.e. $Friend(f(c_2), c_2)$	

Therefore success.

Example 7.5.2 *Using* backward rules *for classical logic provide formal proofs for the following:*

$$\exists x.(A(x) \to A(a) \wedge A(b))$$

7.5. WORKED EXAMPLES

Solution

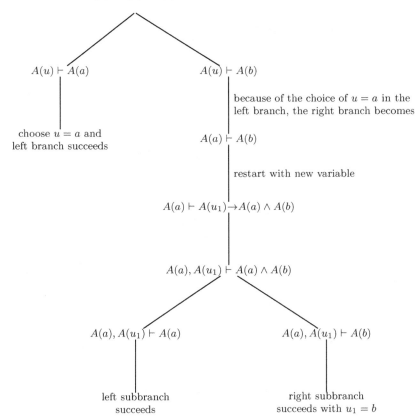

Using the soundness theorem, Theorem 8.3.6, we get that the success of the goal $A(u) \to A(a) \wedge A(b)$ with $\theta_1(u) = a, \theta_2(u) = b$ means that

$$\models (A(a) \to A(a) \wedge A(b)) \vee (A(b) \to A(a) \wedge A(b))$$

8

INTRODUCTION TO PROLOG

PROLOG is a high level programming language based on a subset of predicate logic that we introduced in Chapter 6. The term PROLOG stands for Programming in Logic. PROLOG is a declarative language, as opposed to an imperative language such as C++ or JAVA.

From the programming point of view, you can think of a PROLOG program as a collection of *facts* and *rules*. The facts represent basic information about the world, whereas the rules are special statements that tell the computer how to derive information from the facts, behaving pretty much as implications. Whenever a *query* is made to a PROLOG program, the system looks at the facts and rules and tries to find an answer for the query. Later on we shall see that behind the scenes the system gives the answer by trying to prove the unsatisfiability of a set of special formulae called clauses.

We are going to assume a working environment with PROLOG. A popular and free implementation is called SWI-Prolog freely available from http://www.swi-prolog.org/. SWI-Prolog is available for a variety of platforms.

8.1 Getting started with PROLOG's syntax

SWI-Prolog is close to ISO-Prolog standard syntax, which is based on the Edinburgh Prolog syntax. We are not going to fully present this syntax here, but we will discuss it briefly so you can get started writing your first programs. You can find a full description of the syntax in the ISO standard document. Notice that there may be variations in the terminology used to describe the several components. The idea here is to concentrate on the underlying concepts.

Components of the language

The syntax of the language is built around a data structure called a term. A term is defined recursively in terms of a Number, an Atom, a Variable or a Compound term.

Term ::= Number | Atom | Variable | Compound_Term

A Compound term is formed by an Atom, followed by "(", a nonempty sequence of terms, separated by commas and ended by ")":

Compound_Term ::= Atom(Term1, ..., Termk) for some $k \geq 1$.

In a compound term, the initial atom is called the functor and the number of its arguments is called its arity. A term is said to be ground if it has no variable in it. Sometimes compound terms are also called structures.

Do not feel discouraged by the apparent complexity of these definitions. You will see that in practice they are much simpler than they seem.

Let us now turn to each one of the basic components mentioned above.

Numbers

A number in PROLOG is made up of a sequence of digits and special characters such as +, -, E, .. All implementations of PROLOG will contain the type Integer which is formed by an optional sign symbol (+ or -) and a sequence of digits. Most will also have a Float type, such as the SWI-Prolog. A full description of the numeric types supported by SWI-Prolog can be found in its documentation.

Examples: 276, -127, 120.45, -1.2E4.

In the examples above, the expression -1.2E4 represents the number -12000 (since, $-1.2E4 = -1.2 * 10^4$).

Atoms

An atom is a sequence of letters starting with a lower-case letter and possibly including underscore characters in the middle, or a sequence of arbitrary characters in single quotes.

Examples: likes, w%@!, day_of_the_week, etc.

8.1. GETTING STARTED WITH PROLOG'S SYNTAX 245

Atoms and numbers are also called constants.

Variables

A variable is a sequence of letters, digits or underscore characters starting with a capital letter or by the underscore character _. There is also a special variable typed _ (an underscore on its own) and referred to as the anonymous variable. _ is used when the value of the variable is of no particular interest. Multiple occurrences of the anonymous variable in a PROLOG expression are assumed to be all distinct. In other words, their values do not need to be the same across the expression.

Another important thing about variables in PROLOG is that they are assumed to be universally quantified, you will learn more about this in the Classical Logic part of the Course.

Examples: X, Employee, Student, S123, S_1, _, etc.

8.1.0.1 Revisiting terms

We are now in a position to give some examples of terms:

Examples of terms: likes(john,mary), f(X,g(_,1.20)), +(A,1.23)

f(X,g(_,1.20)) is a compound term: it is formed by the atom "f", followed by the special character "(", a sequence of two terms (X and g(_,1.20)) separated by a comma and ended by ")".

Special Terms

Some terms are more conveniently expressed in infix form. For instance, the term +(2,3) is a legal PROLOG term. Its functor is the atom +, which corresponds to the arithmetic operation +. It is easier to write this expression simply as 2 + 3, where the functor + appear as a binary operator. Other operators in PROLOG include \+, -, /, *, is, =, \=, ., ;, ,, -> and :-. You will learn more about the latter when you start writing your own programs.

There are also a number of other more interesting terms.

Lists

A *list* is a term constructed from the atom [], representing the empty list, and the functor |. | takes as arguments a term and a list, respectively. A list of k elements $e1, \ldots, ek$ corresponds to [e1|[e2,...,ek]]. We will have a special section on how to manipulate lists later.

Predication

A predication is either an atom or a compound term.

Examples:
    father(X,Y), member(X,[X]), factorial(0,1), happy.

PROLOG Clauses

In its simplest form, a PROLOG clause is an expression of the form

    head :- body1, body2, ..., bodyk.

where $k \geq 0$, head is a predication and body1, body2, ...bodyk are predications or possibly multiple applications of the operator not (\+) to predications. head is the head of the clause and the expression body1, body2, ..., bodyk is its body. Later on, we will see that the body may contain more complex structures formed with the help of operators such as ; and ->.

When $k = 0$, i.e., the body is empty, the clause is called a *fact* and the symbol :- is omitted. Notice that the negation operator is not allowed in the head of a clause and, therefore, negation of facts cannot appear in a PROLOG program. The intuitive idea is to express only the things that are true in the world and assume that everything else is false.[1] There is also a special clause called the *empty clause* which is used to represent contradiction and denoted by □.

Examples:
    member(X,[X]).
    member(X,[_|T]) :- member(X,T).
    concession(X) :- student(X), not receives_grant(X).

Notice that in PROLOG clauses must be terminated by a full stop (.) and should be grouped by the predication in the head.

Examples: not married(john, mary). is not a valid clause, because not married(john, mary) is not a predication and hence cannot be used in the head of a clause, however

    not_married(X,Y) :- not married(X,Y).

is a valid clause.

In logical terms, the comma "," in the body of the clauses corresponds to the conjunction connective, the operator :- corresponds to the converse of the implication ($\leftarrow$) and the not operator corresponds

---

[1]This is known in PROLOG as *negation by failure*, which we will explain in detail later.

8.1. GETTING STARTED WITH PROLOG'S SYNTAX

to a weaker form of classical negation known as *negation by failure*. We can rephrase the clauses in the examples above as the following formulae in classical logic:

(a) $\forall x.member(x, [x])$

(b) $\forall x.\forall \_.\forall t.(member(x, t) \rightarrow member(x, [\_|t]))$

(c) $\forall x.((student(x) \land \neg receives\_grant(x)) \rightarrow concession(x))$

The clauses above can be intuitively read in the following way.

(a*) "Every thing is a member of a list whose only element is that thing."

(b*) "For all things x and lists y and t, if the tail of the list y is t and x is a member of t, then x is also a member of y.[2]

(c*) "All students x who are not known to receive a grant are entitled to concessions."

The last phrase was stated in terms of "not known to" instead of simply "not" because of the way negation as failure works.

It is also possible to have disjunction in the body of clauses. This is done by using ";" instead of ",". For instance, the clause

```
parent_of(X,Y) :- mother_of(X,Y); father_of(X,Y).
```

is equivalent to the two clauses below

```
parent_of(X,Y) :- mother_of(X,Y).
parent_of(X,Y) :- father_of(X,Y).
```

A query to `parent_of` succeeds if either `mother_of(X,Y)` succeeds or `father_of(X,Y)` succeeds. ";" binds less tightly than ",", but it is possible to override this through the use of parentheses. For example, the clause

```
concession(X) :- unemployed(X), \+married(X); student(X).
```

is equivalent to the two clauses below:

---

[2] Note that in this statement, the reference to Y is symbolic and implicit in the definition of the clause. The requirement that "the tail of Y is T" is made implicitly upon unification with the head of the clause. We will discuss how these bindings are made in a special section for lists later.

```
concession(X) :- unemployed(X), \+married(X).
concession(X) :- student(X).
```

If the intention is to make the disjunction between \+married(X) and student(X) only, then the clause must be rewritten as

```
concession(X) :- unemployed(X), (\+married(X); student(X)).
```

In any case, avoiding the use of ; improves readability of your programs.

Goal Clauses

In its simplest form, a goal clause is an expression of the type

:- body1, body2, ..., bodyk.

where $k \geq 0$ and body1, body2, ...bodyk are predications or possibly multiple applications of the operator not ('\+') to predications. As with normal clauses, a goal clause may have more complex structures formed with the help of operators such as ";" and "->" and parentheses. Goal clauses are used to submit queries to PROLOG programs. When submitting the queries, the symbol ':-' must be omitted, but keep in mind that this is a reminder of the logical meaning of a goal clause:

A clause of the kind :- p(x). is equivalent to the sentence $\forall x. \neg p(x)$, since the term p(x) is in the left-hand side of the implication symbol, which in PROLOG is represented in its reverse form ($\leftarrow$) as :-.[3] We will see that what we are in fact doing is trying to prove a contradiction from our original program and the goal clause. The bindings for the variables in the goal clause that we receive as output of the program are simply a by-product of this process.

Program

A program in PROLOG is a list of clauses.

8.2 Trying it all out

We are now in a position to try a few simple programs in PROLOG. You can edit your PROLOG programs in any text editor that can save files as text only. The instructions given here are for a typical linux system, but they can be easily applied for other platforms.

---

[3]Remember that an implication A→B is equivalent to the formula ¬A∨ B.

8.2. TRYING IT ALL OUT

It will be easier if you keep all of your PROLOG files in the same directory. For that, create a new directory from a shell window with the command mkdir:

1. mkdir ~/prolog

Change to that directory with the command cd:

1. cd ~/prolog

Start your favourite editor program, for example Emacs, with the example below. Type

1. emacs example-1.pl &

Now you are ready to type and save the simple program below:

Program 8.1. Illustrating the use of facts and rules in PROLOG.

```
studies(mark,logic).
studies(lucy,geometry).
studies(lucy,english).
studies(paul,geometry).
studies(paul,logic).
subject(john,english).
subject(dov,logic).
subject(mary,geometry).
teaches(X,Y) :- subject(X,Z), studies(Y,Z).
```

We now have to start SWI-Prolog and load the file you just saved. To start PROLOG type "pl" in the shell window and press the return key. If all goes well, you should be able to see something like the content below in your screen (depending on version and program used).

Screen output 8.8.2. Sample session in SWI-Prolog

```
Welcome to SWI-Prolog (Multi-threaded, 64 bits, Version 7.2.3)
Copyright (c) 1990-2015 University of Amsterdam, VU Amsterdam
SWI-Prolog comes with ABSOLUTELY NO WARRANTY. This is free software,
and you are welcome to redistribute it under certain conditions.
Please visit http://www.swi-prolog.org for details.

For help, use ?- help(Topic). or ?- apropos(Word).

?-
```

The symbol ?- is the PROLOG prompt and it means that PROLOG is waiting for queries to be typed in. When you type in a query, PROLOG tries to evaluate it and answers true or false according to the result of the evaluation of the query. You will see, that depending on the query, PROLOG might show you additional information regarding the query's evaluation.

At the prompt, type consult('example-1.pl')., or alternatively, ['example-1.pl']. (note the full stop at the end, as with any PROLOG clauses, the goal clauses must be ended by a full stop). If the extension of the file to be read is ".pl", it can actually be omitted. Note the correct syntax and make sure you use single quotes. SWI-Prolog should report:

Screen output 8.8.3. Consulting a sample file in SWI-Prolog

```
?- ['example-1.pl'].   true.
```

The response true means that the query (i.e., consulting the file) succeeded, and PROLOG is ready to receive queries to our initial program.

8.2. TRYING IT ALL OUT

As you can see from the source code, the program has seven facts and one rule. The facts state information about students and the subjects they currently take (in the predicate `studies`). The association is up to the programmer's interpretation. One can choose either argument for the student or the subject. This means one must be consistent in the representation. In this example, the first argument indicates the student and the second argument indicates the subject. The predicate `subject` specifies the association between a lecturer and a subject. In this example, the lecturer is in the first argument and the subject in the second. The rule `teaches(X,Y) :- subject(X,Z), studies(Y,Z).` specifies a relationship between lecturers and subjects. It states that a lecturer X teaches a student Y if (`:-`) X teaches a subject Z that Y studies. The variables in a PROLOG clause are universally quantified.

Let us now try a simple query. Type `studies(mary,maths).` at the prompt and press ENTER. PROLOG will reply with "`false`". The reason for this is simple: it is impossible to derive `studies(mary,maths)`" from the facts and rules given. PROLOG tries to match the queries against the head of the clauses and replies `true` when a match can be found. The matching process is what we have seen in Chapter 7 as unification. If instead we submit the query `studies(mary,geometry)`, PROLOG will reply `true`, since this is one of the facts in the program.

The use of variables in a query is even more interesting. For example, try the query `studies(lucy,X)`.[4] This can be understood as "give me the values X for which a successful unification with `studies(lucy,X)` can be performed in the program". In other words, you will be asking "what does Lucy study?" PROLOG will respond as follows:

```
?- studies(lucy,X).
X = geometry
```

This means that PROLOG found a solution to your query, by replacing X by `geometry`. You can just press return if you are happy or you can tell PROLOG to look for other solutions. If you press ENTER, PROLOG will return to the prompt and stop. If you type ";" PROLOG will try for alternative answers to your query. In this case, there is also the alternative solution `english`. The answers will be given in the order in which the unifications are performed. In this case, in the order of the appearance of facts in the program.

---

[4] Do not forget to always terminate a query with full stop, which from now on we shall omit for brevity.

The following diagram explains the behaviour of PROLOG in this situation:

```
studies(mark,logic).
  ↑     ↑
  ok    fails, since lucy and mark do not unify
studies(lucy,X).

studies(lucy,geometry).
  ↑     ↑     ↑
  ok    ok    ok, with X=geometry
studies(lucy,X).
```

"ok" above indicates a successful unification. For instance, **studies** matches **studies**, the open parentheses match, but **lucy** does not match mark, so the unification fails in the first clause. The unification succeeds in the second clause, because the variable **X** is free and it can assigned to the constant **geometry**. The two clauses then unify. Once **X** is bound to **geometry**, it would be bound for every other occurrence of **X** in the current clause and it would not unify with anything else.

PROLOG then waits for some input. If you just press ENTER, it reports "true". If you type ";" and press ENTER, it will try to find alternative solutions to the original query by proceeding to the next clause in the program **studies(lucy,english)**. This clause also unifies with X=english, giving the next and last answer to the original query.

```
?- studies(lucy,X).
X = geometry ;
X = english.

?-
```

A similar pattern would happen with queries that match the head of a rule. Say, for instance, that you make the query: **teaches(mary,lucy)**.

```
teaches(X,Y) :- subject(X,Z), studies(Y,Z).
  ↑    ↑           ok, with X=mary, Y=lucy
  ok
teaches(mary,lucy).
```

At this stage, the variable **X** is bound to **mary** and the variable **Y** is bound to **lucy**. The original query will succeed if the query **subject(mary,Z), studies(lucy,Z)**. succeeds. This *goal* succeeds if both components succeed for the same value of Z. It should be easy to see that `subject(mary,Z)` will unify with `subject(mary,geometry)`, binding Z to geometry. This binding is passed on to the next literal in the goal, which now becaomes `studies(lucy,geometry)`. This sub-goal also succeeds and the original query will succeed giving answer `true`.

In general, if you submit to PROLOG a goal of the form

$$?\text{- p,q1,}\ldots\text{,qk.}$$

PROLOG will look in the program for a clause whose head matches p. The exact nature of this matching mechanism will be explained later. Now, suppose your program has a clause like the one below:

$$\text{p :- r1,}\ldots\text{,rj.}$$

After resolving the first sub-goal p, the rest of the original goal will become

$$?\text{- r1,}\ldots\text{,rj, q1,}\ldots\text{,qk.}$$

This makes sense, if you think that your program stated that p could be proved by showing that `r1,...,rj` can be proved. Thus, if you were originally trying to prove `p,q1,...,qk`, then you could also do it by proving `r1,...,rj` and `q1,...,qk`. In general, the actual mechanism is a bit more complex, because the terms `p,q1,...,qk` might share some variables, whose values, if bound, must be the same throughout the goal clause.

8.3 PROLOG's strategy for proving goals

Consider the following PROLOG program, with data about a relation r and two rules stating how the *transitive closure of r* can be computed. The data is expressed in PROLOG as facts of the form

$$\text{r(x,y).}$$

where x and y are some values for which $r(x,y)$ holds, whereas the transitive closure of r, in symbols r^*, is given by rules of the form

$$\text{r\_star(X,Y):- } \cdots$$

In practice, you ask PROLOG queries about `r_star(x,w)` to find out whether the pair (x,w) is in the transitive closure of r.

Program 8.4. Computing the transitive closure of a relation r.

```
r(a,b).
r(b,c).
r(c,d).
r_star(X,Y) :- r(X,Y).
r_star(X,Y) :- r(X,Z), r_star(Z,Y).
```

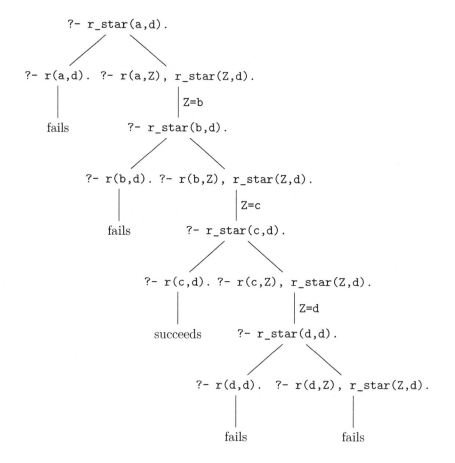

Figure 8.1: Sample search space for a query to Program 8.5.

Figure 8.1 illustrates the search space for answers to the query
$$?\text{- } r\_star(a,d).$$
and Program 8.5. The tree in the figure shows PROLOG's strategy for

8.3. PROLOG'S STRATEGY FOR PROVING GOALS

finding answers to a given query.

1. Clauses in the program are selected from top to bottom
2. Predicates in the goal are selected from left to right

The tree is traversed in a depth-first order. You should pay special attention to this strategy, for misunderstandings usually cause unexpected behaviour, such as loops, in your programs, as explained next. Suppose you need to write a program which, given a number X, writes all numbers from X down to 0. Let us call the rule that implements this procedure loop. One of the simplest ways of implementing this is by defining loop recursively, in such a way that given X

(C1) The program will do nothing if X is less than or equal to 0.

(C2) Otherwise, the program writes X on the screen, subtracts 1 from X giving Y and then calls itself recursively for the value Y.

The base of the recursion is Condition (C1) above, because it indicates that no further computation is needed. Now suppose that even though you know that PROLOG selects the clauses in the program from top to bottom, you decide to write the program implementing (C1) and (C2) above in the following order:

Program 8.5. Sample program that loops indefinitely.

```
loop(X):- write(X), nl, Y is X - 1, loop(Y).
loop(X):- X < 0.
```

The first rule in Program 8.5 implements the recursive call to loop and the second rule is the base of the recursion. If you submit any goal of the form ?- loop(A)., where A is a number, say ?- loop(100)., the program will loop, because PROLOG will look for a rule with the atom loop in the head. The first such rule is the one in the first line of Program 8.5, which tells it to write the current value of X on the screen,[5] then subtract 1 from X, then unify Y with the result,[6] and finally call loop recursively for Y. The recursive call will itself use the first rule matching the predicate in the query, and the first rule will be called over and over for ever smaller values. As a result, the program

---

[5]This is done by the special predicate write(···).
[6]Since Y is free, this always succeeds and Y is bound to the value $X - 1$.

never terminates.[7] A correct version of the program can be found in Program 8.6. Of course you make further improvements.

Program 8.6. The corrected version of Program 8.5.

```
no_loop(X):- X < 0, !.
no_loop(X):- write(X), nl, Y is X - 1, no_loop(Y).
```

In Program 8.6, PROLOG first checks whether the number is smaller than 0 (the first rule selected). If this is the case, X<0 succeeds and the predicate ! is evaluated. This is a special predicate called "cut" and its objective is to tell PROLOG not to look for alternative solutions past this point. If the number is greater or equal to 0, then the call to the first rule will fail and PROLOG will look for an alternative solution. The head of the second rule also unifies with the query, so it is then invoked. The number is written, decreased and the process restarts. Let us now give some further explanation about some of the other predicates used in this program:

- As we mentioned, the predicate write(X), simply outputs the value bound to X to the screen;

- The predicate nl writes a blank line on the screen (i.e., subsequent output will be written to a new line);

- The predicate is binds Y to the value obtained from calculating X - 1.

It can be easily seen that no_loop is called recursively in the second rule for the newly calculated value Y. At each "iteration", the value of X is decreased by 1 until it eventually it becomes smaller than 0. At that point, the first rule can be used since X<0 succeeds, the cut is invoked and the program terminates. One rule of thumb for writing your programs is to place terminating conditions in recursive definitions at the top of the associated set of clauses of a PROLOG program.

8.4 Working with Lists

As mentioned in Section 8.1.0.1, a list in PROLOG is a term constructed from the atom [], which represents the empty list, and the binary op-

---

[7]You can interrupt the computation with CONTROL-C and then selecting a for "abort".

8.4. WORKING WITH LISTS

erator "|". The operator | takes as arguments a term and a list, respectively.

The built-in predicate is_list(X) is very useful when you do not know in advance whether the argument passed to a predicate is a list. It will succeed if X is a list.

Lists in PROLOG can be nested and can also contain elements of different types. The terms below are all valid examples of lists.

Examples:
   [1,5,7]
   [prolog, is, a, nice, language]
   [[1,5,7], [who, are, you], 2016]

8.4.1 Terminology and Notation

We already mentioned that the special term [] is used to represent the empty list. The first element in a list is called the *head of the list*. The *tail* of the list is the list formed by whatever remains in the list when its element is removed. Apart from the empty list, every list has a head and a tail.

Examples:

List	head	tail
[1,5,7]	1	[5,7]
[prolog, is, a, nice, language]	prolog	[is, a, nice, language]
[[1,5,7], [who, are, you], 1998]	[1,5,7]	[[who, are, you], 1998]
[1]	1	[]

The head and tail notions only apply to non-empty lists. The head of a list is always an element, that may sometimes also be a list, whereas the tail of a list is *always* another list formed by the original list without its head element. For instance, the head of the list [1,5,7] is the element 1, whereas its tail is the list [5,7]. The head of the list [[1,5,7], [who, are, you], 1998] is the list [1,5,7], and its tail is the list [[who, are, you], 1998].

The following notation is used to split up a non-empty list into its head and tail or to build a list up from an element and a list:

[Head | Tail]

The vertical bar splits the list into elements on the left-hand side and a single list on the right side. For instance, [1,5,7] can me matched against [Head | Tail]. Head will be instantiated with the element 1 and Tail will be instantiated with the list [5,7]. Similarly, when [1] is matched against [Head | Tail], Head will be instantiated with the element 1 and Tail will be instantiated with the empty list! More than one variable can be placed on the left-hand side of the bar. For example, the list ['element one','element two',element three'], can be matched to the list [A1,A2|A3]. Notice the difference in meaning of the objects in either side of |: A1 is instantiated to 'element one', A2 is instantiated to 'element two' but A3 is instantiated to the *list* ['element three'].

The bar can also be used to build up lists. For instance, the list [1|[2,3,4]] is equivalent to the list [1,2,3,4]. This is especially useful when building lists from values received as arguments to a PROLOG clause.

8.4.2 Unifying lists

As with any other term in PROLOG a list can be unified with any free variable. However, unifying two lists is a bit more complicated, since lists can have elements of any type and can contain variables as well.

We have seen some examples about how lists unify. The general process is done recursively and involves the steps below

1. First try to unify the two heads

2. If the heads match, try to unify the tails recursively

As the tail of a list is itself a list, so the process can be applied recursively in 2. above. By applying common sense, it is possible to answer quite complicated questions about list unification using the steps above.

Exercise 8.4.1 *(List Unification) Try to unify the lists in columns A and B below. If unification is possible, write the corresponding variable instantiations in column C.*

A	B	C
[X,Y,Z]	[1,2,7]	
[X,Y]	[1,[3,4]]	
[X,Y,Z]	[a,[b],c]	
[X,Y]	[a,b],[3,4]]	
[[X,Y],Z]	[1,[2,3]]	
X	[]	
[X]	[]	
[X,[1,2],[a,B]]	[a,[R,S],[a,c]]	

8.4. WORKING WITH LISTS

8.4.3 Useful predicates about lists

One of the most useful predicates about lists is the one which tests membership of an element with respect to a list. PROLOG has the built-in predicate for this.

- member(X,L)

A call to member(X,L) will succeed if X is a member of the list L. The definition of member is quite interesting in itself. You must get used to manipulating lists, so here goes a sample definition:

Program 8.7. Sample definition of the predicate member

```
member(X,[X|_]).
member(X,[_|T]):- member(X,T).
```

The code above be read as "X is a member of a list L, if X is the head of L, or if it is an element in the tail of L".

Other useful pre-defined predicates include:

- append(X,Y,L)

A call to append(X,Y,L) succeeds whenever L can be unified to the concatenation of the lists X and Y. If L is a free variable and X and Y are lists, a call to append(X,Y,L) will instantiate L to the list obtained from concatenating X and Y.

Exercise 8.4.2 *(List concatenation)*
Will the following goal succeed? If it does succeed, give the values for each of the variables in the clause.

```
?- append([X|Xs],[3|Ys],[1,2,3]).
```

More list manipulation predicates:

- length(L,N)

The above predicate succeeds whenever N can be instantiated to the length of the list L. Notice that since PROLOG has no explicit notion of input and output parameters in clauses, the following goal will succeed:
```
?- length(X,3).
```

The variable X will be instantiated to a "hollow" list with three (non-instantiated) elements.

There are many other useful pre-defined predicates for list manipulation. The reader is referred to the user manual of her preferred PROLOG distribution.

Exercise 8.4.3 *(Finding the last element of a list)*

Write a PROLOG program to extract the last element of a list. Call the predicate to perform this operation `last_elem`, whose syntax should obey the following: `last_elem(L,X)` succeeds if and only if the list L has at least one element and X can be unified with the last element of L.

Note: SWI-Prolog already has pre-defined predicate to do this (it is called `last(L,X)`). However, the objective of this exercise is for you familiriase yourself with list manipulation in PROLOG.

Exercise 8.4.4 *(Adding up the values of a numerical list)*

Write a PROLOG program to display the total accumulated value of a list of integers. You should call your main predicate `acc_list(L,T)`, where L is the list of integers and T is the sum of all elements in L.
Example: The goal

```
?- acc_list([1,2,4],X).
```

must succeed and return the value X=7, whereas the goal

```
?- acc_list([0,1,3],3).
```

must fail.

Exercise 8.4.5 *(Insert an element in an ordered list)*

Write a PROLOG program to insert a number in an ordered list. You should call the predicate that accomplishes the insertion `sort_ins`. The predicate should have three arguments. The first is the number to insert; the second one is the ordered list where the number will be inserted; and the third is new list with the value inserted in the correct location. Assume that the list given as input is in ascending order. You can use the operator =<, where X=<Y returns true if X is smaller or equal to Y.

Example: The goal

```
?- sort_ins(4,[1,2,7],X).
```

must succeed and return the value X=[1,2,4,7].

8.5 Operators in PROLOG

We have already informally seen some operators in PROLOG. For instance, we have seen the addition + operator and the comparison operator =<. In this section, we will analyse these operators in more detail and show you how you define operators of your own.

Operations in PROLOG cannot be used directly (as a predicate), except for those whose evaluation provides a true/false answer, such as relational operators. This is the reason why we use the special predicate is in order to obtain the result of an arithmetic operation. For instance, instead of simply writing write(X+Y), we need to use the predicate is in the goal below in order to have the value resulting from the addition of X to Y written to the screen:

$$\cdots, \text{Z is X+Y, write(Z)}, \cdots$$

If we use $\cdots$, write(X+Y), $\cdots$, PROLOG will succeed, but will simply write "3+4" on the screen.

For the same reason, it does not make sense to write something like

$$\cdots, \text{X+Y}, \cdots$$

because X+1 evaluates to a number and prolog expects a `true/false` result when evaluating the body of a clause.

Comparisons operations that produce a true/false result can be used directly. For example,

$$\cdots, \text{X=<Y}, \cdots$$

must be used, instead of $\cdots$, Z is X =< Y, Z,$\cdots$

This is as it should be, since the predicate is can only be used with arithmetic expressions. Both is and relational operators such as =< will either succeed or fail, provided their arguments have the correct binding.

Operators are just functors, with their own particular syntax. Remember from Section 8.1.0.1 that a functor is the initial atom in a compound term. In fact, X+Y is more formally written as the compound binary term +(X,Y). Since the operation + is more commonly used in infix form, it is thus defined in PROLOG. More formally, operators can be of three forms:

(a) **infix**, when the operator appears in the middle of its two arguments. Such as the arithmetic operators below.

   Examples:

   – +, as in 3+4

- -, as in 4-5
- *, as in 2*7
- /, as in 3/6

(b) **prefix**, when the operator precedes its argument.

Examples:

- -2

And finally,

(c) **postfix**, when the operator appears after its argument.

Examples:

- The operator ! (factorial) as in 4![8]

8.5.1 Operator Precedence and Associativity

In the description of operators, atoms of the form `fx`, `xfx`, `xfy`, `yfx` are usually used. `f` indicates the position the operator assumes with respect to its arguments, whereas `x` and `y` indicate how the operator associates with other operators whenever parentheses are not explicitly given. For instance, `fx` represents a *postfix* operator, since the argument (represented by "x") appears after `f`. Similarly, `xf` would be associated to a *prefix* operator, whereas `xfx` would be associated with an *infix* operator.

The actual role played by `x` and `y` is a bit more complicated. An `x` indicates that any embedded operators must have a strictly lower precedence class than the class held by this operator. On the other hand, a `y` indicates that any embedded operators can be of a lower or same precedence class of the current operator. Precedence classes are given as a number between 0 and 1200. The special class 0 is used to remove a previously made declaration of the operator. We will see the classes associated with each operator in the table that follows. We hope that the meaning of `x` and `y` will become clearer.

---

[8]The factorial operator is not actually pre-defined in most PROLOG installations. As an exercise, write a simple program to calculate the factorial of any natural number.

8.5.2 Some built-in PROLOG operators

PROLOG provides a number of useful pre-defined operators. We will divide these into three types: *arithmetic*, *relational* and *extra-logical*. Arithmetic operators are associated with a mathematical operation such as addition, subtraction, multiplication, division, etc. There are also operators that can be used to evaluate whether numbers are positive or negative, to convert in between data types, etc. The relational operators provide a means of comparing terms in the language whereas the extra-logical operators help to control the way PROLOG evaluates queries.

8.5.2.1 Arithmetic operators

In modern PROLOG distributions, these operators are usually used with the special operator `is` which can be used with a free or bound left operand. PROLOG will try to unify the left operand to the right operand. If this is possible, then `is` will succeed and instantiate variables as necessary. If your objective is simply to test equality, then the operator `=:=` can also be used instead. This means that

$$X \text{ is } 2+0$$

will succeed and instantiate `X` to 2, but

$$X =:= 2+0$$

will fail, because the free variable `X` is not equal to the number 2 (although it can be instantiated to 2 as was the case with the use of `is`).

As a rule of thumb, if your intention is to obtain the result of an expression, use a free variable on the left-hand side of the operator `is`. The list of operators below is not comprehensive, check the manual of your distribution for more details.

Operator	Precedence	Associativity	Meaning
^, **	200	xfy	exponentiation
mod	300	xfx	rest of integer division
*	400	yfx	multiplication
/	400	yfx	division
+	500	fx	positive sign
-	500	fx	negative sign
+	500	yfx	addition
-	500	yfx	subtraction
is	700	xfx	evaluation

Understanding the associativity description given above

Say you give the expression -2*3 to PROLOG. How should the expression be interpreted? There are two options: (-2)*3 and -(2*3). In fact, every valid expression, regardless of the use of parentheses has a unique interpretation, much in the same way as the unique interpretation for logical formulas with the logical connectives ∧, ∨, →, etc.. In this example, the correct interpretation is -(2*3). The reason can be understood as follows: - as a prefix operator has associativity fx and precedence 500, which means that it is a prefix operator. It allows embedded operators on its right-hand side, as long as they have strictly lower precedence than 500. The operator * has precedence 400, which means that it must be evaluated before -.[9] The result is therefore -(2*3). Similarly, the expression -2+3, has the unique interpretation (-2) + 3, for the following reason: - and + have the same precedence orders. However, - will only allow on its right-hand side expressions with operators strictly less than 500, which is not the case of the expression 2+3. Therefore, it must be applied to the number 2. This must also agree with the definition of +, which is the case here. The associativity of + is yfx, and therefore, it accepts on its left-hand side expressions with operators with precedence at most 500. It follows that - can be used on the left-hand side of +. This gives the interpretation (-2)+3.

In practice, you can override these definitions by using parentheses in a sensible way. If you want to be clear, use -(2*3), because this format does not assume knowledge of the reader about the arithmetic operators' precedence order. Obviously, PROLOG's arithmetic operators are defined in a such a way to agree with their common use in Mathematics: ^ binds more tightly than * and /, which in turn bind more closely than + and -.

8.5.2.2 Relational operators

The operators in the table below are self-explanatory. The only ones worth mentioning are the ones involved with equality and equivalence. These will be discussed next in a series of examples. The \+ operator (also known as "not") will be discussed in the next section.

---

[9]Lower numbers indicate higher priority. The lower the precedence, the earlier an operator must be evaluated.

8.5. OPERATORS IN PROLOG

Operator	Precedence	Associativity	Meaning
<	700	xfx	less than
>	700	xfx	greater than
=<	700	xfx	less than or equal
>=	700	xfx	greater than or equal
=	700	xfx	succeeds if the operands are unifiable
\=	700	xfx	not the case that operands are unifiable
==	700	xfx	equivalent expressions
\==	700	xfx	expressions are not equivalent
=:=	700	xfx	equivalent arithmetic expressions
=\=	700	xfx	non-equal expressions
\+, not	900	fy	fails if the argument succeeds, succeeds otherwise

- `Term1 = Term2` succeeds if `Term1` and `Term2` can be unified, i.e., if they can be made equal by some substitution of variables.

- `Term1 \= Term2` succeeds, if `Term1 = Term2` fails.

- `Term1 =:= Term2` succeeds, if `Term1` and `Term2` evaluate to equal numbers. Unlike = and ==, =:= forces the evaluation of the expression on either side first.

- `Term1 =\= Term2` succeeds if `Term1` and `Term2` evaluate to non-equal numbers.

- `Term1 == Term2` succeeds if `Term1` and `Term2` are equivalent. Variables are only equivalent if they are bound to the same value.

- `Term1 \== Term2` succeeds if `Term1 == Term2` fails.

Examples: (Assume that `X` and `Y` are free variables)

- `X = 2+3` succeeds and instantiates `X` to the expression `2+3`.

- `X =:= 2+3` fails as `X` is not bound to an expression or number.

- X = 2+3, X =:=5 succeeds, as X becomes the expression 2+3 by the first clause, and then can be evaluated to a value which is equal to 5. Note that this would fail if any value other than 5 is used.

- X == 5 fails and so does X == Y. 2+3 == 3+2 fails, but 2+3 == 2+3 succeeds.

- X =\= 2+3 generates an exception, as X does not evaluate to any number.

- 4+2 =\= 5-3 succeeds.

- X \= 5 fails, as X is unifiable to 5.

- X is 3, X \= 5, succeeds, as X is 3 succeeds and instantiates X to 3, which then cannot be unified with 5.

- X \== 5 succeeds and so does X \== Y unless they have been previously bound to the same term.

8.5.2.3 The \+ operator

The \+ operator can be used with any term that you would normally use in a goal clause or in the body of a program clause. In some systems, \+ is available only as **not**. In SWI-Prolog both notations are accepted, the form **not** being kept for compatibility reasons. We will refer to both notations simply as **not**, instead of the more formal definition of \+, namely, "not provable using the current program".

A goal of the form

$$?\text{- }\text{\textbackslash}+(\texttt{any\_goal}).$$

will succeed if `any_goal` fails, and will fail if `any_goal` succeeds.

Examples: The following queries will be answered as follows when submitted to the program below.

```
age(mary,25).
age(paul,33).
older(X,Y):- age(X,AgeX), age(Y,AgeY), AgeX > AgeY.

?- age(mary,27).
false.
```

8.5. OPERATORS IN PROLOG

```
?- not(age(mary,27)).
true.

?- not(older(mary,paul)).
true.
```

But be careful with the intended meaning of your query, especially if it has free variables. For instance, if you ask

```
?- not(older(X,Y)).
```

The answer will be `false`, instead of what some might expect "all values of X, such that X is not older than Y". The reason is simple to understand: `not(older(X,Y))` fails if `older(X,Y)` succeeds, which is the case here. The correct interpretation to the query is "is it true that for no values of X and Y, it is the case that X is older than Y?" This is false. If you take X=paul and Y=mary, then `older(paul,mary)` succeeds, and therefore, `not(older(X,Y))` fails.

8.5.3 Declaring your own operators

You can declare your own operators with the predicate `op(Precedence, Type, Name)`. `op` takes three arguments: `Precedence` is a number between 0 and 1200 stating the precedence of the operator as explained before. `Type` is one of the atoms `xfy`, `xfx`, `fx`, etc, and `Name` is the "name" of the new operator. Obviously, just declaring an operator is not enough, you must also give a meaning to it.

If the operator is to be used as an arithmetic expression, you must use the SWI-Prolog predicate `arithmetic_function(nfunction/Arity)`, where `nfunction` is the "name" of the new arithmetic function and `Arity` is the number of arguments it takes. If the operator is to be used in comparisons, you can define it in a PROLOG clause in the usual way. Just remember that no matter what type you give to the operator, they are internally treated as "functor(... arguments ...)" by PROLOG and hence that is how their declaration must be made.

The example below should help you to understand the concepts:

Example: `op(1000,xfx,minus)` declares a new infix operator called `minus`, with precedence 1000 and of associativity `xfx`. Here we see a novelty! Some clauses in the program body have no head. This makes PROLOG evaluate them as soon as the program is loaded into memory.

In your program:

```
:-initialization(op(1000,xfx,precedes)).
:-initialization(op(1000,xfx,minus)).
:-arithmetic_function(minus/2).
minus(X,Y,Z) :- Z is X-Y.
precedes(X,Y):- X < Y.
```

The first two lines in the program declare two new operators: `minus` and `precedes`, both of which are infix. The predicate `initialization` tells PROLOG that `op(1000,xfx,precedes)` is an initialisation directive. `precedes` is a relational operator, and therefore will evaluate to either true or false. That is exactly how `precedes(X,Y)` is declared: it succeeds if `X < Y` succeeds and fails otherwise.

`minus(X,Y)` is an arithmetic function and returns the result of subtracting the second argument from the first. Therefore, it is declared with the `arithmetic_function` predicate. The call `arithmetic_function(minus/2)` declares a clause with head `minus` and three arguments (the two stated in /2 plus a third one which is the result of the operation). In the actual definition of `minus`, we must use three arguments, the third one being instantiated to the value we want `minus` to return as defined in the program line

```
minus(X,Y,Z) :- Z is X-Y.
```

You can now use your newly created operators. For instance, the following clauses are valid PROLOG statements (the answer given by PROLOG is also given).

```
?- 2 precedes 3.
true.

?- 3 precedes 3.
false.

?- X is (3 minus 2).
X = 1.
```

8.6 Backtracking

We have already experienced backtracking indirectly as a result of evaluating queries in some programs. The following examples will illustrate

8.6. BACKTRACKING

the general idea.

Consider the following PROLOG program:

```
father(paul,susan).
mother(mary,susan).
parent(X):- father(X,_).
parent(X):- mother(X,_).
```

If we ask the query

?- parent(mary).

PROLOG will look for answers in the program. The first definition of parent is in the rule

parent(X):- father(X,_).

It is possible to unify X with mary, so the new goal now becomes

?- father(mary,_).

This goal now fails, because PROLOG cannot find any clauses which can unify with it. However, there is an alternative definition in the code for the original goal parent(X). Since the first attempt of proof failed, PROLOG now *backtracks* and tries the alternative definition:

parent(X):- mother(X,_).

Again, it is possible to match X with mary, so the new goal now becomes

?- mother(mary,_).

_ unifies with susan and the goal succeeds. The execution pattern can be seen in the tree of Figure 8.2.

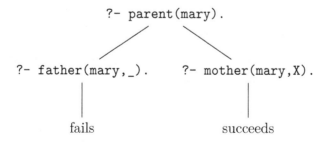

Figure 8.2: Backtracking in PROLOG.

You can find a comprehensive example for the PROLOG program and goal clauses given below in Figure 8.3. Whenever a failed branch is found, the program backtracks to the level immediately above containing alternative solutions. These solutions are searched all the way down until a solution is found or the branch fails.

Program 8.8. Sample PROLOG Program

```
local_port(london,lhr).
local_port(paris,cdg).
local_port(paris,orly).
local_port(amsterdam,sch).
local_port(cannes,nice).
direct(lhr,cdg).
direct(lhr,orly).
direct(lhr,sch).
direct(nice,orly).
link(X,X).
link(X,Y):- direct(X,Z), link(Z,Y).
link(X,Y):- direct(Y,Z), link(Z,X).
connect(A,B):- local_port(A,X), local_port(B,Y),
link(X,Y).
```

Goal:

?- connect(london,cannes).

8.7 Cut

The operator !, and read as "cut", is a built-in PROLOG predicate that always succeeds. Cut has however the effect of limiting the backtracking performed by PROLOG. As a result, not all possible solutions to a given predicate/query are searched for.

This can be best seen by means of an example. Consider the PROLOG program and the goal given below.

8.7. CUT

Program 8.9. Sample PROLOG program used to illustrate Cut

```
father(X,Y):- parent(X,Y), male(X).
parent(paul,susan).
parent(mary,susan).
male(paul).
female(mary).
female(susan).
```

Goal
?-father(X,susan).

The search tree for this program can be seen below.

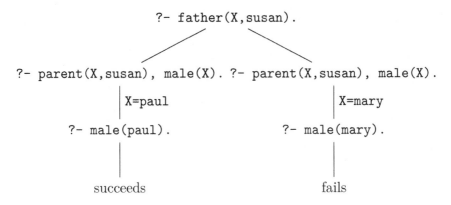

Now observe the use of cut in the modified version of Program 8.9 program with the goal clause below:

Program 8.10. Modified version of Program 8.9 including cut.

```
father(X,Y):- parent(X,Y), male(X), !.
parent(paul,susan).
parent(mary,susan).
male(paul).
female(mary).
female(susan).
```

Cut prevents backtracking to the sub-goals on its left-hand side. Execution proceeds as usual with the literals following the cut. For example, if you submit the goal ?-father(X,susan). to Program 8.10, then, PROLOG would interrupt the search after the successful call to male(paul) and as a result, the failed branch terminating at male(mary) would never be explored. This is because once male(paul) succeeds, the next predicate to evaluate it !, which succeeds and prevents any backtracking in any of the clauses appearing before it. The use of cut in the example above is safe, because the intended meaning of the predicate father is to provide one single solution. Moreover, the branches pruned by cut would have failed anyway.

However, it is possible to use cut in a less elegant and potentially dangerous way as demonstrated in the program found on the left-hand side of the table below. A version demonstrating the correct use of cut can be found on the right-hand side of the table.

Wrong use	Correct use
min(X,Y,X):- X < Y, !.	min(X,Y,X):- X<Y.
min(_,Y,Y).	min(X,Y,Y):- Y=<X.

The program on the left-hand side will provide the right answers whenever the third argument is free. For example, consider the goals

$$?- min(4,5,X).$$

and

$$?-min(7,2,X).$$

Both will succeed providing the correct answers X=4 and X=2, respectively. However, the program on the left will give an incorrect answer whenever the value of the first argument is less than the value of the second argument, but the second and third arguments are equal. For instance, the goal

$$?- min(2,7,7).$$

will succeed, when this clearly should not be the case.

8.8 PROLOG and classical logic

This section will explain the relationship between the execution of a PROLOG program and the underlying logical background supporting it. We shall see that when we submit a query to a PROLOG program, what we actually do is to try to prove that the query (with associated variable bindings) follows from the formulae representing the program.

In this book, we have seen a few techniques that can be used to check whether a conclusion (e.g., a query), follows from a given set of assumptions (e.g., a program):

1. The truth-tables of Chapter 1.

2. The natural deduction technique using forward rules of Chapter 3.

3. And the backward reasoning method presented in Chapter 4.

If we see the clauses of a PROLOG program as formulae of predicate logic, the execution of the program can be seen as a form of theorem proving and the answers to the queries can be regarded as the by-product of the proof process. In PROLOG a query to a program is implicitly negated. PROLOG tries to find a refutation from the query and the set of assumptions (the program). This amounts to showing that the program and the negation of the query are together inconsistent (the is what the empty clause represents). There is a procedural interpretation to all this which is actually very similar to technique 3. mentioned above, except that the goal-directed technique is more powerful, since it captures the full expressive power of classical logic. As we have seen, the negation in PROLOG is a weaker variant of classical logic negation known as *negation by failure*.

It is important to emphasize that the goal is implicitly negated and we will see that the translation from clauses to first-order formulae shows this clearly. The reason for this is that it is possible to check whether the argument $\{A_1, A_12, \ldots, A_n\} \vdash B$ is valid, by checking whether the set $\{A_1, A_2, \ldots, A_n, \neg B\}$ is inconsistent. The formulae $\{A_1, A_2, \ldots, A_n\}$ constitute our program and B is the goal we try to prove.

8.9 Clauses as first-order logic formulae

Consider the following PROLOG program:

```
mother(susan,john).
father(paul,john).
mother(mary,edward).
parent(X,Y):- mother(X,Y).
```

```
parent(X,Y):- father(X,Y).
sibling(X,Y):- parent(Z,X), parent(Z,Y).
```

We have seen that all clauses in a PROLOG program are implictly universally quantified; that the symbol `:-` can be read as "if" ($\leftarrow$) and that the comma "," represents conjunction. This gives the clause

```
parent(X,Y):- mother(X,Y).
```

the interpretation

"X is a parent of Y if X is the mother of Y"

By the assumption that all variables in a clause are implicitly universally quantified, the same clause can be translated into the following predicate logic formula:

$$\forall x.\forall y.(mother(x,y) \rightarrow parent(x,y))$$

We have intentionally swapped the order of the atoms in the clause. This is just to reflect the meaning of the clause which can of course also be read as

"if X is the mother of Y, then X is a parent of Y"

Or even more precisely

"for all X and for all Y, if X is the mother of Y then X is a parent of Y"

Obviously the translation of the clause to the formula

$$\forall x.\forall y.(parent(x,y) \leftarrow mother(x,y))$$

is equally acceptable, as long as the logical language has the connective $\leftarrow$. We have not used "$\leftarrow$" directly in the description of our first-order language and, therefore, we prefer the other form.

Clauses with "," in the body can be translated in a similar way. All we need to do is to replace "," with $\wedge$. For instance, the clause

```
sibling(X,Y):- parent(Z,X), parent(Z,Y).
```

could be translated into the formula

$$\forall x.\forall y.\forall z[(parent(z,x) \wedge parent(z,y)) \rightarrow sibling(x,y)]$$

Translating operators is similar. Relational operators can be translated directly, by considering the operator as a predicate symbol that is evaluated to *true* or *false*. However, arithmetic operators require the introduction of function symbols and pattern matching with lists requires additional constraints. We will not cover translation of operators in this book.

8.9. CLAUSES AS FIRST-ORDER LOGIC FORMULAE

Translating goal clauses

The translation of goal clauses follows the same pattern. You just have to take into account the meaning of the "←" connective, which is represented by :- in a PROLOG clause, and by ?- in a goal clause. Consider the following goal clause

    ?- sibling(X,Y).

This is a clause with an empty head, which you think of as $\bot$. With this in mind, everything becomes very similar to the concepts given in the preceding chapters if thsi book. We have seen that the goal-directed technique for theorem proving represents negation via $\bot$. For instance, the $\neg A$ was rewritten as $A \to \bot$. The clause in the goal ?- sibling(X,Y) can therefore be translated to the predicate formula

$$\forall x.\forall y.(sibling(x,y) \to \bot)$$

which is equivalent to

$$\forall x.\forall y.(\neg sibling(x,y))$$

Compound goal clauses have similar translations. Consider the clause

    ?-parent(Z,X), parent(Z,Y).

Following the same reasoning, we can re-write the clause as

$$\forall x.\forall y.\forall z.((parent(z,x) \land parent(z,y)) \to \bot)$$

which is equivalent to the formula

$$\forall x.\forall y.\forall z \neg(parent(z,x) \land parent(z,y))$$

There is a special reason for writing formulae in this way. It is always possible to translate them to a disjunction of literals with *at most* one positive literal. The theorem prover working behind the scenes takes advantage of this fact using the rule of *resolution* that we say in Chapter 7. Since any clause in the program has exactly one positive literal and goal clauses have none, there is a unique way to resolve them.

8.10 Theorem proving in PROLOG

Let us look into more detail at what happens when a query is submitted to a PROLOG program and what the whole process means in logical terms. Consider again Program 8.1 and let us analyse the execution from a different perspective:

```
studies(mark,logic).
studies(lucy,geometry).
studies(lucy,english).
studies(paul,geometry).
studies(paul,logic).
subject(john,english).
subject(dov,logic).
subject(mary,geometry).
teaches(X,Y) :- subject(X,Z), studies(Y,Z).
```

Query: ?- studies(mark,X).

Looking at the program, we can easily see that the answer to the query will be true with X=logic. What we need to understand in more detail now is how this follows from classical logic. First of all, notice that the answer X=logic makes sense, since the query ?- studies(mark,X). is universally quantified, and one of the possible values for X is the term logic (which is in the domain of the program). By substituting logic for X in the clause above, our problem becomes to prove that studies(mark,logic) follows from the set of formulae represented by the program.[10] PROLOG performs the proof by showing that the set representing the program together with the goal clause is inconsistent.

This is Program 8.1 seen as a set of predicate logic formulae:
$\Delta = \{$ $studies(paul, geometry),$
          $studies(mark, logic),$
          $studies(lucy, geometry),$
          $subject(mary, geometry),$
          $\forall x. \forall y. \forall z.((subject(x,z) \land studies(y,z)) \to teaches(x,y))$ $\}$

The goal clause seen as a predicate logic formula is

$$G = \forall x. \neg studies(mark, X)$$

---

[10]This is obvious, since studies(mark,logic) is in the program and classical consequence is *reflexive*.

8.10. THEOREM PROVING IN PROLOG

$\Delta \cup G$ is obviously inconsistent. If you substitute, for instance, *logic* for x in G, we obtain the set

$\Delta = \{$ $studies(paul, geometry),$
           $studies(mark, logic),$
           $studies(lucy, geometry),$
           $subject(mary, geometry),$
           $\forall x. \forall y. \forall z ((subject(x,z) \land studies(y,z)) \rightarrow teaches(x,y))$
           $\neg studies(mark, logic)$ $\}$

which is clearly inconsistent. You can think of the answer `logic` as the value whose substitution for `X` in the query `?-studies(mark,X).` gives an inconsistency (there could be other values, which can be ontained by successively asking for more answers with `;`).

It is even easier to interpret the whole process as a goal-directed technique such as the one we have seen in Chapter 4. Consider the query

                     `?- teaches(X,lucy).`

with intended meaning "Who teaches Lucy?" and look at the set Δ representing the program.

By replacing *lucy* for y in the last formula, we obtain the formula

$$\forall x. \forall z((subject(x,z) \land studies(lucy,z)) \rightarrow teaches(x,lucy))$$

In a goal-directed reasoning technique we can think backwards, so we can prove $teaches(x, lucy)$ if we prove that the following goal follows

                 `?- subject(X,Z), studies(lucy,Z).`

By similar reasoning, we can now replace *mary* for x and *geometry* for z in the goal clause above. We need to remember that the `X` of the original goal is the same x used in this substitution. If success is obtained, these substitutions must be reported back to the user, and they become the answers to the original query. The acual names of the variables are irrelevant otherwise. The `X` from the query and the x from the formula in the program incidentally coincided here, but this need not be the case. In practice, what really matters is the position the variable occupies in the predicate (since the clauses are universally quantified). The substitutions mentioned before result in the following new goal:

       `?- subject(mary,geometry), studies(lucy,geometry).`

`subject(mary,geometry)` can be proved, since it is in the database (4th clause). The new goal then becomes

?- `studies(lucy,geometry).`

which also succeeds, since that literal is also in the database (3rd clause). The original goal succeeds and the by-product is the substitution used in the proof.

The universal qualification explains why the actual names of the variables in a PROLOG program are not important across different clauses, but rather the positions they occupy in the predicates. Look at the two clauses below:

(A) `teaches(X,Y) :- subject(X,Z), studies(Y,Z).`
(B) `teaches(A,B) :- subject(A,C), studies(B,C).`

(A) and (B) correspond to the predicate logic formulae (A*) and (B*) below:

(A*) $\forall x.\forall y.\forall z.((subject(x,z) \land studies(y,z)) \to teaches(x,y))$
(B*) $\forall a.\forall b.\forall c.((subject(a,c) \land studies(b,c)) \to teaches(a,b))$

which are logically equivalent.

8.11 From PROLOG to classical logic

The translation to classical logic is quite straightforward. We outline the main steps to translate a PROLOG program into a set of first-order formulae.

1. All variables in a program are universally quantified.

   Given a program clause of the form

   `head_pred :- body_pred1, ..., body_predk.`

1. Create a universally quantified prefix with all the variables appearing in the clause.

2. Create a conjunction of all the literals appearing in the body of the clause.

3. The final formula will be an implication with the prefix found in (1.); the antecedent as the conjunction found in (2.); and consequent `head_pred`.

8.11. FROM PROLOG TO CLASSICAL LOGIC

PROLOG facts translate roughly as they already appear in the program, with the exception that some might contain variables too. In this case, these variables must appear in the translation as universally quantified, in the same way that was done for the rules.

Given a goal clause of the form

?- body_pred1, ..., body_predk.

1. Create a universally quantified prefix with all the variables appearing in the goal.

2. Create a conjunction of all the literals appearing in the goal.

3. The final formula will be the negation of a formula with the prefix found in (1.) and the body found in (2.).

We now conclude this chapter with some examples of translations of program and goal clauses to predicate logic formulae.

Examples:

Rules:

Rule	Translation
bird(X) :- penguin(X).	$\forall x.(penguin(x) \rightarrow bird(x))$
loves(X,Y) :- loves(Y,X).	$\forall x.\forall y.(loves(y,x) \rightarrow loves(x,y))$
depends(X,Y):- calls(X,Z), depends(Z,Y).	$\forall x.\forall y.\forall z((calls(x,z) \land depends(z,y)) \rightarrow depends(x,y))$

Facts:

Fact	Translation
student(john).	$student(john)$
thing(X).	$\forall x.thing(x)$

Goal clauses:

Goal	Translation
:- husband(X,Y).	$\forall x.\forall y \neg husband(x,y)$
:- employee(X), employee(Y), married(X,Y).	$\forall x.\forall y \neg (employee(x) \land employee(y) \land married(x,y))$

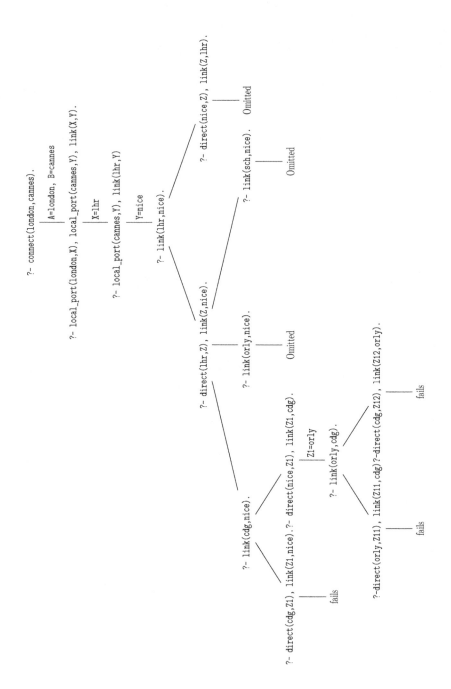

Figure 8.3: Backtracking in PROLOG.

ANSWERS TO THE EXERCISES

Exercise 1.1.1

The truth values of the formulae, when p has the value $\top$, and q has the value $\bot$, are:

1. $\neg((p)\wedge(q))$ is $\top$
2. $(p)\to(p)$ is $\top$
3. $(p)\to(\neg(p))$ is $\bot$
4. $((q)\vee(\neg(q)))\wedge(p)$ is $\top$

Exercise 1.1.3

You were asked to find the truth tables for the following formulae:

1. $\neg A \vee B$ has the same table as $A \to B$.
2. $\neg(\neg A \wedge \neg B)$ has the same table as $A \vee B$.
3. $A \vee \neg A$ has $\top$ in every row in its table.
4. $(A \wedge B \to C) \to [A \to (B \to C)]$ also has $\top$ in every row in its table.
5. $(A \to B) \wedge (B \to A)$ has the following table:

A	B	$(A\to B)\land(B\to A)$
T	T	T
T	⊥	⊥
⊥	T	⊥
⊥	⊥	T

6. $[A\to(B\land C)]\to[(A\to B)\land(A\to C)]$ again has T in every row in its table.

Exercise 1.1.4

1. The wff for the first table is $(p\land q)\lor(\neg p\land\neg q)$, another possibility being $(p\to q)\land(q\to p)$. Both formulae have this table. The wff for the second table of this exercise is

$$[p\land q\land r]\lor[\neg p\land\neg q\land r]\lor[\neg p\land q\land\neg r]\lor[p\land\neg q\land\neg r]$$

2. Let p stand for 'the television is switched on', q stand for 'the television has power', and r for 'there is a picture'. The data to be checked for consistency is

$$p\land(p\land q\to r)\land\neg r$$

We need to draw up a truth table to see whether we can find truth values for the propositions so that the formula is true. The table is

p	q	r	$p\land q\to r$	$\neg r$	$p\land(p\land q\to r)\land\neg r$
T	T	T	T	⊥	⊥
T	T	⊥	⊥	T	⊥
T	⊥	T	T	⊥	⊥
T	⊥	⊥	T	T	T
⊥	⊥	T	T	⊥	⊥
⊥	⊥	⊥	T	T	⊥
⊥	⊥	T	T	⊥	⊥
⊥	⊥	⊥	T	T	⊥

As the third row has a T for the full conjunction, the sentences can be all true together.

Exercise 1.2.2

1. The definitions using the Sheffer stroke '|' are:

$$\neg p \stackrel{\text{def}}{=} p|p$$

$$p \vee q \stackrel{\text{def}}{=} \neg p | \neg q \equiv (p|p)|(q|q)$$
$$p \wedge q \stackrel{\text{def}}{=} \neg(\neg p \vee \neg q) \equiv (p|q)|(p|q)$$
$$p \rightarrow q \stackrel{\text{def}}{=} \neg p \vee q \equiv p|(q|q)$$

2. All except the third formula are tautologies.

Exercise 1.2.5

To write the disjunctive normal form equivalent to

$$A \equiv [(p \rightarrow q) \vee r] \wedge \neg q \rightarrow p$$

we can use two methods. One, which we have already learnt, is to find the truth table for A and then find the wff for the table. Two, use equivalences to transform A into its normal form. We start with the first method. Here is the table for A.

p	q	r	$\neg q$	$p \rightarrow q$	$(p \rightarrow q) \vee r$	$((p \rightarrow q) \vee r) \wedge \neg q$	A
T	T	T	$\bot$	T	T	$\bot$	T
$\bot$	T	T	$\bot$	T	T	$\bot$	T
T	$\bot$	T	T	$\bot$	T	T	T
$\bot$	$\bot$	T	T	T	T	T	$\bot$
T	T	$\bot$	$\bot$	T	T	$\bot$	T
$\bot$	T	$\bot$	$\bot$	T	T	$\bot$	T
T	$\bot$	$\bot$	T	$\bot$	$\bot$	$\bot$	T
$\bot$	$\bot$	$\bot$	T	T	T	T	$\bot$

The wff of this table is:

$$[p \wedge q \wedge r] \vee [\neg p \wedge q \wedge r] \vee [p \wedge \neg q \wedge r] \vee [p \wedge q \wedge \neg r] \vee [\neg p \wedge q \wedge \neg r] \vee [p \wedge \neg q \wedge \neg r]$$

As we can see the wff is quite long. We can find a shorter formula if we look at the table for $B \equiv \neg A$. This table has only two Ts, in the last row (where A has $\bot$) and in the fourth row. The formula for the table for $B \equiv \neg A$ is therefore $B \equiv (\neg p \wedge \neg q \wedge r) \vee (\neg p \wedge \neg q \wedge \neg r)$, B is equivalent to $\neg p \wedge \neg q$ and so A is therefore $A \equiv \neg B \equiv \neg(\neg p \wedge \neg q)$ and hence $A \equiv p \vee q$, a much simpler formula.

We now demonstrate the second method, using equivalences. We know that $A \rightarrow B$ is the same as $\neg A \vee B$. Hence $[(p \rightarrow q) \vee r]$ is the same as $[\neg p \vee q \vee r]$. The expression is now

$$[\neg p \vee q \vee r] \wedge \neg q \rightarrow p$$

We also know that
$$(A\lor B)\land C \equiv (A\land C)\lor(B\lor C)$$
Hence
$$[\neg p\lor q\lor r]\land\neg q \equiv (\neg p\land\neg q)\lor(q\land\neg q)\lor(r\land\neg q)$$
Since $q\land\neg q$ is a contradiction, we have for any x
$$x\lor(q\land\neg q) \equiv x$$
Thus we are left with
$$[\neg p\lor q\lor r]\land\neg q \equiv (\neg p\land\neg q)\lor(r\land\neg q)$$
The expression becomes $A \equiv [(\neg p\land\neg q)\lor(r\land\neg q)]\to p$. Since $x\to y \equiv \neg x\lor y$ we get $A \equiv \neg[(\neg p\land\neg q)\lor(r\land\neg q)]\lor p$. Since $\neg(x\land y) \equiv \neg x\lor\neg y$ and $\neg(x\lor y) \equiv \neg x\land\neg y$, we get

$$\begin{aligned}A &\equiv [\neg(\neg p\land\neg q)\land\neg(r\land\neg q)]\lor p\\ &\equiv [(p\lor q)\land(\neg r\lor q)]\lor p\end{aligned}$$

Distributing, we get

$$\begin{aligned}A &\equiv [p\land(\neg r\lor q)]\lor[q\land(\neg r\lor q)]\lor p\\ &\equiv (p\land\neg r)\lor(p\land q)\lor(q\land\neg r)\lor(q\land q)\lor p\\ &\equiv (p\land\neg r)\lor(p\land\neg q)\lor(q\land\neg r)\lor q\lor p\end{aligned}$$

This has the same truth table as $p\lor q$.

Exercise 1.2.7

1. A logically implies B iff $A\to B$ is a tautology. $A\land\neg B$ is not consistent iff $\neg(A\land\neg B)$ is a tautology. We must therefore show the equivalence $(A\to B) \equiv \neg(A\land\neg B)$, which we can do using the equivalences $p\to q \equiv \neg p\lor q$ and $p\lor q \equiv \neg(\neg p\land\neg q)$.

2. Yes.

Exercise 1.2.9

1. The answer depends on whether n is even, in which case the set of sentences can be consistent, or odd, in which case it is impossible for the set to be consistent.

ANSWERS TO THE EXERCISES

Exercise 1.2.10

1. The sentences translate as follows:

   (a) $p \rightarrow q \wedge \neg s$

   (b) $r \wedge q \rightarrow \neg s$

   (c) We cannot express that Mary loves herself. The atomic sentences given for this exercise do not contain an appropriate proposition. We would need to add a new atom, t say, to stand for 'Mary loves Mary.'

   Now we can translate as $p \wedge q \rightarrow t$.

2. Using the equivalence rules to push all negations onto the atoms:

   (a) $\neg((a \rightarrow b) \vee (a \rightarrow c) \wedge \neg a)$
   $\equiv \neg(a \rightarrow b) \wedge \neg((a \rightarrow c) \wedge \neg a)$ ($\wedge$ binds tighter than $\vee$)
   $\equiv a \wedge \neg b \wedge (\neg(a \rightarrow c) \vee \neg a)$
   $\equiv a \wedge \neg b \wedge (a \wedge \neg c \vee \neg a)$

   (b) $\neg(a \wedge \neg b) \rightarrow a$
   $\equiv \neg\neg(a \wedge \neg b) \vee a$
   $\equiv a \wedge \neg b \vee a$

3. Simply draw up the truth tables for the left and right sides of each equivalence, and it is easy to see that the sides take on the same truth values as each other, throughout the truth table.

4. The arguments translate as follows:

   (a) p = it is warm
   q = it is raining
   r = we go outside

   $p \vee q$
   $\underline{\neg p \vee \neg r}$
   r

   The truth table for $(p \vee q) \wedge (\neg p \vee \neg r) \rightarrow r$ shows that it is not a tautology, and hence that this argument is not logically valid.

   (b) p = the king is in the room
   q = the courtiers laugh
   r = the jester is in the room
   s = the king laughs

$$p \to (q \to s)$$
$$r \to q$$
$$\underline{r \to \neg s}$$
$$\neg p \vee \neg r$$

The truth table for $(p \to (q \to s)) \wedge (r \to q) \wedge (r \to \neg s) \to (\neg p \vee \neg r)$ shows that it is a tautology, and hence that this argument is also logically valid.

(c) p = we are hungry
        q = the food is very hot
        r = we eat slowly

$$\neg p \wedge q \to r$$
$$\neg p \to q \vee r$$
$$\underline{\neg q}$$
$$p$$

This argument is not valid. If you draw up the truth table for $(\neg p \wedge q \to r) \wedge (\neg p \to q \vee r) \wedge (\neg q) \to p$, the row with r true and p and q both false would have false for the implication.

(d) Let us denote 'either A or B' by $A \oplus B$. Its truth table

A	B	$A \oplus B$
T	T	⊥
T	⊥	T
⊥	T	T
⊥	⊥	⊥

shows that $A \oplus B \equiv (A \wedge \neg B) \vee (\neg A \wedge B)$.

Now let

     j = Jones did not meet Smith last night,
     s = Smith was a murderer,
     l = Jones is telling a lie,
     m = the murder happened after midnight,

and consider the argument in question:

$$\frac{j \to s \oplus l, \neg s \to j \wedge m, m \to s \oplus l}{s}$$

This argument is not valid, because we can take the values

j	s	l	m
T	⊥	T	T

ANSWERS TO THE EXERCISES

Exercise 2.1.1

1. $d_1(x) = x \to x$
   $d_3(x) = \neg(x \to x)$
   $d_2(x)$ does not exist because every formula gets the value 0 or 1 when each of its arguments is 0 or 1. (The latter is easily proved by induction.)

2.

A	1	1	1	1/2	1/2	1/2	0	0	0
B	1	1/2	0	1	1/2	0	1	1/2	0
$\neg A$	0	0	0	1/2	1/2	1/2	1	1	1
$\neg A \vee B$	1	1/2	0	1	1/2	1/2	1	1	1
$A \vee \neg A$	1	1	1	1/2	1/2	1/2	1	1	1
$\neg B$	0	1/2	1	0	1/2	1	0	1/2	1
$\neg A \wedge \neg B$	0	0	0	0	1/2	1/2	0	1/2	1
$A \to B$	1	1/2	0	1	1	1/2	1	1	1
$B \to A$	1	1	1	1/2	1	1	0	1/2	1
$(A \to B) \wedge (B \to A)$	1	1/2	0	1/2	1	1/2	0	1/2	1

This gives the answers for (a), (b), (c), (e). The answers for (d), (f) are given by Table 8.1.

3. (a) First let us show that for any k,

$$(*) \quad (\neg x)^k \to x = \min(1, (k+1)x)$$

This comes by induction:
If $k = 0$, we have $\neg x^0 \to x = x = \min(1, x)$.
If $(k+1)x \geq 1$, then also $(k+2)x \geq 1$, and we have $(\neg x)^{k+1} \to x = \neg x \to ((\neg x)^k \to x) = \neg x \to 1$ (by the induction hypothesis) $= 1 = \min((k+2)x, 1)$.
If $(k+1)x < 1$, we have

$$(\neg x)^{k+1} \to x = \neg x \to (k+1)x = \min(1, 1 - \neg x + (k+1)x)$$
$$= \min(1, (k+2)x)$$

Now if $k \geq n - 1$, it follows that $(k+1)x \geq 1$, unless $x = 0$ (recall that $x \in \{0, 1/n, \ldots, 1\}$). Hence $(\neg x)^k \to x$ is either 0 or 1.

A	B	C	$A\wedge B$	$B\wedge C$	$B\to C$	$A\to(B\to C)$	$A\wedge B\to C$	(d)	$A\to B$	$A\to C$	$(A\to B)\wedge(B\to C)$	$A\to B\wedge C$	(f)
1	1	1	1	1	1	1	1	1	1	1	1	1	1
1	1	1/2	1	1/2	1/2	1/2	1/2	1	1	1/2	1/2	1/2	1
1	1	0	1	0	0	0	0	1	1	0	0	0	1
1	1/2	1	1/2	1/2	1	1	1	1	1/2	1	1/2	1/2	1
1	1/2	1/2	1/2	1/2	1	1	1	1	1/2	1/2	1/2	1/2	1
1	1/2	0	1/2	0	1/2	1/2	1/2	1	1/2	0	1/2	0	1
1	0	1	0	0	1	1	1	1	0	1	0	0	1
1	0	1/2	0	0	1	1	1	1	0	1/2	0	0	1
1	0	0	0	0	1	1	1	1	0	0	0	0	1
1/2	1	1	1/2	1	1	1	1	1	1	1	1	1	1
1/2	1	1/2	1/2	1/2	1/2	1	1	1	1	1	1/2	1	1
1/2	1	0	1/2	0	0	1/2	1/2	1	1	1/2	0	1/2	1
1/2	1/2	1	1/2	1/2	1	1	1	1	1	1	1	1	1
1/2	1/2	1/2	1/2	1/2	1	1	1	1	1	1	1	1	1
1/2	1/2	0	1/2	0	1/2	1	1	1	1	1/2	1/2	1/2	1
1/2	0	1	0	0	1	1	1	1	1/2	1	1/2	1/2	1
1/2	0	1/2	0	0	1	1	1	1	1/2	1	1/2	1/2	1
1/2	0	0	0	0	1	1	1	1	1/2	1/2	1/2	1/2	1
0	1	1	0	1	1	1	1	1	1	1	1	1	1
0	1	1/2	0	1/2	1/2	1	1	1	1	1	1/2	1	1
0	1	0	0	0	0	1	1	1	1	1	0	1	1
0	1/2	1	0	1/2	1	1	1	1	1	1	1	1	1
0	1/2	1/2	0	1/2	1	1	1	1	1	1	1	1	1
0	1/2	0	0	0	1/2	1	1	1	1	1	1/2	1	1
0	0	1	0	0	1	1	1	1	1	1	1	1	1
0	0	1/2	0	0	1	1	1	1	1	1	1	1	1
0	0	0	0	0	1	1	1	1	1	1	1	1	1

Table 8.1: Partial answer to Exercise 2.1.1 part 2.

(b) A possible example is $[(\neg x)^{n-1} \to x] \lor \neg[(\neg x)^{n-1} \to x]$, with x atomic.

As we have proved in the solution of 3(a), $(\neg x)^{n-1} \to x$ is either 0 or 1 in L_n; but in L_{n+1} we have $(\neg x)^{n-1} \to x = \min(1, nx)$, and it can get also the value $n/(n+1)$ (when $x = 1/n+1$). So

$$[(\neg x)^{n-1} \to x] \lor \neg[(\neg x)^{n-1} \to x] = 1$$

in L_n, but in L_{n+1} we have

$$[(\neg(1/n+1))^{n-1} \to 1/n+1] \lor \neg[(\neg(1/n+1))^{n-1} \to 1/n+1] = (n/n+1) \lor (1/n+1) = n/n+1.$$

4. First let us prove by induction that

$$(\sharp) \quad x^k \to y = \min(1, k(1-x) + y)$$

(in L_n). If $k = 0$ we have: $x^0 \to y = y = \min(1, y)$. If ($\sharp$) is true for k, we have

$$x^{k+1} \to y = x \to \min(1, k(1-x)+y) = \min(1-x+\min(1, k(1-x)+y), 1).$$

Now if $k(1-x) + y > 1$, it follows that

$$(k+1)(1-x) + y > 1, \text{ and } x^k \to y = 1$$

In this case $x^{k+1} \to y = \min(2 - x, 1) = 1$.
If $k(1-x) + y < 1$ we get

$$x^{k+1} \to y = \min(1 - x + k(1-x) + y, 1) = \min((k+1)(1-x) + y, 1)$$

Anyway ($\sharp$) holds for $n = k+1$.
Now in L_n we have that

$$x^n \to y = \min(1, n(1-x) + y) = \begin{cases} y, & \text{if } x = 1 \\ 1, & \text{if } x \neq 1 \end{cases}$$

(since $1 - x \geq 1/n$ in the latter case).
Thus $((x^n \to y) \to x) \to x = (y \to 1) \to 1 = 1 \to 1 = 1$ if $x = 1$, and

$$((x^n \to y) \to x) \to x = (1 \to x) \to x = x \to x = 1 \text{ if } x \neq 1$$

On the other hand in L_{n+1} we have (again by ($\sharp$)):

$$(((\tfrac{1}{n+1})^n \to 0) \to \tfrac{1}{n+1}) \to \tfrac{1}{n+1} = (\tfrac{n}{n+1} \to \tfrac{1}{n+1}) \to \tfrac{1}{n+1} =$$
$$\tfrac{2}{n+1} \to \tfrac{1}{n+1} = \tfrac{n}{n+1}$$

5. Again we can use the equality (∗) from the solution of 3(a).

   Consider the formula $A(x) = (\neg x) \Leftrightarrow [(\neg x)^{k-2} \to x]$. We have that $A(x) = 1$ iff $\neg x = (\neg x)^{k-2} \to x$ iff $1 - x = \min(1, (k-1)x)$.

   Now we notice that $1 - x = (k-1)x$ iff $1 = kx$ iff $x = 1/k$.

   Thus $x = 1/k$ implies $A(x) = 1$, and also $(k-1)x < 1$ together with $x \neq 1/k$ imply $A(x) \neq 1$. Finally, if $(k-1)x \geq 1$ (i.e. $x \geq 1/(k-1)$) and $1 - x = \min(1, (k-1)x) = 1$, we obtain that $x = 0$, which is a contradiction.

   Therefore $A(x)$ has the property required.

6. Assume that m is a divisor of n, and that a formula $A(p_1, \ldots, p_k)$ with atoms $p_1, \ldots, p_k$ is not an L_m-tautology. Then $A(x_1, \ldots, x_k) \neq 1$ for some values $x_1, \ldots, x_k \in \{0, 1/m, \ldots, (m-1)/m, 1\}$. But since m divides n, all the numbers $1/m, \ldots, (m-1)/m$ are among $\{1/n, \ldots, (n-1)/n\}$, and thus $x_1, \ldots, x_k \in \{0, 1/n, \ldots, 1\}$.

   Then $A(x_1, \ldots, x_k) \neq 1$ also in L_n because the functions $\neg, \to, \vee, \wedge$ in L_n are computed by the same rules as in L_m. Therefore A is not an L_n-tautology.

   ('If') Assume that $m \leq n$. Consider an arbitrary formula A without negation, such that A is not an L_m-tautology. Let $p_1, \ldots, p_k$ be all atoms occurring in A; then $A(x_1, \ldots, x_k) \neq 1$ for some values $x_1, \ldots, x_k \in \{0, 1/m, \ldots, 1\}$ of these atoms. Further on, we can assume that $x_1 \leq x_2 \leq \cdots \leq x_k$ (otherwise we make the list of atoms in a corresponding order).

   Now we map the values $\{0, 1/m, 2/m, \ldots, 1\}$ into $\{0, 1/n, \ldots, (n-1)/n, 1\}$ in such a way that $1 \mapsto 1, (1 - 1/m) \mapsto (1 - 1/n), \ldots, 0 \mapsto (1 - m/n)$. This is done by the function

   $$f(x) = 1 - \frac{m}{n} + \frac{m}{n}x$$

   Indeed, one can observe that $f(1) = 1$ and f increases all distances by (m/n) times. Obviously f is monotonic, and thus $f(x \vee y) = f(x) \vee f(y)$, $f(x \wedge y) = f(x) \wedge f(y)$.

Moreover, f preserves the implication. For, if $x \to y = 1$ then $x \le y$, and so $f(x) \le f(y)$,

$$f(x) \to f(y) = 1 = f(x \to y)$$

If $x \to y \ne 1$, then

$$x > y, x \to y = 1 - x + y,$$
$$f(x \to y) = 1 - \tfrac{m}{n} - \tfrac{m}{n}x + \tfrac{m}{n}y$$

On the other hand, $f(x) > f(y)$, thanks to monotonicity, and thus

$$f(x) \to f(y) = 1 - f(x) + f(y)$$
$$= 1 - (1 - \tfrac{m}{n} + \tfrac{m}{n}x) + (1 - \tfrac{m}{n} + \tfrac{m}{n}y)$$
$$= 1 - \tfrac{m}{n}x + \tfrac{m}{n}y.$$

Hence $f(x \to y) = f(x) \to f(y)$.

Since f preserves $\vee, \wedge, \to$, we obtain that

$$f(A(x_1, \ldots, x_k)) = A(f(x_1), \ldots, f(x_k))$$

and this value is not 1, because $A(x_1, \ldots, x_k) \ne 1$ and f is monotonic. Therefore, $A(f(x_1), \ldots, f(x_k)) \ne 1$ in L_n, which means that A is not an L_n-tautology.

Thus (in the language $\{\vee, \wedge, \to\}$) every L_m-non-tautology is an L_n-non-tautology, i.e. $L_n \subseteq L_m$.

('Only if') Exercise 4 shows that $L_n \not\subseteq L_{n+1}$ for formulae without negation.

Hence $L_n \not\subseteq L_m$ for any $m > n$ because otherwise we had

$$L_n \subseteq L_m \subseteq L_{n+1}$$

by the 'If' part proved above.

Exercise 2.1.3

$A_1, \ldots, A_k \vdash_{L_n} B \to C$ iff for any assignment h,

1. $h(A_1) + \cdots + h(A_k) \le (k-1) + h(B \to C)$.

$A_1, \ldots, A_k, B \vdash_{L_n} C$ iff for any h,

2. $h(A_1) + \cdots + h(A_k) + h(B) \leq k + h(C)$.

If $h(B) \geq h(C)$ then (1) is equivalent to (2), because $k - 1 + h(B \to C) = k - 1 + (1 - h(B) + h(C)) = k - h(B) + h(C)$.
If $h(B) < h(C)$ then (2) is true because it is obtained by addition of inequalities $h(A_1) \leq 1, \ldots, h(A_k) \leq 1, h(B) \leq h(C)$.
 (1) is also true in this case, because $h(B \to C) = 1$, and $h(A_1) + \cdots + h(A_k) \leq \underbrace{1 + \cdots + 1}_{} = k = (k-1) + h(B \to C)$.
So again (1) and (2) are equivalent.

Exercise 2.2.4

This is shown by induction on A (for any t, s).

- If A is an atom, A is persistent by definition;

- Suppose $A = B \wedge C$ and B, C are persistent, $t \leq s, h(t, A) = 1$. Then $h(t, B) = h(t, C) = 1$. By definition, $h(s, B) = h(s, C) = 1$ by persistence, and thus $h(s, A) = 1$ by Definition 2.2.3.

- Suppose $A = B \vee C$, with B, C persistent, $t \leq s$, $h(t, A) = 1$. Then $h(t, B) = 1$ or $h(t, C) = 1$ by definition; hence $h(s, B) = 1$ or $h(s, C) = 1$ by persistence, and finally, $h(s, A) = 1$ by Definition 2.2.3.

- Suppose $A = B \to C$, with B, C persistent, $t \leq s$, $h(t, A) = 1$. Take any $r \geq s$; then $r \geq t$ by transitivity. Now if $h(r, B) = 1$, we have $h(r, C) = 1$ (since $h(t, A) = 1, r \geq t$). Thus $h(s, A) = 1$ because r is arbitrary. (Note that we have not used persistence of B, C at all.)

- Obviously $\bot$ is persistent.

- Suppose $A = \neg B$, with B persistent, $t \leq s, h(t, A) = 1$. Take any $r \geq s$; then $r \geq t$ by transitivity, and hence $h(r, B) = 0$ (since $h(t, A) = 1$).

  This implies persistence of A (again we did not use persistence of B).

ANSWERS TO THE EXERCISES 293

Exercise 2.2.6

1. $H_\emptyset(A \twoheadrightarrow A) = 1$ iff for any t, $H_t(A) = 1$ implies $H_{\emptyset \otimes t}(A) = 1$. The latter is true because $\emptyset \otimes t = t$.

   So $A \twoheadrightarrow A$ is valid.

2. $H_\emptyset(A \twoheadrightarrow (A \twoheadrightarrow A)) = 1$ iff for any t, $H_t(A) = 1$ implies $H_t(A \twoheadrightarrow A) = 1$, iff for any t, r, $H_t(A) = 1$, $H_r(A) = 1$ imply $H_{t \otimes r}(A) = 1$.

   This is not necessarily true, even in the case (a). For example, we can take A atomic, $I = \{\emptyset, a, b, 1\}$, with the operation $\otimes$ given by the following multiplication table:

$\otimes$	$\emptyset$	a	b	1
$\emptyset$	$\emptyset$	a	b	1
a	a	a	1	1
b	b	1	b	1
1	1	1	1	1

   (This corresponds to the operation '∪' on subsets of a two-element set.) Now put
   $$H_a(A) = H_b(A) = 1$$
   and
   $$H_\emptyset(A) = H_1(A) = 0$$
   The above argument shows that
   $$H_\emptyset(A \twoheadrightarrow (A \twoheadrightarrow A)) = 1$$
   because $H_{a \otimes b}(A) = 0$.

3. $H_\emptyset[(A \twoheadrightarrow (B \twoheadrightarrow C)) \twoheadrightarrow (B \twoheadrightarrow (A \twoheadrightarrow C))] = 1$ iff for any s, $H_s(A \twoheadrightarrow (B \twoheadrightarrow C)) = 1$ only if $H_s(B \twoheadrightarrow (A \twoheadrightarrow C)) = 1$.

   In the case (a) this is true. For, assume that

   (a_1) $H_s(A \twoheadrightarrow (B \twoheadrightarrow C)) = 1$,
   and show that

   (a_2) $H_s(B \twoheadrightarrow (A \twoheadrightarrow C)) = 1$,
   i.e. that for any t, $H_t(B) = 1$ only if $H_{s \otimes t}(A \twoheadrightarrow C)$.

   So assume also

(a_3) $H_t(B) = 1$

and show that

(a_4) $H_{s\otimes t}(A \twoheadrightarrow C) = 1$.

To check this, assume finally that

(a_5) $H_u(A) = 1$

and show that

(a_6) $H_{(s\otimes t)\otimes u}(C) = 1$.

By our assumptions we get

(a_7) $H_{s\otimes u}(B \twoheadrightarrow C) = 1$ (by (a_1), (a_5)),

(a_8) $H_{(s\otimes u)\otimes t}(C) = 1$ (by (a_7), (a_3)).

If the operation $\otimes$ is commutative and associative, (a_8) is the same as (a_6). In the case (b) we can find a model where the formula 3 fails, i.e. we can take a three-element set with the following multiplication table:

$\otimes$	$\varnothing$	u	t
$\varnothing$	$\varnothing$	u	t
u	u	u	u
t	t	t	t

Associativity follows because

$$(u \otimes x) \otimes y = u \otimes y = u = u \otimes (x \otimes y),$$
$$(t \otimes x) \otimes y = t \otimes y = t = t \otimes (x \otimes y),$$
$$(\varnothing \otimes x) \otimes y = x \otimes y = \varnothing \otimes (x \otimes y).$$

Now let $H_u(C) = 1, H_t(C) = 0$,

$$H_x(A) = 1 \text{ iff } x = u$$
$$H_y(B) = 1 \text{ iff } y = t$$

Then $H_\varnothing[(A \twoheadrightarrow (B \twoheadrightarrow C)) \twoheadrightarrow (B \twoheadrightarrow (A \twoheadrightarrow C))] = 0$ because

(b_1) $H_\varnothing(A \twoheadrightarrow (B \twoheadrightarrow C)) = 1$,

(b_2) $H_\varnothing(B \twoheadrightarrow (A \twoheadrightarrow C)) = 0$,

and $\varnothing \otimes \varnothing = \varnothing$.

To show (b_1), assume

(b_3) $H_x(A) = 1$.

Then $x = u$ (by definition of H), and we have to check that

ANSWERS TO THE EXERCISES 295

(b_4) $H_u(B \twoheadrightarrow C) = 1$

(because $\varnothing \otimes u = u$).

This means: for any y, $H_y(B) = 1$ only if $H_{u \otimes y}(C) = 1$. The latter is true since $u \otimes y = u$.

To prove (b_2) it suffices to notice that $H_t(B) = 1$, but $H_t(A \twoheadrightarrow C) = 0$. This happens since $H_u(A) = 1$, but $H_{t \otimes u}(C) = 0$.

4. First we show validity in the case (b), which means: for any s, $H_s(A \twoheadrightarrow (A \twoheadrightarrow B)) = 1$ implies $H_s(A \twoheadrightarrow B) = 1$.

So assume that

(4.1) $H_s(A \twoheadrightarrow (A \twoheadrightarrow B)) = 1$, and that

(4.2) $H_r(A) = 1$. Then we get

(4.3) $H_{s \otimes r}(A \twoheadrightarrow B) = 1$ (by (4.1), (4.2)),

(4.4) $H_{(s \otimes r) \otimes r}(B) = 1$ (by (4.2), (4.3)),

(4.5) $H_{s \otimes r}(B) = 1$ since $(s \otimes r) \otimes r = s \otimes r$.
  Thus (4.2) implies (4.5), i.e. $H_s(A \twoheadrightarrow B) = 1$.

On the other hand, the formula can be refuted in the case (a). To see this, we take the model $\langle I, \otimes, \varnothing \rangle$ in a two-element set $\{\varnothing, r\}$ with the following multiplication table:

$\otimes$	$\varnothing$	r
$\varnothing$	$\varnothing$	r
r	r	$\varnothing$

and with the assignment (for A, B atomic)

$$H_r(A) = H_\varnothing(B) = 1, H_r(B) = H_\varnothing(A) = 0$$

Now the formula 4 becomes false at $\varnothing$ because

(4.6) $H_\varnothing(A \twoheadrightarrow (A \twoheadrightarrow B)) = 1$,

(4.7) $H_\varnothing(A \twoheadrightarrow B) = 0$,
  and $\varnothing \otimes \varnothing = \varnothing$.
  (4.7) holds because $H_r(A) = 1$ but $H_r(B) = 0$. To verify (4.6) assume that

(4.8) $H_s(A) = 1$, and show that

(4.9) $H_{\emptyset \otimes s}(A \twoheadrightarrow B) = 1$.
By definition, (4.8) can be true only for $s = r$, so (4.9) is equivalent to

(4.10) $H_r(A \twoheadrightarrow B) = 1$,
which means that for any u, $H_u(A) = 1$ only if $H_{r \otimes u}(B) = 1$. This is true because $H_u(A) = 1$ implies $u = r, r \otimes u = \emptyset$.

5. (a) The formula is valid. Assume that

(5.1) $H_s(A) = 1$,
and prove that $H_s((A \twoheadrightarrow A) \twoheadrightarrow A) = 1$. For this, assume also

(5.2) $H_r(A \twoheadrightarrow A) = 1$,
and show that

(5.3) $H_{s \otimes r}(A) = 1$.
From (5.1) and (5.2) it follows that

(5.4) $H_{r \otimes s}(A) = 1$, which is equivalent to (5.3) by commutativity.

(b) The formula is non-valid. Consider the following model:

$\otimes$	$\emptyset$	s	t
$\emptyset$	$\emptyset$	s	t
s	s	s	t
t	t	s	t

with the assignment (for A atomic):

$$H_s(A) = 1, H_t(A) = H_\emptyset(A) = 0$$

Now $H_s((A \twoheadrightarrow A) \twoheadrightarrow A) = 0$ follows from $H_{s \otimes t}(A) = H_t(A) = 0$ and $H_t(A \twoheadrightarrow A) = 1$. The latter holds because $H_x(A) = 1$ only if $x = s$, only if $H_{t \otimes x}(A) = H_s(A) = 1$.
Hence $H_\emptyset(A \twoheadrightarrow ((A \twoheadrightarrow A) \twoheadrightarrow A)) = 0$.

Associativity for our operation can be checked in a similar way as in item 3 above.

Exercise 3.1.5

One way (there may be others) of proving the validity of each of the arguments is given:

ANSWERS TO THE EXERCISES

1. $\dfrac{\dfrac{p \wedge q}{p} \quad (\wedge\text{E})}{p \vee q \quad\quad (\vee\text{I})}$

2. $\dfrac{p \to q \quad \dfrac{\neg q}{p \to \neg q} \;(\to\text{I2})}{\neg p \quad\quad (\neg\text{I})}$

3. $\dfrac{p \vee r \quad r \to q \quad \dfrac{\neg p}{p \to q} \;(\to\text{I1})}{q \quad\quad\quad (\vee\text{E})}$

Exercise 3.2.3

We prove the two tautologies using the forward rules.

1. We must show $\dfrac{\text{nothing}}{(p \vee q) \wedge \neg p \to q}$ and so by ($\to$I) we have to show $\dfrac{(p \vee q) \wedge \neg p}{q}$ which is given by ($\wedge$E) and ($\vee$E1).

2. We must show $\dfrac{\text{nothing}}{(\neg q \to \neg p) \to (p \to q)}$ and so by ($\to$I) we have to show $\dfrac{\neg q \to \neg p}{p \to q}$ and by ($\to$I) again, we have to show $\dfrac{\neg q \to \neg p,\; p}{q}$ which we do by showing

   (a) $\neg q \to p$ from $\{\neg q \to \neg p, p\}$
   (b) $\neg q \to \neg p$ from $\{\neg q \to \neg p, p\}$

   and using the negation rule ($\neg$E) to get q from $\{\neg q \to \neg p, p\}$. To show $\neg q \to p$, we use ($\to$I2). To show $\neg q \to \neg p$, we observe that it is one of the assumptions.

Exercise 3.2.4

1. (a) Rule $(\to \text{I1}) : \neg A \vdash A \to B$.
   - (1) $\neg A$ data
   - (2) $A \to B$ subcomputation

		$\underline{B}$
(2.1)	A	assumption
(2.2)	$\neg B \to A$	subcomputation

		$\underline{A}$
(2.2.1)	$\neg B$	assumption
(2.2.2)	A	from (2.1)

(2.3)	$\neg B \to \neg A$	subcomputation

		$\underline{\neg A}$
(2.3.1)	$\neg B$	assumption
(2.3.2)	$\neg A$	from (1)

(2.4)	B	$(\neg E)$ on (2.2), (2.1)

   (b) Rule $(\to \text{I2}) : B \vdash A \to B$.
   Obviously, $B, A \vdash B$. Then we can use $(\to \text{I})$ to obtain $B \vdash A \to B$.

ANSWERS TO THE EXERCISES   299

(c) Rule $(\to E1) : A \to B \vdash \neg A \lor B$.
  (1) $A \to B$ data
  (2) $A \to \neg A \lor B$ subcomputation

	$\neg A \lor B$
(2.1) A	assumption
(2.2) B	$(\to E)$ on (1), (2.1)
(2.3) $\neg A \lor B$	$(\lor I)$ on (2.2)

  (3) $\neg A \to \neg A \lor B$ subcomputation

	$\neg A \lor B$
(3.1) $\neg A$	assumption
(3.2) $\neg A \lor B$	$(\lor I)$ on (3.1)

  (4) $\neg A \lor B$ $(\neg 2)$ on (2), (3)

2. (a) $p \to (q \to r) \vdash p \land q \to r$.

It is sufficient to show $p \to (q \to r), p \land q \vdash r$, and then apply the rule $(\to I)$.

So we proceed:
  (1) $p \to (q \to r)$ assumption
  (2) $p \land q$ assumption
  (3) p $(\land E)$ on (2)
  (4) q $(\land E)$ on (2)
  (5) $q \to r$ $(\to E)$ on (1), (3)
  (6) r $(\to E)$ on (4), (5)

(b) $p \lor q \to r \vdash (p \to r) \land (q \to r)$.

It is sufficient to show $p \lor q \to r \vdash p \to r$ and $p \lor q \to r \vdash q \to r$, and then apply $(\land I)$. Both proofs are similar, so we do the first:
  (1) $p \lor q \to r$ data
  (2) $p \to r$ subcomputation

	r
(2.1) p	assumption
(2.2) $p \lor q$	$(\lor I)$ on (2.1)
(2.3) r	$(\to E)$ on (1), (2.2)

(c) $\vdash q \to (p \to q)$.

By ($\to$ I2) we have $q \vdash p \to q$. Then we can apply ($\to$ I).

(d) $\vdash p \to p$.

Immediately from $p \vdash p$ by ($\to$ I).

(e) $\neg(p \to q) \vdash p \wedge \neg q$.

(1) $\neg(p \to q)$ data
(2) $\neg p \to \neg(p \to q)$ ($\to$ I2) on (1)
(3) $\neg p \to (p \to q)$ subcomputation

	$p \to q$
(3.1) $\neg p$	assumption
(3.2) $p \to q$	($\to$ I1) on (3.1)

(4) p ($\neg$E) on (2), (3)
(5) $q \to \neg(p \to q)$ ($\to$ I2) on (1)
(6) $q \to (p \to q)$ subcomputation

	$p \to q$
(6.1) q	assumption
(6.2) $p \to q$	($\to$ I2) on (6.1)

(7) $\neg q$ ($\neg$I) on (5), (6)
(8) $p \wedge \neg q$ ($\wedge$I) on (4), (7)

(f) $\neg p \vee q \vdash p \to q$.

This follows by ($\vee$E) since we have

$$\vdash \neg p \to (p \to q) \text{ and } \vdash q \to (p \to q)$$

The second was proved in (c), and the first follows by ($\to$ I1) and ($\to$ I).

Exercise 3.3.1

1. $P \wedge Q \rightarrow R \vdash P \rightarrow (Q \rightarrow R)$

    (1) $P \wedge Q \rightarrow R$ data
    (2) $P \rightarrow (Q \rightarrow R)$ subcomputation

    > $\underline{Q \rightarrow R}$
    >
    > (2.1) P assumption
    > (2.2) $Q \rightarrow R$ subcomputation
    >
    > > $\underline{R}$
    > >
    > > (2.2.1) Q assumption
    > > (2.2.2) $P \wedge Q$ ($\wedge$I) on (2.1),(2.2.1)
    > > (2.2.3) R ($\rightarrow$E) on (1),(2.2.2)

2. $(P \rightarrow Q) \wedge (Q \rightarrow R) \vdash P \rightarrow R$

    (1) $(P \rightarrow Q) \wedge (Q \rightarrow R)$ data
    (2) $P \rightarrow R$ subcomputation

    > $\underline{R}$
    >
    > (2.1) P assumption
    > (2.2) $P \rightarrow Q$ ($\wedge$E) on (1)
    > (2.3) Q ($\rightarrow$E) on (2.1),(2.2)
    > (2.4) $Q \rightarrow R$ ($\wedge$E) on (1)
    > (2.5) R ($\rightarrow$E) on (2.3),(2.4)

3. $P{\to}R$, $Q{\to}S \vdash P{\wedge}Q{\to}R{\wedge}S$

   (1) $P{\to}R$ data
   (2) $Q{\to}S$ data
   (3) $P{\wedge}Q{\to}R{\wedge}S$ subcomputation

		$R{\wedge}S$
(3.1)	$P{\wedge}Q$	assumption
(3.2)	P	($\wedge$E) on (3.1)
(3.3)	R	($\to$E) on (1),(3.2)
(3.4)	Q	($\wedge$E) on (3.1)
(3.5)	S	($\to$E) on (2),(3.4)
(3.6)	$R{\wedge}S$	($\wedge$I) on (3.3),(3.5)

Exercise 3.3.2

1. $\neg(P{\wedge}\neg Q) \vdash P{\to}Q$

   (1) $\neg(P{\wedge}\neg Q)$ data
   (2) $P{\to}Q$ subcomputation

			Q
(2.1)	P		assumption
(2.2)	$\neg Q{\to}(P{\wedge}\neg Q)$		subcomputation
			$P{\wedge}\neg Q$
	(2.2.1)	$\neg Q$	assumption
	(2.2.2)	$P{\wedge}\neg Q$	($\wedge$I) on (2.1),(2.2.1)
(2.3)	$\neg Q{\to}\neg(P{\wedge}\neg Q)$		subcomputation
			$\neg(P{\wedge}\neg Q)$
	(2.3.1)	$\neg Q$	assumption
	(2.3.2)	$\neg(P{\wedge}\neg Q)$	from (1)
(2.4)	Q		from (2.2),(2.3) and ($\neg$E)

2. $\vdash ((P{\to}Q){\to}P){\to}P$

ANSWERS TO THE EXERCISES

(1) $((P \to Q) \to P) \to P$ subcomputation

$\underline{P}$

(1.1) $(P \to Q) \to P$ assumption
(1.2) $\neg(P \to Q) \to P$ subcomputation

$\underline{P}$

(1.2.1) $\neg(P \to Q)$ assumption
(1.2.2) $\neg P \to (P \to Q)$ subcomputation

$\underline{P \to Q}$

(1.2.2.1) $\neg P$ assumption
(1.2.2.2) $P \to Q$ subcomputation

$\underline{Q}$

(1.2.2.2.1) P assumption
(1.2.2.2.2) $\neg Q \to P$ easy to show using (1.2.2.2.1)
(1.2.2.2.3) $\neg Q \to \neg P$ easy to show using (1.2.2.1)
(1.2.2.2.4) Q (1.2.2.2.2),(1.2.2.2.3) and ($\neg$E)

(1.2.3) $\neg P \to \neg(P \to Q)$ subcomputation

$\underline{\neg(P \to Q)}$

(1.2.3.1) $\neg P$ assumption
(1.2.3.2) $\neg(P \to Q)$ from (1.2.1)

(1.2.4) P (1.2.2),(1.2.3) and ($\neg$E)

(1.3) P (1.1),(1.2) and ($\neg$2)

3. ⊢ ¬(P→Q)→P

  (1) ¬(P→Q)→P subcomputation

> $\underline{P}$
>
> (1.1) ¬(P→Q)   assumption
> (1.2) ¬P→(P→Q)   subcomputation
>
>> $\underline{P \to Q}$
>>
>> (1.2.1)   ¬P   assumption
>> (1.2.2)   P→Q   subcomputation
>>
>>> $\underline{Q}$
>>>
>>> (1.2.2.1) P   assumption
>>> (1.2.2.2) ¬Q→P   subcomputation
>>>
>>>> $\underline{P}$
>>>>
>>>> (1.2.2.2.1)   ¬Q   assumption
>>>> (1.2.2.2.2)   P   from (1.2.2.1)
>>>
>>> (1.2.2.3) ¬Q→¬P subcomputation
>>>
>>>> $\underline{\neg P}$
>>>>
>>>> (1.2.2.3.1)   ¬Q   assumption
>>>> (1.2.2.3.2)   ¬P   from (1.2.1)
>>>
>>> (1.2.2.4) Q   (1.2.2.2),(1.2.2.3) and (¬E)
>>
>> (1.2.2.1)   ¬P   assumption
>> (1.2.2.2)   ¬(P→Q)   from (1.1)
>
> (1.3) ¬P→¬(P→Q) subcomputation
>
>> $\underline{P \to Q}$
>>
>> (1.3.1)   ¬P   assumption
>> (1.3.2)   ¬(P→Q)   from (1.1)
>
> (1.4) P   (1.2),(1.3) and (¬E)

ANSWERS TO THE EXERCISES 305

Exercise 4.1.3

To show $(A \to B), \neg B \vdash \neg A$ we show

$$\frac{A \to B, \neg B}{A \to x} \quad \text{and} \quad \frac{A \to B, \neg B}{A \to \neg x}$$

for $x = B$. Therefore, we must show

$$\frac{A \to B, \neg B}{A \to B} \quad \text{and} \quad \frac{A \to B, \neg B}{A \to \neg B}$$

both of which we can obviously do.

Exercise 4.1.14

1. After rewriting, the problem is reduced to $p \to q, q \to r, r \to \bot \vdash p \to \bot$.

   We rewrite the assumption as a set of wffs.

   By **H2**, this is reduced to $p \to q, q \to r, r \to \bot, p \vdash \bot$.
   To show this it is sufficient to show $p \to q, q \to r, r \to \bot, p \vdash r$
   and then apply ($\to$ E).
   The same argument reduces our question to $p \to q, q \to r, r \to \bot, p \vdash q$
   and then to $p \to q, q \to r, r \to \bot, p \vdash p$
   which succeeds.

2. First we do rewriting: $(p \to \bot) \to \bot \vdash p$.
   It is sufficient to show $(p \to \bot) \to \bot \vdash p \to \bot$,
   which reduces (by **H2**) to $(p \to \bot) \to \bot, p \vdash \bot$.
   Now we come into a loop,
   so we restart and ask for $(p \to \bot) \to \bot, p \vdash p$,
   which succeeds.

3. By rewriting this is reduced to $(p \to \bot) \to p \vdash p$.
   Now we can follow the argument of Example 4.1.9, with q replaced by $\bot$.

4. After rewriting we have $(p \to \bot) \to q, p \to \bot \vdash q$.
   This comes by $(\to E)$.
   But to get the result
   automatically, we ask for $(p \to \bot) \to q, p \to \bot \vdash p \to \bot$
   which reduces to $(p \to \bot) \to q, p \to \bot, p \vdash \bot$.
   Now we ask for $(p \to \bot) \to q, p \to \bot, p \vdash p$
   which succeeds.

5. First we rewrite: $q \wedge r \to s, ((p \to \bot) \to r) \to q, p \to \bot \vdash r \to s$.

   By **H2** this is reduced to $q \wedge r \to s, ((p \to \bot) \to r) \to q, p \to \bot, r \vdash s$.

   Now we can ask for $q \wedge r \to s, ((p \to \bot) \to r) \to q, p \to \bot, r \vdash q \wedge r$.

   By **H1** we reduce the problem to
   two questions:
   (a) $q \wedge r \to s, ((p \to \bot) \to r) \to q, p \to \bot, r \vdash q$,
   (b) $q \wedge r \to s, ((p \to \bot) \to r) \to q, p \to \bot, r \vdash r$.
   (b) is trivial.
   (a) can be obtained from
   $$q \wedge r \to s, ((p \to \bot) \to r) \to q, p \to \bot, r \vdash (p \to \bot) \to r,$$
   which is reduced (by **H2**) to
   $$q \wedge r \to s, ((p \to \bot) \to r) \to q, p \to \bot, r, p \to \bot \vdash r,$$
   which succeeds.

6. First we rewrite by **H5**: $(p \to \bot) \to q, p \to r, q \to r \vdash r$.
   Now we may try to get $(p \to \bot) \to q, p \to r, q \to r \vdash p$,
   but then we are stuck,
   so we ask for $(p \to \bot) \to q, p \to r, q \to r \vdash q$,
   which follows from $(p \to \bot) \to q, p \to r, q \to r \vdash p \to \bot$.
   By **H2** this reduces to $(p \to \bot) \to q, p \to r, q \to r, p \vdash \bot$.
   We fail again,
   but now we can restart and $(p \to \bot) \to q, p \to r, q \to r, p \vdash r$.
   ask for
   Then we come to $(p \to \bot) \to q, p \to r, q \to r, p \vdash p$,
   which succeeds.

ANSWERS TO THE EXERCISES

Exercise 4.2.12

1. $p \to q, \neg q \vdash \neg p$
   $|$ rewrite
   $p \to q, q \to \bot \vdash p \to \bot$
   $|$ rule for $\to$
   $p \to q, q \to \bot, p \vdash \bot$
   $|$ rule for atoms
   $|$ using $q \to \bot$
   $p \to q, q \to \bot, p \vdash q$
   $|$ rule for atoms
   $|$ using $p \to q$
   $p \to q, q \to \bot, p \vdash p$
   $|$
   success

2. $\neg p \to q \vdash p \vee q$
   $|$ rewrite
   $(p \to \bot) \to q \vdash (p \to \bot) \to q$
   $|$ rule for $\to$
   $(p \to \bot) \to q, p \to \bot \vdash q$
   $|$ rule for atoms us-
   $|$ ing
   $|$ $(p \to \bot) \to q$
   $(p \to \bot) \to q, p \to \bot \vdash p \to \bot$
   $|$ rule for $\to$
   $(p \to \bot) \to q, p \to \bot, p \vdash \bot$
   $|$ rule for atoms using $p \to \bot$
   $(p \to \bot) \to q, p \to \bot, p \vdash p$
   $|$
   success

3. $\vdash ((a \to b) \to a) \to a$
   $|$ rule for $\to$
   $(a \to b) \to a \vdash a$
   $|$ rule for atoms
   $(a \to b) \to a \vdash a \to b$
   $|$ rule for $\to$
   $(a \to b) \to a, a \vdash b$
   $|$ restart
   $(a \to b) \to a, a \vdash a$
   $|$
   success

Remark: It is more convenient to use restart not with the original goal, but with the second goal (a).

4. $(p \to q) \to q, q \to p \vdash p$

$\quad\quad\quad\quad$ | rule for atoms using
$\quad\quad\quad\quad$ | $q \to p$

$(p \to q) \to q, q \to p \vdash q$

$\quad\quad\quad\quad$ | rule for atoms using
$\quad\quad\quad\quad$ | $(p \to q) \to q$

$(p \to q) \to q, q \to p \vdash p \to q$

$\quad\quad\quad\quad$ | rule for $\to$

$(p \to q) \to q, q \to p, p \vdash q$

$\quad\quad\quad\quad$ | restart

$(p \to q) \to q, q \to p, p \vdash p$

$\quad\quad\quad\quad$ |

$\quad\quad\quad$ success

Remark: We have used restart, because the original goal appeared as an assumption.

5. $\vdash \neg(p \land \neg p)$

$\quad\quad\quad\quad$ | rewrite

$\vdash p \land (p \to \bot) \to \bot$

$\quad\quad\quad\quad$ | rule for $\to$

$p \land (p \to \bot) \vdash \bot$

$\quad\quad\quad\quad$ | rewrite

$p, p \to \bot \vdash \bot$

$\quad\quad\quad\quad$ | rule for atoms

$p, p \to \bot \vdash p$

$\quad\quad\quad\quad$ |

$\quad\quad\quad$ success

ANSWERS TO THE EXERCISES

6. $\vdash \neg(p \to q) \to p$

   | rewrite

   $\vdash ((p \to q) \to \bot) \to p$

   | rule for $\to$

   $(p \to q) \to \bot \vdash p$

   | rule for atoms

   $(p \to q) \to \bot \vdash p \to q$

   | rule for $\to$

   $(p \to q) \to \bot, p \vdash q$

   | restart

   $(p \to q) \to \bot, p \vdash p$

   |

   success

7. $\vdash \neg(p \to q) \to \neg q$

   | rewrite

   $\vdash ((p \to q) \to \bot) \to (q \to \bot)$

   | rule for $\to$

   $(p \to q) \to \bot \vdash q \to \bot$

   | rule for $\to$

   $(p \to q) \to \bot, q \vdash \bot$

   | rule for atoms

   $(p \to q) \to \bot, q \vdash p \to q$

   | rule for $\to$

   $(p \to q) \to \bot, q, p \vdash q$

   |

   success

8. $\vdash b \vee (b \to c)$
   $\Big|$ rewrite
   $\vdash (b \to \bot) \to (b \to c)$
   $\Big|$ rule for $\to$
   $b \to \bot \vdash b \to c$
   $\Big|$ rule for $\to$
   $b \to \bot, b \vdash c$
   $\Big|$ rule for atoms
   $b \to \bot, b \vdash b$
   $\Big|$
   success

9. $\vdash (\neg p \to p) \to p$
   $\Big|$ rewrite
   $\vdash ((p \to \bot) \to p) \to p$
   $\Big|$ rule for $\to$
   $(p \to \bot) \to p \vdash p$
   $\Big|$ rule for atoms
   $(p \to \bot) \to p \vdash p \to \bot$
   $\Big|$ rule for $\to$
   $(p \to \bot) \to p, p \vdash \bot$
   $\Big|$ restart
   $(p \to \bot) \to p, p \vdash p$
   $\Big|$
   success

10. $\vdash (p \to \neg p) \to \neg p$
    $\Big|$ rewrite
    $\vdash (p \to (p \to \bot)) \to (p \to \bot)$
    $\Big|$ rule for $\to$
    $p \to (p \to \bot) \vdash p \to \bot$
    $\Big|$ rule for $\to$
    $p \to (p \to \bot), p \vdash \bot$
    $\Big|$ rewrite
    $p \wedge p \to \bot, p \vdash \bot$
    $\Big|$ rule for atoms
    $p \wedge p \to \bot, p \vdash p \wedge p$
    $\Big|$ rule for $\wedge$
    $p \wedge p \to \bot, p \vdash p$
    $\Big|$
    success

ANSWERS TO THE EXERCISES

12. $\vdash a \to (\neg b \to \neg(a \to b))$

    $\Big|$ rewrite

    $\vdash a \to ((b \to \bot) \to ((a \to b) \to \bot))$

    $\Big|$ rule for $\to$

    $a \vdash (b \to \bot) \to ((a \to b) \to \bot)$

    $\Big|$ rule for $\to$

    $a, b \to \bot \vdash (a \to b) \to \bot$

    $\Big|$ rule for $\to$

    $a, b \to \bot, a \to b \vdash \bot$

    $\Big|$ rule for atoms using $b \to \bot$

    $a, b \to \bot, a \to b \vdash b$

    $\Big|$ rule for atoms using $a \to b$

    $a, b \to \bot, a \to b \vdash a$

    $\Big|$

    success

13. $\vdash \neg p \to ((p \lor q) \to q)$

    $\Big|$ rewrite

    $\vdash (p \to \bot) \to (((p \to \bot) \to q) \to q)$

    $\Big|$ rule for $\to$

    $p \to \bot \vdash ((p \to \bot) \to q) \to q$

    $\Big|$ rule for $\to$

    $p \to \bot, (p \to \bot) \to q \vdash q$

    $\Big|$ rule for atoms using $(p \to \bot) \to q$

    $p \to \bot, (p \to \bot) \to q \vdash p \to \bot$

    $\Big|$ rule for $\to$

    $p \to \bot, (p \to \bot) \to q, p \vdash \bot$

    $\Big|$ rule for atoms using $p \to \bot$

    $p \to \bot, (p \to \bot) \to q, p \vdash p$

    $\Big|$

    success

14. $a \leftrightarrow ((a \land b) \lor (\neg a \land \neg b)) \vdash b$

    $\Big|$ rewrite

    $(a \to (a \land b) \lor (\neg a \land \neg b)) \land ((a \land b) \lor (\neg a \land \neg b) \to a) \vdash b$

    $\Big|$ rewrite

    $a \to (a \land b) \lor ((a \to \bot) \land (b \to \bot)),$
    $((a \land b) \lor ((a \to \bot) \land (b \to \bot)) \to a)$ $\vdash b$

    $\Big|$ rewrite

    $a \to ((a \land b \to \bot) \to (a \to \bot) \land (b \to \bot))$
    $(((a \land b) \to \bot) \to (a \to \bot) \land (b \to \bot)) \to a$ $\vdash b$

$$\begin{array}{c} \Big| \text{ rewrite} \\ a \wedge (a \wedge b \rightarrow \bot) \rightarrow (a \rightarrow \bot) \\ a \wedge (a \wedge b \rightarrow \bot) \rightarrow (b \rightarrow \bot) \qquad\qquad\qquad \vdash b \\ ((a \wedge b \rightarrow \bot) \rightarrow (a \rightarrow \bot)) \wedge ((a \wedge b \rightarrow \bot) \rightarrow (b \rightarrow \bot)) \rightarrow a \\ \Big| \text{ rewrite} \\ a \wedge (a \wedge b \rightarrow \bot) \wedge a \rightarrow \bot \\ a \wedge (a \wedge b \rightarrow \bot) \wedge b \rightarrow \bot \qquad\qquad\qquad \vdash b \\ ((a \wedge b \rightarrow \bot) \wedge a \rightarrow \bot) \wedge ((a \wedge b \rightarrow \bot) \wedge b \rightarrow \bot) \rightarrow a \\ \Big| \text{ rule for atoms using } a \wedge (a \wedge b \rightarrow \bot) \wedge a \rightarrow \bot \\ a \wedge (a \wedge b \rightarrow \bot) \wedge a \rightarrow \bot \\ a \wedge (a \wedge b \rightarrow \bot) \wedge b \rightarrow \bot \qquad \vdash a \wedge (a \wedge b \rightarrow \bot) \wedge a \\ ((a \wedge b \rightarrow \bot) \wedge a \rightarrow \bot) \wedge ((a \wedge b \rightarrow \bot) \wedge b \rightarrow \bot) \rightarrow a \\ \Big| \text{ rule for } \wedge \end{array}$$

Branch I $\qquad$ Branch II
$\mathcal{P} \vdash a \qquad \mathcal{P} \vdash a \wedge b \rightarrow \bot$

($\mathcal{P}$ denotes the set of assumptions in the previous step.)

Now we consider each branch separately.

ANSWERS TO THE EXERCISES

(I)

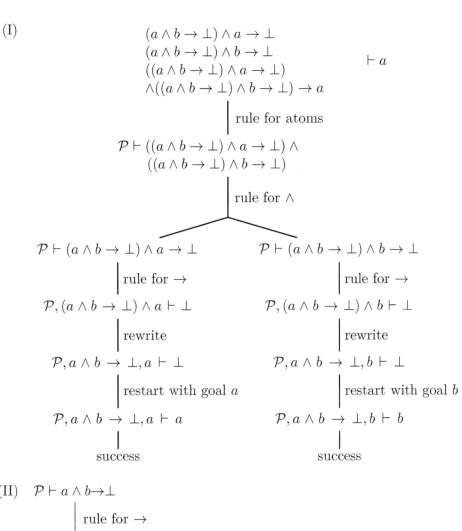

(II) $\mathcal{P} \vdash a \wedge b \to \bot$

$\quad \big|\ \text{rule for} \to$

$\mathcal{P}, a \wedge b \vdash \bot$

$\quad \big|\ \text{rewrite}$

$\mathcal{P}, a, b \vdash \bot$

$\quad \big|\ \text{restart}$

$\mathcal{P}, a, b \vdash b$

$\quad |$

$\quad \text{success}$

Exercise 4.3.8

By induction over the length of computation procedure (cf. the proof of Theorem 6.5.1 below). If x is atomic or $x = \bot$ and $\langle \mathcal{P}, H, x\rangle$ succeeds immediately with restart, then $(1, a) \in \mathcal{P}$ for some $a \in H$. In this case a is a conjunct in $\bigwedge \mathcal{P}$ and a disjunct in $\bigvee H$.

Thus $\bigwedge \mathcal{P} = \top$ only if $a = \top$, only if $\bigvee H = \top$, and so $\bigwedge \mathcal{P} \to x \vee \bigvee H$ is always true.

- If y is atomic or $y = \bot$ and $\langle \mathcal{P}, H, y\rangle$ succeeds with restart, then $\bigwedge \mathcal{P}$ contains the following conjunct A_i:

$$\left(\bigwedge_j (B_j \to x_j)\right) \to x, \text{ with } x = y \text{ or } x = \bot$$

and every $\langle \mathcal{P}'_j, H + y, x_j\rangle$ succeeds with restart (Definition 4.3.3). $\bigwedge \mathcal{P} \equiv A_i \wedge Q$ for some formula Q (Q is the conjunction of all the other available formulae in $\mathcal{P}$; we take $Q = \top$ if A_i is the only one available). From the definition it follows that

$$\bigwedge \mathcal{P}'_j \equiv Q \wedge B_j$$

and thus each formula

$$Q \wedge B_j \to x_j \vee \bigvee H \vee y$$

is a tautology.

We have to show that

$$\bigwedge \mathcal{P} \to y \vee \bigvee H$$

is a tautology. So we choose arbitrary truth assignment, and assume that $\bigwedge \mathcal{P}$ is $\top$. We will prove that $y \vee \bigvee H$ is $\top$. For this, we assume the contrary and come to a contradiction.

(a) $x = y$. Recall that

$$\bigwedge \mathcal{P} \equiv \left[\left(\bigwedge_j (B_j \to x_j)\right) \to x\right] \wedge Q$$

by our assumptions,

ANSWERS TO THE EXERCISES

(a1) $(\bigwedge_j (B_j \to x_j)) \to y$ is $\top$, and also Q is $\top$, $y, \bigvee H$ are $\bot$.

On the other hand, $Q \wedge B_j \to x_j \vee \bigvee H \vee y$ is a tautology, and thus under our assignment we get that $B_j \to x_j$ is $\top$.

This holds for each j, so $\bigwedge_j (B_j \to x_j)$ is $\top$, and then $y = \top$ by (a1), which gives a contradiction.

(b) $x = \bot$. The argument can be the same as in (a), because if $\bigwedge \mathcal{P}$ is $\top$ and $\bigwedge \mathcal{P}$ contains the conjunct $(\bigwedge_j (B_j \to x_j)) \to \bot$, then (a1) holds as well.

- If $\langle \mathcal{P}, H, A \to y \rangle$ succeeds with restart, then by Definition 4.3.3, $\langle \mathcal{P} + A, H, y \rangle$ succeeds with restart. Thus we already know that $\bigwedge (\mathcal{P} + A) \to y \vee \bigvee H$ is a tautology, and it follows from the definition that $\bigwedge (\mathcal{P} + A) \equiv (\bigwedge \mathcal{P}) \wedge A$.

  Now assume that $\bigwedge \mathcal{P}$ is true. Hence we know that $A \to y \vee \bigvee H$ is true, and the latter is equivalent to $(A \to y) \vee \bigvee H$. $((A \to B \vee C) \equiv (A \to B) \vee C$ can be checked by truth tables or by equivalences from 1.2.1.) Therefore $\bigwedge \mathcal{P} \to (A \to y) \vee \bigvee H$ is a tautology.

- If $\langle \mathcal{P}, H, A_1 \wedge A_2 \rangle$ succeeds with restart, then by Definition 4.3.3, $\langle \mathcal{P}, H, A_1 \rangle, \langle \mathcal{P}, H, A_2 \rangle$ also do, and we have $\bigwedge \mathcal{P} \to A_1 \vee \bigvee H$ and $\bigwedge \mathcal{P} \to A_2 \vee \bigvee H$ as tautologies. Assume that $\bigwedge \mathcal{P}$ is true; then $A_1 \vee \bigvee H, A_2 \vee \bigvee H$ are true, i.e. $(A_1 \vee \bigvee H) \wedge (A_2 \vee \bigvee H)$ is true. By equivalences from 1.2.1 it follows that $(A_1 \wedge A_2) \vee \bigvee H$ is true. Therefore $\bigwedge \mathcal{P} \to (A_1 \wedge A_2) \vee \bigvee H$ is a tautology.

Exercise 5.2.2

1.

(1) $\varnothing : A \to A$ subcomputation

	$\{a_1\} : A$	
(1.1)	$\{a_1\} : A$	assumption
(1.2)	$\{a_1\} : A$	from (1.1)

The proof is in linear logic.

2. (1) $\varnothing : (A \to (A \to B)) \to (A \to B)$ subcomputation

$$\boxed{\begin{array}{ll} \hfill \{a_1\} : (A \to B) \\ (1.1) \quad \{a_1\} : A \to (A \to B) \quad \text{assumption} \\ (1.2) \quad \{a_1\} : A \to B \quad \text{subcomputation} \\ \boxed{\begin{array}{ll} \hfill \{a_1, a_2\} : B \\ (1.2.1) \quad \{a_2\} : A \quad \text{assumption} \\ (1.2.2) \quad \{a_1, a_2\} : A \to B \quad \text{by (1.1), (1.2.1)} \\ (1.2.3) \quad \{a_1, a_2\} : B \quad \text{by (1.2.1), (1.2.2)} \end{array}} \end{array}}$$

The proof is in relevance logic. It does not fit for linear logic, because if labels are multisets we obtain (1.2.3) $\{a_1, a_2, a_2\} : B$ in the innermost box, which provides only (1.2) $\{a_1, a_2\} : A \to B$ in the next box, and then (1) $\{a_2\} : (A \to (A \to B)) \to (A \to B)$ in the outermost box. This happens because the assumption (1.2.1) is used twice in this proof.

3. (1) $\varnothing : (C \to A) \to ((B \to C) \to (B \to A))$ subcomputation

$$\boxed{\begin{array}{ll} \hfill \{a_1\} : (B \to C) \to (B \to A) \\ (1.1) \quad \{a_1\} : C \to A \quad \text{assumption} \\ (1.2) \quad \{a_1\} : (B \to C) \to (B \to A) \quad \text{subcomputation} \\ \boxed{\begin{array}{ll} \hfill \{a_1, a_2\} : B \to A \\ (1.2.1) \quad \{a_2\} : B \to C \quad \text{assumption} \\ (1.2.2) \quad \{a_1, a_2\} : B \to A \quad \text{subcomputation} \\ \boxed{\begin{array}{ll} \hfill \{a_1, a_2, a_3\} : A \\ (1.2.2.1) \quad \{a_3\} : B \quad \text{assumption} \\ (1.2.2.2) \quad \{a_1, a_3\} : C \quad \text{by (1.2.2.1), (1.2.1)} \\ (1.2.2.3) \quad \{a_1, a_2, a_3\} : A \quad \text{by (1.2.2.2), (1.1)} \end{array}} \end{array}} \end{array}}$$

The proof is in linear logic.

ANSWERS TO THE EXERCISES

4. Similarly to 3 the proof is in linear logic.
   (1) $\varnothing : (C \to A) \to ((A \to B) \to (C \to B))$ subcomputation

   $\quad\quad\quad\quad\quad\quad\quad\quad\quad\quad\quad\quad\quad$ $\{a_1\} : (A \to B) \to (C \to B)$
   - (1.1) $\{a_1\} : C \to A$ $\quad\quad\quad\quad\quad\quad$ assumption
   - (1.2) $\{a_1\} : (A \to B) \to (C \to B)$ $\quad$ subcomputation

     $\quad\quad\quad\quad\quad\quad\quad\quad\quad\quad\quad\quad$ $\{a_1, a_2\} : C \to B$
     - (1.2.1) $\{a_2\} : A \to B$ $\quad\quad\quad\quad$ assumption
     - (1.2.2) $\{a_1, a_2\} : C \to B$ $\quad\quad$ subcomputation

       $\quad\quad\quad\quad\quad\quad\quad\quad\quad\quad\quad\quad$ $\{a_1, a_2, a_3\} : B$
       - (1.2.2.1) $\{a_3\} : C$ $\quad\quad\quad\quad\quad$ assumption
       - (1.2.2.2) $\{a_1, a_3\} : A$ $\quad\quad\quad$ by (1.2.2.1), (1.1)
       - (1.2.2.3) $\{a_1, a_2, a_3\} : B$ $\quad$ by (1.2.2.2), (1.2.1)

5.
   (1) $\varnothing : (A \to (B \to C)) \to ((A \to B) \to (A \to C))$ subcomputation

   $\quad\quad\quad\quad\quad\quad\quad\quad\quad\quad\quad\quad\quad$ $\{a_1\} : (A \to B) \to (A \to C)$
   - (1.1) $\{a_1\} : A \to (B \to C)$ $\quad\quad\quad\quad$ assumption
   - (1.2) $\{a_1\} : (A \to B) \to (A \to C)$ $\quad$ subcomputation

     $\quad\quad\quad\quad\quad\quad\quad\quad\quad\quad\quad\quad$ $\{a_1, a_2\} : A \to C$
     - (1.2.1) $\{a_2\} : A \to B$ $\quad\quad\quad\quad$ assumption
     - (1.2.2) $\{a_1, a_2\} : A \to C$ $\quad\quad$ subcomputation

       $\quad\quad\quad\quad\quad\quad\quad\quad\quad\quad\quad\quad$ $\{a_1, a_2, a_3\} : C$
       - (1.2.2.1) $\{a_3\} : A$ $\quad\quad\quad\quad\quad\quad$ assumption
       - (1.2.2.2) $\{a_1, a_3\} : B \to C$ $\quad\quad$ by (1.1), (1.2.2.1)
       - (1.2.2.3) $\{a_2, a_3\} : B$ $\quad\quad\quad\quad$ by (1.2.1), (1.2.2.1)
       - (1.2.2.4) $\{a_1, a_2, a_3\} : C$ $\quad\quad$ by (1.2.2.2), (1.2.2.3)

The proof is in relevance logic. It does not fit for linear logic, because the assumption (1.2.2.1) is used twice (cf. the solution to item 2).

6. (1) $A \to (B \to A)$ subcomputation

> $B \to A$
> (1.1) A assumption
> (1.2) $B \to A$ subcomputation
>> A
>> (1.2.1) B assumption
>> (1.2.2) A by (1.1)

The proof is in intuitionistic logic. It cannot be transferred to relevance logic by introducing labels because the label for B in (1.2.1) does not disappear in the outermost box.

7. (1) $\varnothing : ((A \to A) \to A) \to A$ subcomputation

> $\{a_1\} : A$
> (1.1) $\{a_1\} : (A \to A) \to A$ assumption
> (1.2) $\varnothing : A \to A$ subcomputation
>> $\{a_2\} : A$
>> (1.2.1) $\{a_2\} : A$ assumption
>> (1.2.2) $\{a_2\} : A$ by (1.2.1)
>
> (1.3) $\{a_1\} : A$ by (1.1), (1.2)

This proof is *not* in relevance logic, because (1.2) is proved as a 'lemma' with label $\varnothing$. Relevance logic does not allow for such labelling: you should not forget assumptions while making a proof, so $A \to A$ should be proved with a label $\{a_1\}$, which is impossible.

Of course, by deleting all labels we get an intuitionistic proof.

Exercise 5.2.4

1. ('Only if') Assume that $\neg\neg A$ is not in an intuitionistic tautology, and show that A is not a classical tautology. The assumption means that there exists a Kripke model $(T, \leq, h)$, such that $h(t, \neg\neg A) = 0$ for some $t \in T$.

ANSWERS TO THE EXERCISES

Hence (by Definition 2.2.3)
$$h(r, \neg A) = 1 \text{ for some } r \geq t \qquad (1.1)$$

Let $p_1, \ldots, p_n$ be all atoms occurring in A. We construct the sequence $r_0 = r \leq r_1 \leq \cdots \leq r_n$ as follows.

Consider the atom p_1 at r_0. There are two options:

- $h(r_0, \neg p_1) = 1$. Then we put $r_1 = r_0$.
- $h(r_0, \neg p_1) = 0$. Then there exists $r_1 \geq r_0$ such that $h(r_1, p_1) = 1$.

Thus we have found $r_1 \geq r_0$ such that $h(r_1, p_1) = 1$ or $h(r_1, \neg p_1) = 1$, i.e. the truth value of p_1 does not change after r_1.

Similarly, if $r_1, \ldots, r_i$ are already constructed, $i < n$, we construct $r_{i+1} \geq r_i$ such that $h(r_{i+1}, p_{i+1}) = 1$ or $h(r_{i+1}, \neg p_{i+1}) = 1$.

Eventually, we reach $r_n \geq r_0$ such that
$$\text{for every } s \geq r_n, \text{ for any } i \leq n, h(s, p_i) = h(r_n, p_i) \qquad (1.2)$$

That is because the truth value of p_i becomes constant after we reach r_i.

Now we make a classical (two-valued) assignment k for atoms $p_1, \ldots, p_n$:
$$k(p_i) = h(r_n, p_i)$$

We claim that for any B built from atoms $p_1, \ldots, p_n$, for any $s \geq r_n$,
$$k(B) = h(s, B) \qquad (1.3)$$

This is proved by induction on the length of B.

- If B is an atom, this is by (1.2) and the choice of k.
- If $B = C \wedge D$, then
  $k(B) = 1$ iff $k(C) = k(D) = 1$
  iff $(h(s, C) = 1$ and $h(s, D) = 1)$
  (by the inductive hypothesis)
  iff $h(s, C \wedge D) = 1$ (by Definition 2.2.3).
  The case $B = C \vee D$ is considered similarly.
- If $B = C \to D$, then
  $k(B) = 1$ iff $k(C) = 0$ or $k(D) = 1$
  iff $(h(s, C) = 0$ or $h(s, D) = 1)$
  (by the inductive hypothesis).

Now $h(s, C \to D) = 1$ implies $h(s, C) = 0$ or $h(s, D) = 1$ by Definition 2.2.3. The other way round, if $h(s, C) = 0$ or $h(s, D) = 1$ then also $h(s', C) = 0$ or $h(s', D) = 1$ for any $s' \geq s$ (by the inductive hypothesis), and thus $h(s, C \to D) = 1$ by Definition 2.2.3.
Hence $k(B) = 1$ iff $h(s, B) = 1$ in this case too.

- If $B = \bot$, (1.3) is obvious.
- If $B = \neg C$, we can rewrite it as $(C \to \bot)$ and use the argument for implication.

From (1.3) we obtain that
$$h(r_n, \neg A) = k(\neg A)$$
but $h(r_n, \neg A) = h(r, \neg A) = 1$ by persistence (Exercise 2.2.4) and (1.1). Therefore $k(\neg A) = 1$ and so $k(A) = 0$.

Thus A is not a classical tautology.

('If') We will show first that every intuitionistic tautology is a classical tautology. So let B be an intuitionistic tautology, and consider an arbitrary classical assignment k. We have to prove that $k(B) = 1$. For this, we take a Kripke model $(T, \leq, h)$ in which $T = \{t_0\}$ is one element and $h(x) = k(x)$ for any atom x.

We claim that
$$h(t_0, C) = k(C) \text{ for any formula } C \tag{1.4}$$

This is checked easily by induction on C. The base follows from the definition of h, and let us consider the case $C = D \to E$.

$h(t_0, C) = 1$ iff for any $t \geq t_0$ $(h(t, D) = 0$ or $h(t, E) = 1)$
iff $(h(t_0, D) = 0$ or $h(t_0, E) = 1)$
(since $T = \{t_0\}$)
iff $(k(D) = 0$ or $k(E) = 1)$
(by the inductive hypothesis)
iff $k(C) = 1$.

The cases of $\land, \lor$ are even simpler. Now from $\vDash_I B$ and (1.4) we obtain that $k(B) = 1$.

Returning to the original question, we observe that if $\neg\neg A$ is an intuitionistic tautology then $\neg\neg A$ is a classical tautology. But $\neg\neg A \equiv A$ in classical logic, and therefore A is also a tautology.

ANSWERS TO THE EXERCISES

2. Assume that $A \equiv B$ classically. Then both $A \to B, B \to A$ are classical tautologies. By the previous item,

$$\vDash_I \neg\neg(A \to B), \neg\neg(B \to A)$$

Let us consider an arbitrary Kripke model $(T, \leq, h)$, and show that for any $t \in T$

$$h(t, \neg\neg A) = 1 \text{ iff } h(t, \neg\neg B) = 1$$

We show 'only if', and then 'if' follows by symmetry.

So, by assumptions, we have

$$h(t, \neg\neg A) = h(t, \neg\neg(A \to B)) = 1 \qquad (2.1)$$

and we are going to prove that $h(t, \neg\neg B) = 1$, i.e. that for any $s \geq t$

$$H(s, \neg B) = 0 \qquad (2.2)$$

Take an arbitrary $s \geq t$. By (2.1), $h(s, \neg(A \to B)) = 0$. Then for some $r \geq s$

$$h(r, A \to B) = 1 \qquad (2.3)$$

Since $h(t, \neg\neg A) = 1$ by (2.1) and $r \geq t$ by transitivity, it follows that $h(r, \neg A) = 0$ and thus there exists $u \geq r$ such that

$$h(u, A) = 1 \qquad (2.4)$$

Since $u \geq r$ by transitivity, (2.3) and (2.4) yield

$$h(u, B) = 1$$

This proves (2.2) because $u \geq s$.

Exercise 5.2.5

1. We show $A \to A$ is generated.

   (a) $(A \to (B \to C)) \to ((A \to B) \to (A \to C))$, axiom.
   (b) $(A \to (B \to A)) \to ((A \to B) \to (A \to A))$, from (a), substituting A for C.
   (c) $A \to (B \to A)$, axiom.
   (d) $(A \to B) \to (A \to A)$, from (c) and (b), using *modus ponens*.

(e) $(A \to (D \to A)) \to (A \to A)$, from (d), substituting $D \to A$ for B.

(f) $A \to (D \to A)$, axiom.

(g) $A \to A$, from (f) and (e), using *modus ponens*.

2. First let us prove that everything derivable from the intuitionistic axioms is an intuitionistic tautology.

- $A \to (B \to A)$ is a tautology (Example 2.4.2.1(a)).
- To show that $(A \to (B \to C)) \to ((A \to B) \to (A \to C))$ is a tautology, consider a Kripke model $(T, \le, h)$. Let $t \in T$; we have to verify that $h(t, (A \to (B \to C)) \to ((A \to B) \to (A \to C))) = 1$, i.e. that for any $b \ge t$, $h(s, A \to (B \to C)) = 1$ only if $h(s, (A \to B) \to (A \to C)) = 1$.

So assume that

(2.1) $h(s, A \to (B \to C)) = 1$

and show that

(2.2) $h(s, (A \to B) \to (A \to C)) = 1$.

To prove (2.2) we take any $r \ge s$, assume that

(2.3) $h(r, A \to B) = 1$

and show that

(2.4) $h(r, A \to C) = 1$.

Further on, we take any $u \ge r$, assume

(2.5) $h(u, A) = 1$

and show that

(2.6) $h(u, C) = 1$.

In fact, we have that $u \ge r \ge s$, so from (2.5) and (2.1) we get

(2.7) $h(u, B \to C) = 1$,

and from (2.5) and (2.3) we get

(2.8) $h(u, B) = 1$.

(2.8) and (2.7) yield (2.6).

Thus (2.5) implies (2.6) under assumptions (2.1), (2.3), and therefore (2.1) implies (2.2)

- *Modus ponens* derives tautologies from tautologies, because in any Kripke model, if $h(t, A) = -1$ and $h(t, A \to B) = 1$ then $h(t, B) = 1$.

ANSWERS TO THE EXERCISES 323

Second, let us show that

$$((A \to B) \to A) \to A$$

is not an intuitionistic tautology. Take A, B atomic, and consider the following Kripke model:

$h(u, A) = h(u, b) = h(v, B) = 0, h(v, A) = 1$.
Then $h(v, A \to B) = 0$, and thus $h(u, A \to B) = 0$.
This implies $h(u, (A \to B) \to A) = 1$. On the other hand, $h(u, A) = 0$, which yields

$$h(u, ((A \to B) \to A) \to A) = 0$$

Therefore, Peirce's rule is not an intuitionistic tautology, and thus it is not derivable from the intuitionistic axioms.

Exercise 6.1.1

1. The translations into predicate logic of the English sentences are:

   (a) `girl(Carol) ∧ sits-next-to(Brenda, Ann)`
   (b) `¬sits-next-to(Ann,Edward)`
   (c) `boy(David) ∨ girl(David)`

2. The translations into English of the predicate logic formulae are:

   (a) If Carol is a girl then Carol is not a boy.
   (b) Ann and Brenda sit next to each other.
   (c) David is not both a boy and a girl.

Exercise 6.1.2

1. The truth values of the sentences in the given interpretation are as follows:

   (a) `girl(Brenda)` is true and so is `¬boy(Brenda)` and thus `girl(Brenda) →¬boy(Brenda)` is also true

(b) boy(David) is true but girl(David) is false and so the antecedent is false, thus
boy(David) ∧ girl(David) → sits-next-to(David,Edward)
is also true

(c) sits-next-to(Edward,David) is false

(d) best-friend(Edward) is D, and thus sits-next-to(best-friend(Edward), Edward) is true.

2. The truth value of sentence (a) does change if the interpretation is changed so boy maps to $\{B, D, E\}$, because while the truth value of the antecedent is unaffected, the consequent's truth value becomes false, and so the implication as a whole becomes false as well.

3. Five sentences with the predicate boy plus five sentences with the predicate girl plus $5 \times 5 = 25$ sentences with the predicate sits-next-to give us 35 sentences in total.

4. If every atomic sentence has the same truth value in two interpretations, then every sentence has equal truth values in both interpretations. So we have to find the number of all possible combinations of truth values of atomic sentences. Since there are 35 atomic sentences (by the previous exercise), we obtain 2^{35} distinguishable interpretations.

However, this number decreases greatly, if we take only 'natural' interpretations into account, i.e., those in which girl(c) ⇔ ¬boy(c) is true for any constant c and in which sits-next-to is interpreted as a symmetrical relation, such that everyone has a single neighbour, and nobody can be a neighbour of him- or herself. In this case we have to indicate only boys (or girls), i.e. choose a subset of a 10-element set, and also to split this set into five two-element subsets. The first choice can be done by 2^{10} ways, and the number of splittings is

$$\frac{\binom{10}{2} \cdot \binom{8}{2} \cdot \binom{6}{2} \cdot \binom{4}{2} \cdot \binom{2}{2}}{5!}$$

(we choose a two-element subset, then a two-element subset of the remaining eight elements, etc. since the splitting is non-ordered, we have to divide the whole product by 5!). In total this yields

$$2^{10} \cdot \frac{10!}{2^5 \cdot 5!} = 10 \cdot 9 \cdot 8 \cdot 7 \cdot 6 \cdot 2^5 < 2^{20}$$

distinguishable interpretations.

Exercise 6.2.2

1. The translations of the English sentences into predicate logic are:

    (a) $\neg\exists x.\ [\text{man}(x) \wedge \text{woman}(x)]$

    (b) $\exists x.\ [\text{woman}(x) \wedge \text{beautiful}(x)]$

    (c) $\forall x.\ [\text{friend}(x, \text{John}) \to \neg\text{beautiful}(x)]$

    (d) $\text{beautiful}(\text{John}) \wedge \text{man}(\text{John})$

    (e) $\forall x.\ [\neg\text{friend}(x,x) \to \neg\exists y.\text{friend}(y,x)]$, or equivalently
    $\forall y.\forall x.\forall x.\ [\neg\text{friend}(x,\ x) \to \neg\text{friend}(y,\ x)]$

    (f) $\forall x.\ [\text{man}(x) \to \exists y.$
    $[\text{woman}(y) \wedge \text{beautiful}(y) \wedge \text{friend}(x,y)]]$
    $\wedge\ \neg\forall x.[[\text{woman}(x) \wedge \text{beautiful}(x)] \to$
    $\exists y.\ [\text{man}(y) \wedge \text{friend}(x,y)]]$

    The reading of 'but not vice versa' is ambiguous. One can read it as 'but not every beautiful woman is a friend of some man' or another reading is 'but the beautiful woman is not a friend of the man'. A third reading is 'not every friend of some beautiful woman is a man'.

    The third reading can be translated as
    $\neg\forall x.\ \forall y.\ [\text{woman}(y) \wedge \text{beautiful}(y) \wedge \text{friend}(x,y)]$
    $\to \text{man}(x)]$.
    The moral is that natural language is very, very ambiguous.

    (g) $\forall x.\ \forall y.\ \forall z.\ [\text{friend}(x,y) \wedge \text{friend}(y,z) \to \text{friend}(x,z)]$

    (h) $\forall x.\ [[\forall y.\ [\text{friend}(y,x) \to \text{beautiful}(y)]] \to \text{woman}(x)]$

    (i) $\forall x.\ [\text{man}(x) \vee \text{woman}(x)]$

2. Possible translations of the predicate logic sentences into English are:

    (a) There is someone who is a woman if women (do indeed) exist.

    (b) Everyone has someone who would reciprocate his friendship.

    (c) If there is someone who befriends himself then all around are men.

3. In the expression

∃x. [[[∃x. friend(x, y)] → friend(y, y)] ∧
friend(x,y) ∧ woman(x) ∧ [woman(y) → woman(x)] ∧
∀y. friend(y, y)]

x is bound by the outermost ∃x. y is free everywhere except in the last conjunct.

4. The sentences about numbers translate as:

   (a) ∀x.∃y. x<y
   (b) ∀x. (prime(x) → ∃y. (prime(y) ∧ x<y))
   (c) ∃x.∀y. ¬(y<x)
   (d) prime(y)
   (e) We have no symbols for the predicate *divide* so we cannot express it directly. For this particular example, it can be expressed as $(z < 1) \lor ((1 < z) \land (z < 3)) \lor ((3 < z) \land (z < 9))$.
   (f) We have no predicate for *love* and so the sentence cannot be expressed.

5. The barber paradox is a true paradox.

   Let $S(x, y)$ stand for 'x shaves y'. Then the barber formula becomes

   $$\exists x. \forall y. (S(x, y) \Leftrightarrow \neg S(y, y))$$

   This formula is inconsistent.

   Let y be x and we get

   $$S(x, x) \Leftrightarrow \neg S(x, x)$$

   which is impossible.

   The reason that it is not immediately obvious to us that the statement is not consistent is because we think of the barber as shaving other people, not himself. (That is, we do not allow $y = x$.)

Exercise 6.2.3

The outline proofs of the validity of the quantifier equivalences are:[1]

---

[1] If the domain is infinite, we get infinite disjunctions and conjunctions in the proof below, and we should consider the 'proof' as informal and intuitive.

ANSWERS TO THE EXERCISES

1. $\forall \mathbf{x}.[\varphi(\mathbf{x}) \wedge \psi(\mathbf{x})]$
   $\equiv [\varphi(a_1) \wedge \psi(a_1)] \wedge [\varphi(a_2) \wedge \psi(a_2)] \wedge [\varphi(a_3) \wedge \psi(a_3)] \wedge \ldots$
   $\equiv [\varphi(a_1) \wedge \varphi(a_2) \wedge \varphi(a_3) \wedge \ldots] \wedge [\psi(a_1) \wedge \psi(a_2) \wedge \psi(a_3) \wedge \ldots]$
   $\equiv [\forall \mathbf{x}.\varphi(\mathbf{x})] \wedge [\forall \mathbf{x}.\psi(\mathbf{x})]$

2. $\exists \mathbf{x}.\,[\varphi(\mathbf{x}) \vee \psi(\mathbf{x})]$
   $\equiv [\varphi(a_1) \vee \psi(a_1)] \vee [\varphi(a_2) \vee \psi(a_2)] \vee [\varphi(a_3) \vee \psi(a_3)] \vee \ldots$
   $\equiv [\varphi(a_1) \vee \varphi(a_2) \vee \varphi(a_3) \vee \ldots] \vee [\psi(a_1) \vee \psi(a_2) \vee \psi(a_3) \vee \ldots]$
   $\equiv [\exists \mathbf{x}.\,\varphi(\mathbf{x})] \vee [\exists \mathbf{x}.\,\psi(\mathbf{x})]$

3. $\exists x. \forall y.(\varphi(x) \to \varphi(y)) \equiv \exists x.(\varphi(x) \to \forall y\, \varphi(y)) \equiv \forall x.\varphi(x) \to \forall y.\varphi(y)$.

   The latter formula is true because $\forall x.\varphi(x)$ and $\forall y.\varphi(y)$ correspond to the same (maybe infinite) conjunction

   $$\varphi(a_1) \wedge \varphi(a_2) \wedge \ldots$$

Exercise 7.2.5

1. $\exists x.\forall y.p(x,y) \wedge \exists x.\forall y.q(x,y)$ first should be rewritten:

   $$\exists x.\forall y.p(x,y) \wedge \exists z.\forall u.q(z,u)$$

   Then we can apply quantifier equivalences and get

   $$\exists x.(\forall y.p(x,y) \wedge \exists z.\forall u.q(z,u))$$
   $$\equiv \exists x.\forall y.(p(x,y) \wedge \exists z.\forall u.q(z,u))$$
   $$\equiv \exists x.\forall y.\exists z.(p(x,y) \wedge \forall u.q(z,u))$$
   $$\equiv \exists x.\forall y.\exists z.\forall u.(p(x,y) \wedge q(z,u))$$

   Other solutions are:

   $$\exists z.\forall u.\exists x.\forall y.(p(x,y) \wedge q(z,u)),$$
   $$\exists x.\exists z.\forall u.\forall y.(p(x,y) \wedge q(z,u)), \text{ etc.}$$

2. The same method. First we rewrite the formula as

   $$\exists x.\forall y.p(x,y) \to \forall z.\exists u.q(z,u)$$

   Then we get subsequently

   $$\forall z.(\exists x.\forall y.p(x,y) \to \exists u.q(z,u))$$
   $$\equiv \forall z.\exists u.(\exists x.\forall y.p(x,y) \to q(z,u))$$
   $$\equiv \forall z.\exists u.\forall x.(\forall y.p(x,y) \to q(z,u))$$
   $$\equiv \forall z.\exists u.\forall x.\exists y.(p(x,y) \to q(z,u))$$

   There are also other solutions:

   $$\forall x.\forall z.\exists u.\exists y.(p(x,y) \to q(z,u)),$$
   $$\forall z.\forall x.\exists y.\exists u.(p(x,y) \to q(z,u)), \text{ etc.}$$

Exercise 6.3.3

1. The definitions are:

    (a)
    $$\langle \mathcal{M}, V \rangle \models \neg\varphi$$
    iff $\langle \mathcal{M}, V \rangle \models \varphi \rightarrow \bot$
    iff $\langle \mathcal{M}, V \rangle \models \varphi$ implies $\langle \mathcal{M}, V \rangle \models \bot$
    iff $\langle \mathcal{M}, V \rangle \models \varphi$ implies false
    iff $\langle \mathcal{M}, V \rangle \not\models \varphi$

    (b)
    $$\langle \mathcal{M}, V \rangle \models \varphi \vee \psi$$
    iff $\langle \mathcal{M}, V \rangle \models (\varphi \rightarrow \bot) \rightarrow \psi$
    iff $\langle \mathcal{M}, V \rangle \models (\varphi \rightarrow \bot)$ implies $\langle \mathcal{M}, V \rangle \models \psi$
    iff $\langle \mathcal{M}, V \rangle \not\models \varphi$ implies $\langle \mathcal{M}, V \rangle \models \psi$
    iff $\langle \mathcal{M}, V \rangle \models \varphi$ or $\langle \mathcal{M}, V \rangle \models \psi$

    (c)
    $$\langle \mathcal{M}, V \rangle \models \varphi \wedge \psi$$
    iff $\langle \mathcal{M}, V \rangle \models (\varphi \rightarrow (\psi \rightarrow \bot)) \rightarrow \bot$
    iff $\langle \mathcal{M}, V \rangle \not\models \varphi \rightarrow (\psi \rightarrow \bot)$
    iff $\langle \mathcal{M}, V \rangle \models \varphi$ and $\langle \mathcal{M}, V \rangle \not\models \psi \rightarrow \bot$
    iff $\langle \mathcal{M}, V \rangle \models \varphi$ and $\langle \mathcal{M}, V \rangle \models \psi$ and $\langle \mathcal{M}, V \rangle \not\models \bot$
    iff $\langle \mathcal{M}, V \rangle \models \varphi$ and $\langle \mathcal{M}, V \rangle \models \psi$

    (d)
    $$\langle \mathcal{M}, V \rangle \models \forall x.\ \varphi$$
    iff $\langle \mathcal{M}, V \rangle \models (\exists x.\ (\varphi \rightarrow \bot)) \rightarrow \bot$
    iff $\langle \mathcal{M}, V \rangle \not\models \exists x.\ \varphi \rightarrow \bot$
    iff there does not exist $d \in \mathcal{D}$ such that $\langle \mathcal{M}, V_{[x \mapsto d]} \rangle \models \varphi \rightarrow \bot$
    iff there does not exist $d \in \mathcal{D}$ such that $\langle \mathcal{M}, V_{[x \mapsto d]} \rangle \not\models \varphi$
    iff for all $d \in \mathcal{D}$ $\langle \mathcal{M}, V_{[x \mapsto d]} \rangle \models \varphi$

2. The truth values of the formulae are:

    (a) $\langle \mathcal{M}, V \rangle \models \mathtt{p(f(a),b)}$ iff
    $\pi_{\text{pred}}(\mathtt{p})(\pi_{\text{term}}(\mathtt{f(a)}), \pi_{\text{term}}(\mathtt{b})) = \top$ iff
    $\pi_{\text{pred}}(\mathtt{p})(\pi_{\text{func}}(\mathtt{f})(\pi_{\text{cons}}(\mathtt{a})), \pi_{\text{cons}}(\mathtt{b})) = \top$ iff
    $\pi_{\text{pred}}(\mathtt{p})(\pi_{\text{func}}(\mathtt{f})(1), 3) = \top$ iff
    $\pi_{\text{pred}}(\mathtt{p})(2, 3) = \top$, which it is.

ANSWERS TO THE EXERCISES

(b) $\langle \mathcal{M}, V \rangle \models \mathtt{r(a,f(b),c)}$ iff
$\pi_{\text{pred}}(\mathtt{r})(\pi_{\text{term}}(\mathtt{a}), \pi_{\text{term}}(\mathtt{f(b)}), \pi_{\text{term}}(\mathtt{c})) = \top$ iff
$\pi_{\text{pred}}(\mathtt{r})(\pi_{\text{cons}}(\mathtt{a}), \pi_{\text{func}}(\mathtt{f})(\pi_{\text{cons}}(\mathtt{b})), \pi_{\text{cons}}(\mathtt{c})) = \top$ iff
$\pi_{\text{pred}}(\mathtt{r})(1, \pi_{\text{func}}(\mathtt{f})(3), 5) = \top$ iff
$\pi_{\text{pred}}(\mathtt{r})(1,5,5) = \top$, which it is not.

(c) $\langle \mathcal{M}, V \rangle \models \exists \mathtt{x}.\ (\neg \mathtt{q(x)} \to \mathtt{p(f(a),g(x,2))})$ iff
there exists $d \in \mathcal{D}$ such that
$\langle \mathcal{M}, V_{[x \mapsto d]} \rangle \models \neg \mathtt{q(x)} \to \mathtt{p(f(a),g(x,2))}$ iff
there exists $d \in \mathcal{D}$ such that $\langle \mathcal{M}, V_{[x \mapsto d]} \rangle \models \neg \mathtt{q(x)}$ implies
$\langle \mathcal{M}, V_{[x \mapsto d]} \rangle \models \mathtt{p(f(a),g(x,2))}$ iff
there exists $d \in \mathcal{D}$ such that $\langle \mathcal{M}, V_{[x \mapsto d]} \rangle \not\models \mathtt{q(x)}$ implies
$\langle \mathcal{M}, V_{[x \mapsto d]} \rangle \models \mathtt{p(f(a),g(x,2))}$
Since the premise $\neg \mathtt{q(x)}$ is false for $x = 2$, the full expression holds.

(d) In this case the full expression must hold for every choice of x. In particular, for the cases where $\neg \mathtt{q(x)}$ holds we must check the consequent $\mathtt{p(f(a), g(x, 2))}$. The choice of $x = 3$ makes $\neg \mathtt{q(x)}$ true and the consequent false. Hence the sentence does not hold in the model.

Exercise 7.3.4

(a) $s(x, x, x)$,
because $x + x = x$ holds iff $x = 0$.
Another solution is
$$\forall y. s(x, y, y)$$

(b) $p(x, x, x) \land \neg s(x, x, x)$,
because $x \cdot x = x$ holds iff $x = 0$ or $x = 1$.
Another solution is
$$\forall y. p(x, y, y)$$

(c) $\exists y. (\forall z. p(y, z, z) \land s(y, y, x))$,
because $2 = 1 + 1$.
Another solution is:
$$\exists y. (p(x, x, y) \land s(x, x, y)) \land \neg s(x, x, x)$$
because $x \cdot x = x + x$ iff ($x = 0$ or $x = 2$).

(d) $\exists y.s(y,y,x)$.

(e) $\forall y.\forall z.(p(y,z,x) \to p(y,y,y) \lor p(z,z,z)) \land \neg p(x,x,x)$.

Explanation: the first conjunct says that if $y \cdot z = x$ then one of the numbers y, z is either 0 or 1. If we require also that $x \cdot x \ne z$ (i.e. $x \ne 0, x \ne 1$) then the first conjunct is equivalent to
$$\forall y.\forall z.(y \cdot z = x \to y = 1 \lor z = 1)$$
which, taken together with $x \ne 1$, means that x is prime.

(f) One solution is to say that $x \ne 0 \land x \ne 1 \land \cdots \land x \ne 27$.

Another method is to say that
$$\exists y.(27 + y = x \land y \ne 0)$$
or that
$$\exists y.\exists z.(y \ne 0 \land y + z = x \land z = 27)$$
We know how to express the first two conjuncts. $z = 27$ can be expressed, for example, as
$$\exists x.\exists y.(x = 3 \land y = x \cdot x \land z = x \cdot y)$$
Now the problem is to say that $x = 3$. This can be done as follows:
$$\exists y.\exists z.(y + z = x \land y = 1 \land z = y + y)$$
and we can use exercise (c).

Bringing everything together, we obtain the formula
$$\exists y.\exists z.[\neg s(y,y,y) \land s(y,z,x) \land \exists x.\exists y.(p(x,x,y) \land p(x,y,z) \land$$
$$\exists y.\exists z.(s(y,z,x) \land p(y,y,y) \land \neg s(y,y,y) \land s(y,y,z)))]$$
(Note that x has free and bound occurrences here.)

(g) We can express this by saying that $x + 0 = y$, which is equivalent to
$$\exists z.(z = 0 \land x + z = y)$$
i.e. the formula is
$$\exists z.(s(z,z,z) \land s(x,z,y))$$

(h) Similarly, we can say that $\exists z.(z \ne 0 \land x + z = y)$, which is written thus:
$$\exists z.(\neg s(z,z,z) \land s(x,z,y))$$

(i) This is expressed as follows: $\forall x.\exists y.(x < y \land y$ is prime). The expressions for conjuncts are found in (h) and (e).

ANSWERS TO THE EXERCISES

Exercise 7.3.5

(a) No. To show this, we can use quantifier equivalences:

$$\exists x.\forall y.(Q(x,x) \wedge \neg Q(x,y))$$
$$\equiv \exists x.(Q(x,x) \wedge \forall y.\neg Q(x,y))$$
$$\equiv \exists x.Q(x,x) \wedge \forall y.\neg Q(x,y)$$
$$\equiv \exists x.Q(x,x) \wedge \neg \exists y.Q(x,y)$$

If the latter is true, we have for some a in the model:

$$Q(a,a) \wedge \neg \exists y.Q(a,y) = \top$$

But $Q(a,a) = \top$ implies that $\exists y.Q(a,y) = \top$ (namely, one can take $y = a$).

Hence $Q(a,a) \wedge \neg \exists y.Q(a,y) = \top \wedge \bot = \bot$, and this is a contradiction.

(b) Yes. We can take a two-element model

$$D = \{d_1, d_2\}$$
$$\pi_{\text{pred}}(P)(x) = \top \text{ iff } x = d_1$$

Then $P(d_1) \wedge \neg P(d_2) = \top$, implying that $\langle D, \pi_{\text{pred}}, [\ ]\rangle \models (b)$.

(c) Obviously, yes. Take any model in which $\pi_{\text{pred}}(Q)$ sends every pair to $\bot$. Then the premise of the implication is false under any assignment of x, y.

(d) Obviously, yes. Take any model in which $\pi_{\text{pred}}(P)$ sends every element to $\bot$.

Exercise 7.3.6

(a) No. We can make $\exists x.P(x) = \top$ and $\forall x.P(x) = \bot$ in a model where $P(x)$ is sometimes true and sometimes false. For instance, take the model with $D = \{d_1, d_2\}$, $\pi_{\text{pred}}(d_1) = \top$, $\pi_{\text{pred}}(d_2) = \bot$.

(b) No, because $\exists x.P(x) \to \forall x.P(x)$ has a model. It can be the same as for (d) in the previous exercise.

(c) Yes. Assume that $\langle \mathcal{U}, V \rangle \models \exists x.\forall y.Q(x,y)$, and show that $\langle \mathcal{U}, V \rangle \models \forall y.\exists x.Q(x,y)$.

By assumption, $\langle \mathcal{U}, V_{[x \mapsto a]} \rangle \models \forall y Q(x,y)$ for some a, and thus for any b, $\langle \mathcal{U}, V_{[x \mapsto a, y \mapsto b]} \rangle \models Q(x,y)$. Hence for any b, $\langle \mathcal{U}, V_{[y \mapsto b]} \rangle \models \exists x.Q(x,y)$, and therefore $\langle \mathcal{U}, V \rangle \models \forall y.\exists x.Q(x,y)$.

(d) No. Take a model with $D = \{a, b\}$,
$$\pi_{\text{pred}}(Q) : \{(a, a) \mapsto \top, (b, b) \mapsto \top, (a, b) \mapsto \bot, (b, a) \mapsto \bot\}$$

Then
$$\exists y.Q(z, y) = \top \text{ since } Q(a, a) = \top$$
$$\exists y.Q(b, y) = \top \text{ since } Q(b, b) = \top$$

Thus $\forall x.\exists y.Q(x, y) = \top$.

On the other hand,
$$\forall x.Q(x, a) = \bot \text{ since } Q(b, a) = \bot$$
$$\forall x.Q(x, b) = \bot \text{ since } Q(a, b) = \bot$$

Thus $\exists y.\forall x.Q(x, y) = \bot$.

So we obtain that (d) $= \bot$ in this model.

Exercise 7.1.5

1. p(a,f(z)) and p(x,f(b)) unify with most general unifier $\theta = \{$x/a, z/b$\}$.

2. q(f(g(x))) and r(c) do not unify as the predicates differ!

3. r(x,x,f(b)) and r(g(f(b)),g(y),y) unify with most general unifier $\theta = \{$x/g (f(b)),y/f(b)$\}$.

4. p(a,f(g(x))) and p(y,g(z)) do not unify as the functors of the second arguments differ.

5. q(b,x) and q(y,f(x)) do not unify as x occurs within the term f(x).

6. q(f(a,x)) and q(f(y,f(z))) unify with most general unifier $\theta = \{$y/a,x/f(z)$\}$.

Exercise 7.1.6

1. Because the y does not appear in the matrix of the formula, there is no position to place its Skolem function, so it is omitted (if it had appeared, it would have been of the form f(x)), since it is only in the scope of the $\forall$x. The other existentially quantified variable, v, is in the scope of three universal quantifiers, on x, z and u. Thus v's Skolem function g has three arguments:

$$\forall \text{x}.\forall \text{z}.\forall \text{u}. \ [\text{R}(\text{x}, \ \text{z}) \rightarrow \neg \text{Q}(\text{u}, \ \text{g}(\text{x},\text{y},\text{u}))]$$

ANSWERS TO THE EXERCISES

2. The Skolemized forms are:

   (a) ∀x.∀y. [A(x, y) → B(x)]
   (b) ¬A(f) ∨ ¬B(g)
   (c) A(f) ∧ ∀z. B(z, g)
   (d) ∀x.∀y. [A(x, y) → B(x, f(x,y)) ∧ C(g(x,y))]
   (Since u and v do not appear in the matrix, they can be eliminated. The Skolem constants do not depend on them.)

Exercise 7.2.3

A translation into logic of the argument is

    1. ∀x. [interrupt(x) → ¬desirable(x)]
    2. ∃x. [control(x) ∧ interrupt(x)]
    3. ∃x. [control(x) ∧ ¬desirable(x)]

From assumption 2 using ∃ elimination (Skolemization) with the new constant k, we get

    4. control(k) ∧ interrupt(k)

Using ∀ instantiation for k in assumption 1 we get

    5. interrupt(k) → ¬desirable(k)

From sentences 4 and 5, using propositional ∧ elimination and *modus ponens*, we get

    6. ¬desirable(k)

From sentences 4 and 6, using ∧ elimination and ∧ introduction, we get

    7. control(k) ∧ ¬desirable(k)

Finally, ∃ introduction on sentence 7 gives

    3. ∃x. [control(x) ∧ ¬desirable(x)]

Exercise 8.4.1

[X,Y,Z]	[1,2,7]	X=1, Y=2, Z=7
[X,Y]	[1,[3,4]]	X=1, Y=[3,4]
[X,Y,Z]	[a,[b],c]	X=2, Y=[b], Z=c
[X,Y]	[[a,b],[3,4]]	X=[a,b], Y=[3,4]
[[X,Y],Z]	[1,[2,3]]	No match
X	[]	X=[]
[X]	[]	No match
[X,[1,2],[a,B]]	[a,[R,S],[a,c]]	X=a, R=1, S=2, B=c

Exercise 8.4.2

The goal will succeed and the values for the variables are as follows:

```
X = 1
Xs = [2]
Ys = []
```

Exercise 8.4.3

```
last_elem([X],X).
last_elem([_|T],X):-last_elem(T,X).
```

Exercise 8.4.4

```
acc_list([],0).
acc_list([H|T],V):- acc_list(T,V1), V is V1+H.
```

Exercise 8.4.5

```
sort_ins(X,[],[X]).
sort_ins(X,[H|T],[X,H|T]):- X =< H.
sort_ins(X,[H|T],[H|T1]):- X > H, sort_ins(X,T,T1).
```

BIBLIOGRAPHY

[Abramsky et al., 1992–94] S. Abramsky, D. M. Gabbay and T. Maibaum, eds. *Handbook of Logic in Computer Science*, 4 vols. Oxford University Press, Oxford, 1992–1994.

[Aleksander and Burnett, 1987] I. Aleksander and P. Burnett. *Thinking Machines*. Oxford University Press, Oxford, 1987.

[Aleksander and Morton, 1995] I. Aleksander and H. Morton. *An Introduction to Neural Computing*. 2nd Edition. International Thompson Computer Press, London, 1995.

[Antoniou, 1996] G. Antoniou. *Nonmonotonic Reasoning*. The MIT Press, Cambridge, MA, 1996.

[Apt, 1996] K. R. Apt. *From Logic Programming to Prolog*. Prentice Hall, 1996.

[Avron, 1988] A. Avron. *Foundations and proof theory of three valued logic*, LFCS reports, University of Edinburgh, 1988.

[Barendregt, 1984] H. Barendregt. *The Lambda Calculus*. 2nd Edition. North-Holland, Amsterdam, 1984.

[Barwise, 1977] J. Barwise, ed. *Handbook of Mathematical Logic*. North-Holland, Amsterdam, 1977.

[Ben-Ari, 1993] M. Ben-Ari. *Mathematical Logic for Computer Science*, Prentice Hall, 1993.

[Besnard, 1989] P. Besnard. *Default Logic*. Springer-Verlag, Heidelberg, 1989.

[Bizam and Herczeg, 1978] G. Bizam and J. Herczeg. *Sokszin Logika* (175 logikai feladat). Müszaki könyvkiado, Budapest, 1975 (in Hungarian); Russian translation: Mir, Moscow, 1978.

[Blumenthal, 1980] L. M. Blumenthal. *A Modern View of Geometry*. Dover, New York, 1980.

[Bradley and Swartz, 1979] R. Bradley and P. Swartz. *Possible World: An Introduction to Logic and its Philosophy*. Blackwell, Oxford, 1979.

[Brewka, 1991] G. Brewka. *Nonmonotonic Reasoning: Logical Foundations of Commonsense*. Cambridge University Press, Cambridge, 1991.

[Brewka et al., 1997] G. Brewka, J. Dix and K. Konolige. *Nonmonotonic Reasoning: An Overview*. CSLI, 1997.

[Broda et al., 1994] K. Broda, S. Eisenbach, H. Khoshnevisian and S. Vickers. *Reasoned Programming*. Prentice Hall, 1994.

[Burke and Foxley, 1996] E. Burke and E. Foxley. *Logic and its Applicatons*. Prentice Hall, 1996.

[Chellas, 1980] B. F. Chellas. *Modal Logic, An Introduction*. Cambridge University Press, Cambridge, 1980.

[Clarke and Gabbay, 1987] M. Clarke and D. M. Gabbay. An intuitionistic basis for non-monotonic reasoning. In *Automated Reasoning for Non-standard Logic*, ed. P. Smets, pp. 163–179. Academic Press, London, 1987.

[Copeland, 1993] J. Copeland. *Artificial Intelligence*. Blackwell, Oxford, 1993.

[Copi, 1986] I. M. Copi. *Introduction to Logic*, 7th Edition. Macmillan, London, 1986.

[Fagin et al., 1995] R. Fagin, J. Y. Halpern, Y. Moses and M. Y. Vardi. *Reasoning about Knowledge*. The MIT Press, Cambridge, MA, 1995.

BIBLIOGRAPHY

[Field and Harrison, 1988] A. J. Field and P. G. Harrison. *Functional Programming*. Addison Wesley, Reading, 1988.

[Fitting, 1996] M. Fitting. *First Order Logic and Automated Theorem Proving*, 2nd Edition. Springer-Verlag, Berlin, Heidelberg, New York, 1996.

[Flach, 1994] P. Flach. *Simply Logical*. Wiley, 1994.

[Gabbay, 1976] D. M. Gabbay. *Investigations in Modal and Tense Logics*. D. Reidel, Dordrecht, 1976.

[Gabbay, 1982] D. M. Gabbay. Intuitionistic basis for non-monotonic logic. In *Proceedings of CADE-6*, LNCS vol. 138, pp. 260–273. Springer-Verlag, Berlin, Heidelberg, New York, 1982.

[Gabbay, 1985] D. M. Gabbay. Theoretical foundations for non-monotonic reasoning. In *Expert Systems, Logic and Models of Concurrent Systems*, ed. K. Apt, pp. 439–459. Springer-Verlag, Berlin, Heidelberg, New York, 1985.

[Gabbay, 1988] D. M. Gabbay. *The Tübingen Lectures*, Technical report 90-55, pp. 1–127. University of Tübingen, 1988.

[Gabbay, 1996] D. M. Gabbay. *Labelled Deductive Systems, Vol. 1, Basic Theory*. Oxford University Press, Oxford, 1996.

[Gabbay and Guenthner, 1977–99] D. M. Gabbay and F. Guenthner, eds. *Handbook of Philosophical Logic*, vols 1–4. Kluwer, Dordrecht, 1979–1986; 2nd Edition 10–12 vols. to appear during 1998–1999.

[Gabbay and Hunter, 1991] D. M. Gabbay and A. B. Hunter. Making inconsistency repsectable, part 1. In *Fundamentals of Artificial Intelligence Research (FAIR '91)*, eds. Ph. Jorrand and J. Kelemen, LNAI Vol. 535, pp. 19–32. Springer-Verlag, Berlin, Heidelberg, New York, 1991.

[Gabbay and Reyle, 1984] D. M. Gabbay and U. Reyle. N-Prolog: An extension of Prolog with hypothetical implications 1. *Journal of Logic Programming*, **1**, 319–355, 1984.

[Gabbay *et al.*, 1993–95] D. M. Gabbay, C. Hogger and J. A. Robinson, eds. *Handbook of Logic in Artificial Intelligence and Logic Programming*, 5 vols., Oxford University Press, Oxford, 1993–1996.

[Gabbay et al., 1994] D. M. Gabbay, I. Hodkinson and M. Reynolds. *Temporal Logic, Vol. 1.* Oxford University Press, Oxford, 1994.

[Gabbay et al., 1995] D. M. Gabbay, L. Giordano, A. Martelli and N. Olivetti. Hypothetical updates, priority and inconsistency in a logic programming langauge. *Logic Programming and Non-monotonic Reasoning*, eds. V. W. Marek, A. Nerode and M. Truszczynski, LNCS Vol. 928, pp. 203–216. Springer-Verlag, Berlin, Heidelberg, New York, 1995.

[Gabbay and Olivetti, 2000] D. M. Gabbay and N. Olivetti. *Goal-Directed Proof Theory.* ISBN: 978-0792364733. Springer, 2000.

[Gallier, 1986] J. H. Gallier. *Logic for Computer Science: Foundations of Automatic Theorem Proving.* Harper and Row, New York, 1986.

[Gibbins, 1988] P. Gibbins. *Logic with Prolog.* Oxford University Press, Oxford, 1988.

[Gillies, 1996] D. Gillies. *Artificial Intelligence and Scientific Method.* Oxford University Press, Oxford, 1996.

[Gottwald, 1993] S. Gottwald. *Fuzzy Sets and Fuzzy Logic.* Vieweg, Braunschweig, 1993.

[Goubault-Larrecq and Mackie, 1997] J. Goubault-Larrecq and I. Mackie. *Proof Theory and Automated Deduction*, Applied Logic Series. Kluwer, Dordrecht, 1997.

[Hankin, 1994] C. L. Hankin. *Lambda Calculi: A Guide for Computer Scientists.* Oxford University Press, Oxford, 1994.

[Hein, 1995] J. L. Hein. *Discrete Structures. Logic and Computability.* Jones and Bartlett, Sudbury, Massachusetts, 1995.

[Hindley and Seldin, 1986] J. R. Hindley and J. P. Seldin. *Introduction to Combinators and λ-calculus.* Cambridge University Press, Cambridge, 1986.

[Hodges, 1977] W. Hodges. *Logic.* Pelican, Harmondsworth, 1977.

[Hogger, 1994] C. J. Hogger. *Essentials of Logic Programming.* Oxford University Press, Oxford, 1994.

[Hughes and Cresswell, 1996] G. E. Hughes and M. J. Cresswell. *A New Introduction to Modal Logic*. Routledge, London, 1996.

[Hunter, 1996] A. Hunter. *Uncertainty in Information Systems*. McGraw Hill, New York, 1996.

[Kahane, 1990] H. Kahane. *Logic and Philosophy*, 6th Edition. Wadsworth, Belmont, California, 1990.

[Klir and Folger, 1988] G. J. Klir and T. A. Folger. *Fuzzy Sets, Uncertainty and Information*. Prentice Hall, London, 1988.

[Kowalski, 1979] R. A. Kowalski. *Logic for Problem Solving*. North-Holland, Amsterdam, 1979.

[Kowalski and Sergot, 1986] R. A. Kowalski and M. Sergot. A logic based calculus of events. *New Generation Computing*, **4**, 1986.

[Kraus et al., 1990] S. Kraus, D. Lehmann and M. Magidor. Nonmonotonic reasoning, preferential models and cumulative logics. *Artificial Intelligence*, **44**, 167–208, 1990.

[Kuliakowski, 1984] C. Kuliakowski. *A Practical Guide to Designing Expert Systems*. 1984.

[Lalement, 1993] R. Lalement. *Computation as Logic*. Prentice Hall, 1993.

[Lambert and Ulrich, 1980] K. Lambert and W. Ulrich. *The Nature of Argument*. Macmillan, London, 1980.

[Lavrov and Maksimova, 1995] I. A. Lavrov and L. L. Maksimova. *Problems in Set Theory, Mathematical Logic and Theory of Algorithms*, 3rd Edition. Nauka, Moscow, 1995. (in Russian).

[Łukasiewicz, 1990] W. Łukasiewicz. *Non-monotonic Reasoning: Formalisation of Commonsense Reasoning*. Ellis Horwood (now Prentice Hall), Chichester, 1990.

[Malinowski, 1993] G. Malinowski. *Many Valued Logics*. Oxford University Press, Oxford, 1993.

[Marek and Truszczynski, 1993] V. W. Marek and M. Truszczynski. *Nonmonotonic Logic*. Springer-Verlag, Berlin, Heidelberg, New York, 1993.

[Mendelson, 1964] E. Mendelson. *Introduction to Mathematical Logic*. Van Nostrand, 1964; 3rd Edition, Wadsworth, Belmont, California, 1987.

[Metcalfe et al., 2009] G. Metcalfe and N. Olivetti and D. M. Gabbay. *Proof Theory for Fuzzy Logics*. Springer Science+Business Media B.V., 2009.

[Müller et al., 1997] J. P. Müller, M. J. Wooldridge and N. R. Jennings, eds. *Intelligent Agents III*, LNAI, vol. 1193, Springer-Verlag, Berlin, Heidelberg, New York, 1997.

[Nidditch, 1962] P. H. Nidditch. *Propositional Calculus*. RKP, London, 1962.

[Novikov, 1964] P. S. Novikov. *Elements of Mathematical Logic*. Oliver and Boyd, Edinburgh, 1964.

[Nwana and Azarmi, 1997] H. S. Nwana and N. Azarmi, eds. *Software Agents and Soft Computing*. LNAI, vol. 1198, Springer-Verlag, Berlin, Heidelberg, New York, 1997.

[Potter et al., 1991] B. Potter, J. Sinclair and D. Till. *An Introduction to Formal Specification and Z*. Prentice Hall, London, 1991.

[Reeves and Clarke, 1990] S. Reeves and M. Clarke. *Logic for Computer Science*. Addison Wesley, Reading, 1990.

[Russell and Norvig, 1995] S. Russell and P. Norvig. *Artificial Intelligence: A Modern Approach*. Prentice Hall, London, 1995.

[Sandewall, 1994] E. Sandewall. *Features and Fluents: Volume 1: A Systematic Approach to the Representation of Knowledge about Dynamical Systems*. Oxford University Press, Oxford, 1994.

[Shoham, 1988] Y. Shoham. *Reasoning about Change*. The MIT Press, Cambridge, MA, 1988.

[Spivak, 1995] A. V. Spivak. *Mathematical festival*. Moscow Centre of Continuous Mathematical Education, Moscow, 1995 (in Russian).

[Truss, 1991] J. K. Truss. *Discrete Mathematics for Computer Scientists*. Addison Wesley, Reading, 1991.

[van Dalen, 1994] D. van Dalen. *Logic and Structure*, 3rd Edition. Springer-Verlag, Berlin, Heidelberg, New York, 1994.

[Walton, 1989] D. Walton. *Informal Logic*. Cambridge University Press, Cambridge, 1989.

[Woods and Walton, 1982] J. Woods and D. Walton. *Argument, The Logic of the Fallacies*. McGraw-Hill, New York, 1982.

INDEX

Symbols
$(\to E)$, 94
$(\to E1)$, 94
$(\to I)$, 94
$(\to I1)$, 94
$(\to I2)$, 94
$(\neg 2)$, 94
$(\neg E)$, 94
$(\neg E1)$, 94
$(\neg I)$, 94
$(\neg I1)$, 94
$(\vee E)$, 94
$(\vee E1)$, 94
$(\vee I)$, 94
$(\wedge E)$, 94
$(\wedge I)$, 94
Γ-computation, 139
H1, 109
H2, 109
H3, 109
H4, 109
H5, 109
H6, 109
H7, 109
H8, 109

A
a logic, 130
affine planes, 190
annotated computation, 90
annotated database, 123
argument, 16
    valid, 17, 19
arity, 169, 170, 184
associated answer substitution θ, 229
assumption, 17
atomic sentences, 169
Avron, 135

B
$\mathcal{B}$-clauses, 184, 206
backward computation with diminishing resource, 123
backward reasoning, 158
binding, 209
body, 111
bound, 176
bounded restart, 125

box, 91
    consequence, 92
    display, 91
    soundness of consequence, 92

C

classical logic, 125, 158
classical propositional logic, 3, 7
clausal form, 111
completeness, 138
completeness of the backwards computation rules, 144
computation state, 124
computation trees without restart, 144
conclusion, 16
consequence relation, 16, 18, 130
consequence relations
    associated with Hilbert systems, 135
consistency, 205
consistency of a set of formulae, 19
consistent, 13
constant symbols, 169, 184
contradictions, 12
cut elimination, 148
cut rule, 18, 94
cut theorem, 136

D

data, 17
database, 228
decidability, 237
deduction theorem for a Hilbert system, 163
diminishing resource, 122, 158
diminishing resource computation procedures, 124

domain, 171, 175, 186
dual, 180

E

English sentences, 170
equivalences, 13
Euler's problem, 189
existential introduction, 180
existential variable, 216, 227

F

fails
    with bounded restart, 124
    with restart, 124
forward and backward predicate rules, 205
forward reasoning, 158
forward rules, 69
free, 176
function
    over the domain, 172
function name, 170
function symbols, 185
functional relationships, 170

G

Glivenko Theorem, 133
goal, 16
goal-directed algorithm, 122
grafting, 146

H

head, 111
Herbrand theorem, 234
heuristics, 96
    for negation, 105
Hilbert axioms, 133
Hilbert formulation
    for classical logic, 134

INDEX

for intuitionistic logic, 134
for predicate logic, 234
of a logic, 132
of three-valued logic, 135
Hilbert systems, 130

I

immediate failure, 124
independence of Hilbert axioms, 158
indices, 89
infinite-valued logic L_∞, 41
interpolation, 137, 156
interpretation, 185
introduction to PROLOG, 243
intuitionistic
    implicational tautologies, 133
    logic, 46, 50, 125, 158
    logic, fragment of $\{\wedge, \to, \bot\}$, 155
    many-valued logic, 47

K

Kripke models for intuitionistic logic, 47, 132

L

L_n-contradiction, 41
L_n-tautology, 41
logical system, 36
Łukasiewicz many-valued logics, 40

M

many-valued logic, 40, 158
many-valued logic L_n or L_∞, 158
many-valued resource logic, 49
matrix, 182, 206, 207
metatheorems, 129
modalities, 45

model, 185
modus ponens, 17, 19
monotonicity, 18, 137

N

names, 171
natural deduction rules, 72
natural numbers, 171
Nidditch, 134
non-classical logics, 39
normal form, 14
    conjunctive, 15
    disjunctive, 15

O

Olivetti, 158

P

Peirce's rule, 133, 134
predicate
    symbols, 184
predicate $\mathcal{B}$-clauses, 227
predicate $\mathcal{B}$-computation, 229
predicate interpretation, 171–173
predicate logic, 167
    semantics, 185–187
    syntax, 184–185
predicates, 169
prefix, 182, 206
prenex normal form, 182–184, 206
proof of completeness, 143
proofs, 81
    formal descriptions of, 89
propositional $\mathcal{B}$-clauses, 140

Q

quantifier
    equivalences, 178–182
    inequivalences, 181

rules, 178–182
quantifier ∃, 185
quantifiers, 173–184
    existential, 175
    scope, 176–178
    universal, 174
query, 227

R

ready for computation form, 122
reasoning backward, 227
reasoning forwards, 221
recursively enumerable, 237
reflexivity, 18
relationships, 171
resolution, 235
resource logics, 48
restart rule, 103, 104, 112, 158
rewriting assumptions, 106
rules, 19
    backward, 110
    for conjunction, 72
    for disjunction, 73
    for implication, 74
    for negation, 75
    proofs involving negation, 84
    summary of computation, 109
    summary of natural deduction, 94

S

satisfiability, 205
satisfiable, 13
semi-decidable, 237
set of terms, 185
Sheffer stroke, 14, 282
Skolem functions, 207, 217
Skolemisation, 217
Smetanich, 62

Smetanich tautologies, 62
soundness, 138
    of the backwards computation rules, 142
    of the diminishing resource computation procedure, 128
    of the predicate $\mathcal{B}$ computation, 233
strict normal form, 15
strong cut, 156
sub-query, 211
substitution, 137, 209
    composition of, 209
    functional, 209
    idempotent, 209
substitution instance, 209
succeeds, 124
    with bounded restart, 124
    with restart, 124
successful computation tree, 144, 148
summary of systems and completeness theorems, 158

T

tautology, 12, 205
temporal many-valued logic, 44, 158
temporal tautologies, 45
term mapping, 186
term stack, 212
The $\{\wedge, \rightarrow, \bot\}$ fragment of intuitionistic logic, 158
theorems of logic, 12
transitivity, 137
translate, 170
tree cut, 147
tree for restart, 148

tree properties, 145
truth, 12
truth tables, 8

U

unification, 210–211
    algorithm, 212, 213
unifier, 212
    most general, 212
universal elimination, 179
universal prenex form, 207, 210
universal quantifiers, 207
    omission of, 210
universal variable, 216, 227
using sub-computation, 77

V

validity, 12
variable
    in query, 215
    meaning of in clauses, 215
variable assignment, 186, 208
variables, 173, 185
    reasoning with, 206

W

well-formed formulae, 185
wff, 185

Lightning Source UK Ltd.
Milton Keynes UK
UKOW05f0747071016

284703UK00006B/98/P